Plumbing

NVQ and Technical Certificate Level 3

Mike Phoenix

John Thompson

on behalf of JTL

JTL
Connecting to the future

www.jtltraining.com

www.heinemann.co.uk

✓ Free online support
✓ Useful weblinks
✓ 24 hour online ordering

01865 888058

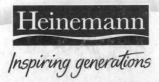

Heinemann

Inspiring generations

Heinemann is an imprint of Pearson Education Limited, a company incorporated in England and Wales, having its registered office at Edinburgh Gate, Harlow, Essex, CM20 2JE.
Registered company number: 872828

Heinemann is a registered trademark of Pearson Education Limited

Text © JTL, 2005

First published 2005

10 09 08
10 9 8 7 6

British Library Cataloguing in Publication Data is available from the British Library on request.

ISBN: 978 0 435 401 95 5

Typeset and illustrated by HL Studios, Long Hanborough, Oxford

Original illustrations © Harcourt Education Limited, 2005

Printed in China (SWTC/06)

Cover design by GD Associates

Cover photo: © Gareth Boden/Harcourt Education

Picture research by Ginny Stroud-Lewis

Acknowledgements

Every effort has been made to contact copyright holders of material reproduced in this book. Any omissions will be rectified in subsequent printings if notice is given to the publishers.

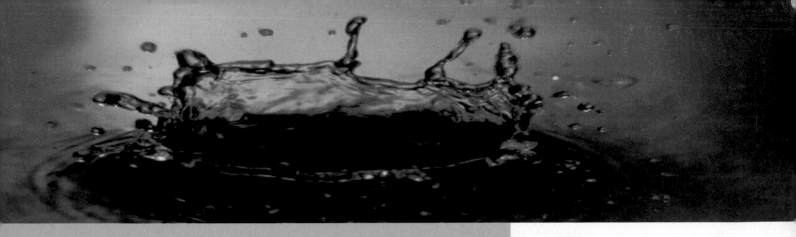

Contents

Acknowledgements

JTL would like to express its appreciation to all those members of staff who contributed to the development of this book, ensuring that the professional standards expected were delivered and generally overseeing the high quality of the final product. Without their commitment and support – as much to each other as to the project – this project would not have been completed successfully. Particular thanks go to Mike Phoenix, Keith Arrell, Keith Powell, Dave Bowers, Frank Kirkham, Mike Bullock and Kevan Thomas. Our thanks also to the staff at Building Training Services, Caerphilly, and Oxford College for their patience and assistance at photoshoots, and to John Thompson for his help in getting this project off the ground.

The authors and publishers would like to thank the following for permission to reproduce their material:

Saniflo
WRAS
CE Mark reproduced by courtesy of EUROPA/New Approach (Council Decision 93/465/EEC)

PICTURE ACKNOWLEDGEMENTS

The authors and publishers would like to thank the following for permission to reproduce photographs:

All pictures **Harcourt Education Ltd/Gareth Boden** apart from the following:

Alamy Images 82; **Alamy Images/Liam Bailey** 161; **Alamy Images/Michael Booth** 423, 434, 426; **Alamy Images/F1 Online** 71; **Alamy Images/Andrew Linscott** 79; **Alamy Images/Bob Pardue** 99; **Alamy Images/Photofusion Picture Library** 261 (right); **Alamy Images/Trevor Smithers ARPS** 368 (top); **Alamy Images/Janine Wiedel Photolibrary** 76; **Keith Arrell/JTL** 117; **Art Directors and Trip** 17, 26, 28, 104, 170, 180, 211, 238, 271 (left); **Construction Photography** 432; **Corbis** 77; **Corbis/Brand X** 108; **DIY Photo Library** 431; **Jake Fitzjones/Redcover** 368 (bottom); **Getty Images/Stone** 107, 115 **Getty Images/Taxi** 15; **Getty Photodisk** 182; **Harcourt Education Ltd/Ginny Stroud-Lewis** 206, 367, 426; **JTL** 121; **Santon** 121 (bottom), 129 (bottom right); 398; **Kevan Thomas/JTL** 120, 133, 174 (bottom), 134, 151, 153, 248.

Introduction

What is this book?

This book has been designed with you in mind. It has a dual purpose:

1   To lead you through the Level 3 MES Plumbing qualifications, providing background information and technical guidance.

2   To provide a future reference book that you will find useful to dip into long after you have gained your qualification.

The book has been specially produced to assist you in completing your Plumbing Technical Certificate 6129 Scheme and NVQ qualification. The schemes are applicable in England, Wales and Northern Ireland and are designed to set a quality standard for learning and training while attending college or learning centre, or for the mature experienced worker wishing to update qualifications.

The book details the various sections that you will undertake to complete all the Level 3 Units from the Plumbing 6129 Scheme, all of which support the full NVQ. Each chapter concludes with job knowledge tests, which are essential parts of the overall qualification.

The book is also a key reference document for you to refer to in support of your plumbing work. Your tutor will further explain the content of the various sections to you when you attend an approved assessment centre.

The chapters are in a similar format to the units of the City & Guilds assessment, making it easier to complete a section with a successful assessment before moving on to the next.

Qualifications

1   **Technical Certificate:** this is the job knowledge and training part of the qualification for new entrants.

2   **NVQ:** this in full means National Vocational Qualification which is the collection of evidence that you have done the work in the real workplace.

Together they give the full qualification for an operative to be able to work with a company under supervision.

For those of you who have been engaged in the trade for some time you can simply do the Technical Certificate tests to accompany the NVQ workplace evidence.

Either way, this book has the content you require .

Note:

This publication is designed to complement studies towards the Plumbing NVQ qualifications and Technical Certificates. It should be noted that its use and interpretation is primarily for the purpose of training and should not be used as the sole, definitive guide to actual installations.

Specific reference should always be made to current Regulations, British Standards and Codes of Practice to ensure that all installations and work complies with the latest requirements

How this book can help you

There are other key features of this book which are designed to help you make progress and reinforce the learning that has taken place. Such features are:

- **Photographs:** easy-to-follow sequences of key operations.

- **Illustrations:** clear drawings, many in colour, showing essential information about complex components and procedures.

- **Margin notes:** short helpful hints to aid good practice.

- **Tables, bullet points and flowcharts:** easy-to-follow features giving information at a glance.

- **On the job scenarios:** typical things that happen on the job. What would you do?

- **Did you know?** useful information about things you have always wondered about.

- **End of section job knowledge checks:** test yourself to see if you have absorbed all the information. Are you ready for the real test?

- **Glossary:** clear definitions and explanations of those strange words and phrases.

Why choose plumbing?

Plumbing is a very satisfying and rewarding industry to join with a variety of work on a range of systems. There is always a demand for plumbers who have the skills, knowledge and qualifications.

Why this book?

This book is structured to give all the basic information required to gain the qualification and set you on course towards an exciting career in a buoyant industry…Well done for choosing such a good resource!

There is a similar qualification at Level 2 and another book like this to match! You may have already used it, of course, as a starting point.

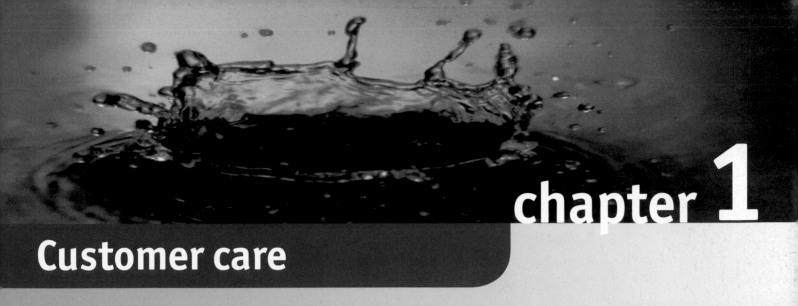

chapter 1

Customer care

OVERVIEW

Customer care plays a very important part in running a successful business and making it grow. This chapter outlines the key approaches that a Level 3 qualified plumbing operative needs to ensure proper customer care. It discusses how to put effective communications in place and looks at how poor communication can cause breakdowns in the working relationship between customer and plumbing company. This chapter will cover:

- **Presenting a quality image**
  - Key customers
  - Basic customer care
  - On-site customer care
  - Customer complaints
  - Improving customer care

- **Customer-service policies**
  - Formal customer-service policy
  - Putting a customer-service policy into practice

Figure 1.1 What is your image?

Presenting a quality image

At the end of this section you should be able to:

- identify good practice when dealing with customers
- state the requirements for taking responsibility for jobs in domestic properties
- detail the acceptable procedures for dealing with customer complaints.

The easiest and most cost-effective way of getting work (and profit) is through repeat business or recommendations from customers. If a business gets a bad reputation, recommendations soon turn negative. With a poor reputation, the business will have to rely on getting work from new customers (using things like advertising). If the number of customers drops because of a negative image, a business could ultimately be forced to cease trading.

It is as important to look after customers as it is to install a good system or do a good service job.

Key customers

As an employee of a business, the first step in customer care is to recognise who your customers are.

A private customer is where your company will have been invited to do a job directly by the customer. Usually they will not have any technical knowledge of the work to be carried out, and will put their trust in you and your company. However, they will have certain expectations of the work. Trying to meet their expectations before they need to state them is beginning to care for the customer.

Your company may do contract work for organisations such as property developers, housing associations or local authorities. Do not assume that the customer is different for this type of work – your customer is the organisation. That organisation will have various staff representatives who take the lead in running the job, such as a site agent or clerk of works. These people can be thought of as your front-line customer. They too will have expectations of what they need from you and your company, which may not be very different from those of a private householder. However, the customer's representative may have an in-depth technical knowledge of the service you are providing.

Finally, you may work for a plumbing company that is part of a larger building services or construction company. In this situation it may be easy to forget customer care issues because the customer feels very distant from the work. Your customer is the parent company representative, who, in these days of competitive contracting, can go outside your company for services if customer care is lacking.

Basic customer care

To give effective customer care as an employee, you need to know that there are things that you can and can't influence on a job. Things you can't influence will be your employer's responsibility.

You are the front line when it comes to customer-care issues, as it will normally be you who is on site and who deals with the customer. There are certain things that are important for good customer care.

Good customer relations are essential to good business

Personal presentation of staff

A customer's opinion of you and your company is often made in the first few seconds of seeing you. A well-presented, tidy and clean image goes a long way towards starting a good professional working relationship with a customer. You might, for example, change your boots or shoes before you go into a fully carpeted house to do some maintenance work. Most customers understand that plumbing can be a dirty job but there is no excuse for gear that goes filthy and unwashed from week to week.

If a company van is part of your contract it should be kept clean and tidy. The customer may see the tidiness of your vehicle as an indication of the way in which you will work on their property.

Communicating with customers

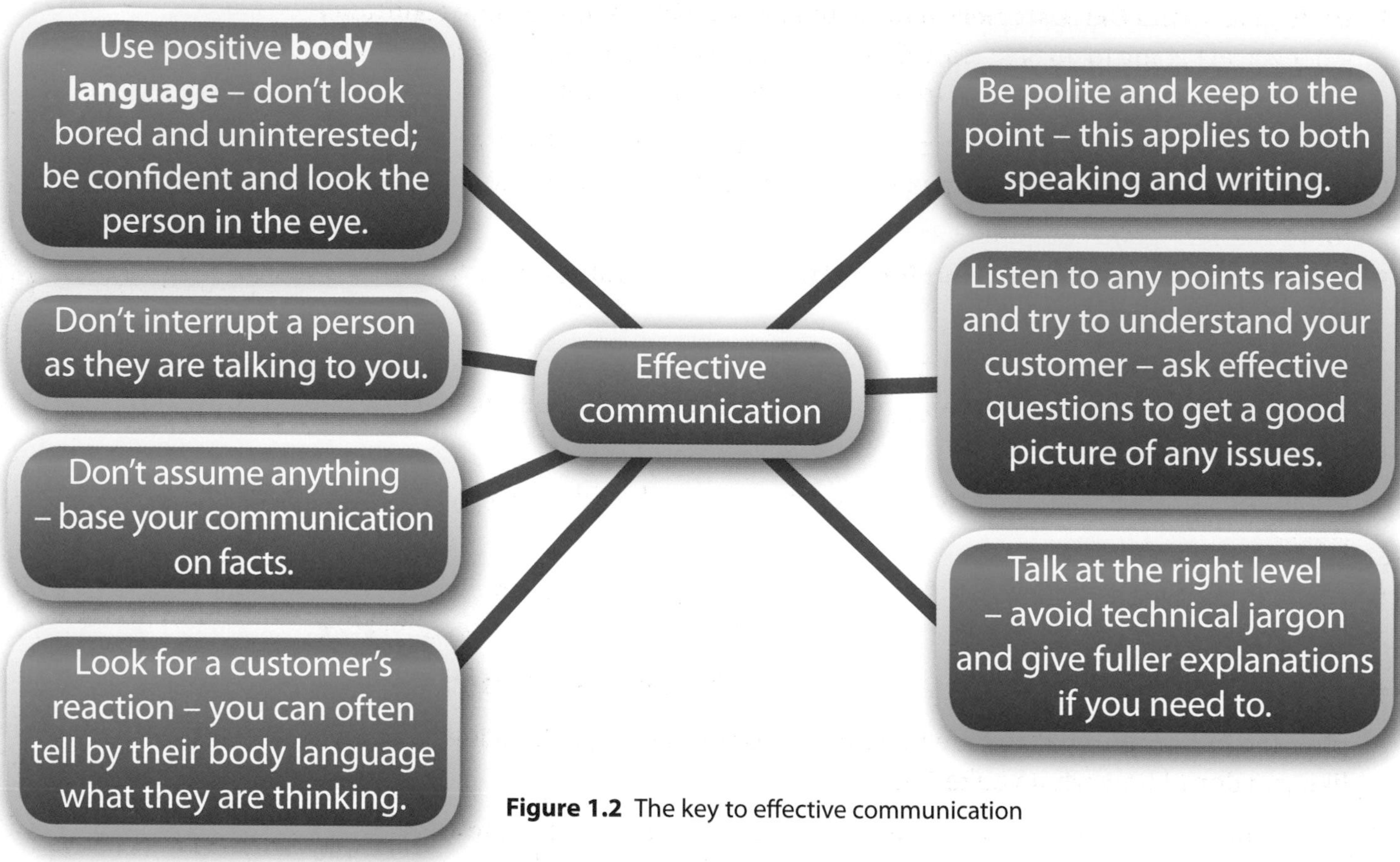

Use positive **body language** – don't look bored and uninterested; be confident and look the person in the eye.
Don't interrupt a person as they are talking to you.
Don't assume anything – base your communication on facts.
Look for a customer's reaction – you can often tell by their body language what they are thinking.
Effective communication
Be polite and keep to the point – this applies to both speaking and writing.
Listen to any points raised and try to understand your customer – ask effective questions to get a good picture of any issues.
Talk at the right level – avoid technical jargon and give fuller explanations if you need to.

Figure 1.2 The key to effective communication

Remember

Don't assume a customer will like the position of a radiator just because you do

Remember

Written communication means there is a permanent record. If you use verbal communication, you may not be able to prove your discussions ever took place

It is important to use the right type of communication for the right circumstance.

Use verbal communication for:	Use written communication for:
confirming the location of components before they're installed	giving a job **quotation**
establishing the initial job requirement	confirming the work (this is done by the customer)
things that need to be carried out before you commence the work, e.g. emptying the airing cupboard	major alterations to the original quotation or **specification**
solving simple problems and straightforward complaints.	confirming quotation/specification alterations (this is done by the customer)
	commissioning reports or **job records** (usually kept by the customer for future use)
	dealing with more complicated problems and complaints.

Table 1.1 Verbal communication should be used for fact-finding and simple issues. Written communication should be for more formal activities

You needn't be the boss to have a customer-caring attitude. You can show you care if you:

- **understand what the customer's real needs are**. A small amount of time confirming what you are going to do on a job before it starts could save a lot of wasted time later. Misunderstandings happen – try to take care of any before you start.

- **respect customer concerns**. Do not put customers down. Try to solve any issues raised. If you can't deal with it make sure it gets passed back to your boss as quickly as possible.

- **show you are committed to your job**. Remember punctuality and tidiness and always be honest.

Remember

A problem ignored usually grows

You are employed to give a service that the customer either can't or doesn't want to do themselves. As a specialist, you are expected to know your job thoroughly. To show your competence you should:

- **know your subject**. But be prepared to seek advice and guidance if you need support.

- **communicate well**. Show you have a clear understanding of your trade, but if there is anything you are unsure of don't bluff your way through – get back to them later.

- **be positive about your subject**. This shows you are in control of what you do.

- **plan the job properly**. If you are making things up as you go along it will seem like you don't know what you're doing.

Showing confidence in yourself is all about having a positive attitude to work.

- **Learn to deal with problems that don't go your way**. Never reach the argument stage and avoid shouting and confrontation. Hand serious problems to your employer to deal with.

- **Develop work routines**. Apply yourself to your work consistently – if a job works well one way, work that way again in future. But be prepared to try better ways of doing the job.

- **Do what you say you're going to do**. Follow things up – nothing annoys customers more than empty promises.

- **Never set unrealistic targets**. If you know you can't meet them when you start, how will you be able to meet them later?

On-site customer care

Starting the job

It is important for you to know the extent of your job responsibility. At Level 3 you are being prepared to take responsibility for running a domestic job under your own initiative. Before you start a job you need to work out with your boss what your responsibilities are when dealing with customers and the types of issues that should be handed over to him/her. A bigger company may have a customer-service policy that you work to.

The level of responsibility that you have may well be different from business to business. For example, in a company specialising in service and maintenance you may be expected to give a price for the job and collect the money before you leave.

With service and maintenance work it is normal for a customer to require a reasonable **estimate** of the cost before work begins. Your company should have procedures in place for you to get the price of any parts. They will also have a standard **cost of labour calculation**.

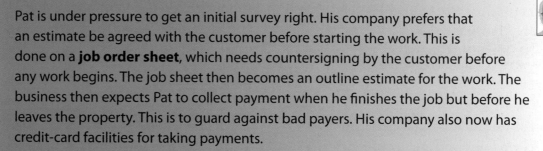

On the job: Pricing the job

Pat is under pressure to get an initial survey right. His company prefers that an estimate be agreed with the customer before starting the work. This is done on a **job order sheet**, which needs countersigning by the customer before any work begins. The job sheet then becomes an outline estimate for the work. The business then expects Pat to collect payment when he finishes the job but before he leaves the property. This is to guard against bad payers. His company also now has credit-card facilities for taking payments.

1   Why is an estimate preferred to a quotation?

2   What are the benefits to the company of this approach?

If your company specialises in installation, then it is likely you will install full systems. When working on these jobs it is important to have clear communication right from the start. You may not have a full system layout plan. Much of the specification and design will have been done in your employer's head. It is not standard with private customers to give a highly detailed job specification. However, details of pipe runs and so on will have been discussed at the initial survey visit. This survey should end with a quotation that outlines limited details of the job (such as the specification and placement of major components).

Paperwork is a part of any plumbing project

Before starting major jobs you need a good briefing from your boss, or the person who did the survey/pricing, on what was originally agreed with the customer. Your supervisor/employer will go through the details of the system layout, telling you things such as where pipe runs are to go, how you are going to tackle the job and what rooms you'll need to go in. It's vital that this job planning is done before the work begins. This is when any misunderstandings about the job can be sorted out. It is much easier to deal with problems before installation or maintenance has taken place.

Did you know?

Some companies have a policy of cleaning up right down to doing the dusting!

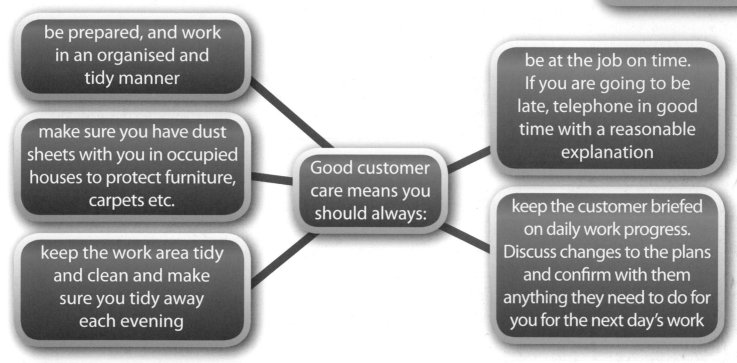

be prepared, and work in an organised and tidy manner

make sure you have dust sheets with you in occupied houses to protect furniture, carpets etc.

keep the work area tidy and clean and make sure you tidy away each evening

Good customer care means you should always:

be at the job on time. If you are going to be late, telephone in good time with a reasonable explanation

keep the customer briefed on daily work progress. Discuss changes to the plans and confirm with them anything they need to do for you for the next day's work

Figure 1.3 Good customer care

treat a property like your own home

help yourself to drinks and biscuits

take lots of calls throughout the day on your mobile phone – it gives the impression there's little work going on

Good customer care means you don't:

break anything and try to hide it – your company has insurance to cover accidents

discuss confidential company information with the customer, such as material costs and profit margins

Figure 1.4 Bad customer care

Remember

The customer may have a no-smoking rule

You may also be responsible for any work colleagues who are on a job with you. You should make sure their work standard is acceptable.

While you probably do not have the responsibility for hiring or firing staff, you do have the responsibility for running the job and you are the front line in customer care. Make sure you have steps in place to deal with any problems that may arise. You should be prepared for minor problems (for instance, the customer wasn't happy because your colleague hadn't tidied up) and more serious problems (like your colleague had a blazing row and swore at the customer). Serious problems should go back to your supervisor/employer for action.

Finishing the job

Try to leave the customer's property as you found it – in a tidy state with all rubbish and materials cleared away. As part of the job-completion stage it is very important you commission the installation properly. Time not spent properly here often results in call-backs, and the customer may feel that you and the company were not good enough.

Before you leave, the customer will need a thorough briefing in how to operate any new installations. You need to complete any commissioning records and leave the manufacturer's instructions for everything you've installed on site.

Identify any future service and maintenance needs (you can use this to get future work for the company).

Some companies may ask you to get a satisfaction statement from the customer as part of the job-completion phase. Others may follow up in a few days' time with a phone call. It is common for employers to check up on their employees' work performance.

Customer complaints

By following good customer-care principles, you shouldn't get too many complaints. However, complaints do occur. Your full job description should tell you how much responsibility you have for dealing with customer complaints. (In some companies you may have a lot, for others you may have very few and just report back for a decision on any action.)

Your customer will appreciate your cleaning up after yourself

Reasons customers complain	
1.	The customer didn't get what they had been promised.
2.	A company employee was rude.
3.	The service was regarded as poor – the customer felt no one was going out of their way.
4.	Nobody listened to the concerns or issues the customer raised.
5.	The company employees had a 'can't do' or 'couldn't care' approach.

Table 1.2 Research has shown that these are the top five reasons why customers complain

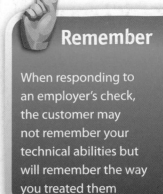

Remember

When responding to an employer's check, the customer may not remember your technical abilities but will remember the way you treated them

The rogue customer

On rare occasions you may have to deal with a rogue customer. A rogue customer may be:

- a customer whose expectations of a job are far higher than the price they will pay (usually due to the quality of materials used). For example, a customer may not accept their new bath because it is not the same as the one in the showroom. They then refuse any further, replacement baths.

- a customer who is determined to get your company to do work without paying for it. For example, a customer may deliberately damage components after installation and then refuse to pay, claiming it was your fault.

Situations like these do happen. As a plumbing employee it is helpful to notice signs that may indicate a rogue customer and report these back to your supervisor/employer. For these circumstances, employers tend to use formal written communication rather than verbal communication. Effective communication needs to take place to attempt to fix the problem, but this should be done by the employer.

Improving customer care

There will always be ways in which a company can improve its customer care. Being in the front line means you may be able to see problems with the care being given. Your company may invite you to suggest ways of improving their customer care.

Many smaller companies will not have procedures for reporting on customer-service issues. If you encounter a situation that you believe could be improved, why not raise it?

Find out

What are some of the problems with customer care that you may encounter on a job?

On the job: A financial dispute

Janie received a phone call one evening from a very good customer whose hot-water cylinder had burst. She visited the site and isolated the hot supply but had no time to do a site inspection.

The next day, she contacted the customer to advise that she could do a site inspection on Saturday. She notified the customer that a replacement cylinder would cost about £170. The customer replied that because there was no hot water he would be away for the weekend but would leave a key next door.

Janie and her apprentice started the work on Saturday and within two hours found the cause of the burst cylinder. Because it was due to a serious fault in the system, Janie went to the next-door neighbour's house to try to contact the customer, but no contact details had been left. As the customer was a very good one, she decided to get on with fixing the system fault. They worked late on Saturday evening and well into Sunday.

Janie called round on Monday morning to make sure the customer was satisfied with the job and to outline the additional work she had done. The customer was extremely grateful. But when Janie sent her invoice for £600 a week later the customer refused to pay and accused her of 'ripping people off'.

1   What should Janie do to get her money paid?

2   Do you think it is likely that Janie will be paid?

3   What could Janie have done to prevent this from happening?

FAQ

If customers are rude to me, why shouldn't I be rude back?

Unfortunately, customers are sometimes rude. This is often because they don't understand what's happened, or why something can't be fixed more quickly or cheaply. Some people think that if they are paying for something they are always right!

Try to avoid conflict by always explaining things as you go along, and if a job is going to take longer or cost more than you first thought, then let the customer know as soon as possible. If people are still rude, then try to keep your cool. If you get angry too, it will just make matters worse. If it's something you really can't deal with, then refer the customer to your boss.

Customer-service policies

At the end of this section you should be able to:

- outline the content of both formal and informal basic customer-service policies
- identify how to put customer-service policies into practice.

There are two approaches that a company can have towards customer care:

- **Formal** – this will usually include a written statement of the company's intentions with regard to customer care. The policy may be supported by a number of procedures outlining how staff are to behave when dealing with different elements of customer care.

- **Informal** – there are often no written procedures, but customer care is still an important part of the business. The employer will usually encourage his/her staff to work to certain standards, even though these are not written down.

The informal approach to customer care is often more simple and tends to be used by smaller businesses; it will have most of the principles of the formal approach.

Find out

What are the benefits of a company having their service procedure in writing?

Formal customer-service policy

The formal customer-service policy is a written statement of standards that the company aims for when dealing with customers. It is made by looking at the work carried out by the business and then working out how best to carry out that work in relation to its customers. A customer-service policy will tend to be a short, straightforward document. It must be communicated effectively to all staff as well as to customers.

A policy will often start with a statement of the company's intentions in relation to customer care. Then it will list aims and targets. It will usually cover:

- general issues, such as staff being courteous etc.
- the timescales over which key activities will be carried out
- the process for customer acceptance of quotations (this could include a commitment to giving an estimate before starting work, how a customer can confirm the work and payment terms)
- the procedure for undertaking small maintenance jobs (day work)
- details of any **cooling-off period** and the refunds policy
- the responsibilities of staff on site and an overview of their general responsibilities while at work
- the complaints procedure (including how to make a complaint, who to make a complaint to, how the complaint will be handled by the company and timescales within which it will be handled)
- guarantee periods (for both materials and labour)
- standards of follow-up service for faults or breakdowns and timescales within which any faults will be looked at.

A customer-care policy is often supported by customer-service procedures. These are based on the key aims of the policy and give more detailed instructions to the staff on how to handle different situations.

It is common for the customer-care policy to be reviewed every year. Some targets may be changed and timescales reduced as the business gets better at meeting customers' needs. The targets become more demanding but this hopefully improves the level of service.

An informal policy will usually include most, if not all, of the principles of a formal policy. It will be verbally communicated to staff and customers. The detail of the policy will then be used by staff as targets to meet customer expectations.

A professional, well-organised company should have a formal customer-care policy

Putting a customer-service policy into practice

Most companies with a formal customer-care policy will give a copy of it to customers when doing a quotation or a work estimate. The details of an informal policy will often be discussed with the customer during the initial survey – and are probably used as a sales approach for getting the work!

All staff in the company need to be aware of what is covered by the policy and how to deal with issues arising from it. Ongoing sessions are often given by management staff or supervisors on what the expectations of the policy are and how staff can meet them. The company may also have procedures in place for checking that staff adhere to the customer-care policy. This will often include contacting customers to see how they feel they have been treated.

FAQs

What do you do if you don't have, or the customer won't sign, a job order sheet?

Your employer may have a policy of working on verbal work instructions. In either case there will be no evidence of an agreement to do the work. This will be a problem if a dispute occurs, especially if it turns into a legal dispute. Because of this, many employers will not continue with service and maintenance work until they get **written confirmation**. Ideally your employer should have a policy that details how it expects you to deal with this situation.

How do you handle a customer complaint?

1   Your company should have procedures for dealing with complaints. This will indentify who handles a complaint (often a senior company official) and how it should be dealt with.

2   It is important that you have full written details of the nature of the complaint, especially if it ends up in legal action – you need records of your actions to back you up.

3   An appointment should be made to see the customer as soon as possible. Try to resolve issues before they get any worse.

4   If the fault is yours or your company's then it must be rectified as soon as possible and at no cost to the customer.

5   If it is not your fault then an explanation of the problem, and why it happened, is needed. A plan of action then needs to be put in place to work out how the problem may be solved. It is not good practice to leave the customer to deal with problems. Try to be of assistance, though additional work will need to be arranged through your boss.

6   Fix the problem as quickly as possible.

7   A few days later a follow-up call should be made to see if everything has been put right.

Knowledge check

1. A written customer-service policy is:
   a. informal
   b. formal
   c. verbal
   d. procedural.

2. During which of the following activities would you use dust sheets?
   a. Replacing a back-boiler.
   b. Replacing bath taps.
   c. Washering a stop valve.
   d. Changing a thermostatic radiator valve head.

3. Which of the following types of clothing are most likely to give a positive image?
   a. Jeans and trainers.
   b. Jacket, shirt and tie.
   c. Clean overalls with company badge and name.
   d. Presenting a positive image is not important.

4. What is a customer-service policy most likely to include?
   a. Work start and finish times.
   b. Cost of a job.
   c. Specification for a job.
   d. Complaints procedure.

5. How can you best determine if a customer is satisfied with your work?
   a. From the volume of complaints received.
   b. By telephone contact after the job.
   c. Whether they write in to complain.
   d. If they come back with repeat business.

6. A customer objects to the proposed position of a WC. What do you do?
   a. Ignore them.
   b. Get on with the installation as the boss wants you on the next job.
   c. Seek advice from your boss on what to do.
   d. Modify the work to suit the customer.

7. A plumber breaks an ornament while working in a house. What should he or she do?
   a. Throw it in the bin and hope no one notices.
   b. Try to repair it.
   c. Report it to the customer.
   d. Leave the site until the boss arrives.

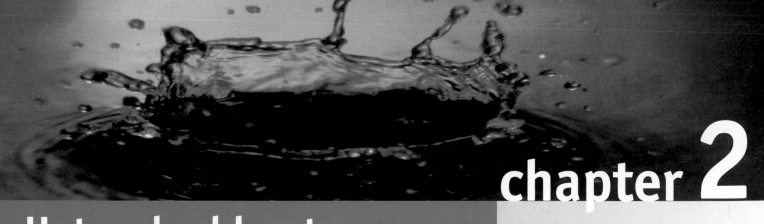

chapter 2

Hot and cold water supply

OVERVIEW

As a plumber the bulk of the work you will carry out will be covered by Regulations, British Standards and Codes of Practice. These include: Building Regulations, Health and Safety legislation, Gas Safety (Installation and Use) Regulations 1998 and many British Standards and Codes of Practice, particularly BS 6700.

In this chapter you will focus on the requirements of the Water Supply (Water Fittings) Regulations 1999, which in the main cover the requirements of cold and hot water supply installations as they apply to the list below. It will also look in more detail at pumped supplies to showers and private water supplies from wells and boreholes.

- Legislation
- Materials and substances in contact with water
- Water fittings
- Design principles and system installation and commissioning
- Prevention of cross connection to unwholesome water
- Backflow prevention and fluid categories
- Cold-water services
- Hot-water services
- Other appliances (WCs, urinals etc.)
- Water for outside use

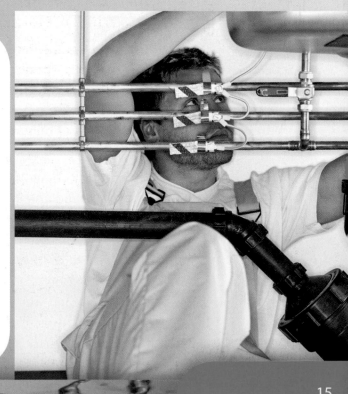

The Regulations themselves can be quite hard to follow. While this text refers to the Regulations, it also includes an interpretation of what they mean. This section also covers some of the content of the following technical certificate units:

- Unit 301 Cold-water Systems
- Unit 302 Domestic Hot-water Systems.

Legislation

At the end of this section you should be able to:

- state the main legislation relevant to the installation and use of cold and hot water services
- explain the main areas of the Water Regulations that relate to the work of the plumber.

This section looks at the legislation that is relevant to the installation and use of cold- and hot-water services. It includes a brief overview of the current Regulations, the Regulations that are relevant to your job, a discussion of some important issues arising from the Regulations, and the European perspective.

Where a particular Regulation is being discussed, the relevant name, Regulation and paragraph is listed beside the text. Copies of the Water Industry Act 1991 and the Water Supply (Water Fittings) Regulations 1999 are published on the website of Her Majesty's Stationery Office: www.hmso.gov.uk/acts/acts1991/Ukpga_19910056_en_1.htm and www.hmso.gov.uk/si/si1999/99114802.htm. Support documents can be found on the DEFRA website: www.defra.gov.uk/environment/water/industry/wsregs99.

Background to the legislation

The control of water-supply installations in England and Wales has been completely revised by the introduction of the Water Supply (Water Fittings) Regulations 1999.

The Secretary of State for the Department of the Environment, Transport and the Regions (DETR) exercised his powers under the Water Industry Act 1991 to enforce Water Regulations that control the installation and use of water fittings, resulting in the making of the Water Supply (Water Fittings) Regulations 1999. The Regulations apply only in England and Wales, but similar requirements have been made by the Scottish Office and Northern Ireland Office.

Previous to the Act, the United Kingdom had a long history of Water Byelaws, which were managed and enforced by local water suppliers. Newly introduced Byelaws expired after ten years, but they would be renewed or updated as necessary. The last renewal of the Water Byelaws was in 1986, before being finally replaced by the new Water Regulations on 1 July 1999.

Abbreviation of relevant regulations and documents are:

WIA – Water Industry Act 1991

WSR – Water Supply (Water Fittings) Regulations 1999

GD – Guidance Document to the WSR

BS – British Standard

The Water Regulations are National Regulations made by the Department of the Environment, Food and Rural Affairs (DEFRA). They apply to all installations in England and Wales that are supplied from a public main by a **water undertaker**. Water undertakers are responsible for the enforcement of the Regulations.

The Regulations have similar aims to the old Byelaws, but are applied differently. They have introduced a significant number of changes in the way water fittings have to be installed and used.

Definition

Water undertaker – a water supply company (this is the technical term used in the Regulations)

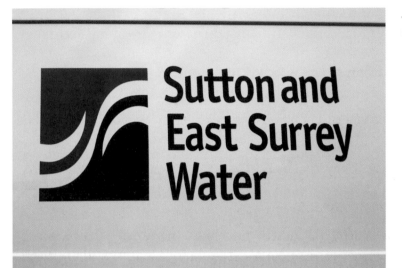

Just one of several water undertakers in the UK

Relevant legislation

The principal legislation governing the creation of the Water Regulations is the Water Industry Act 1991, with sections 73, 74, 75, 84 and 213(2) being particularly relevant. Table 2.1 gives an overview of what these sections discuss.

Did you know?

The Regulations are a means of preventing waste, undue consumption, misuse, contamination and the erroneous measurement of water

Section	Contents of section
73	Offences of contaminating, wasting and misusing water (legal action)
74	Regulations for preventing contamination, waste etc., with respect to water fittings
75	Power to prevent damage, taking steps to prevent contamination, waste etc.
84	Local authority rights of entry etc.
213(2)	Powers to make Regulations

Table 2.1 Relevant sections of the Water Industry Act 1991, which govern the creation of the Water Supply (Water Fittings) Regulations 1999

An extract from Section 74 is reproduced below:

74 Regulations for preventing contamination, waste etc. and with respect to water fittings

(1) The Secretary of State may by Regulations make such provision as he considers appropriate for any of the following purposes, that is to say –

(a) for securing –

  (i) that water in a water main or other pipe of a water undertaker is not contaminated; and

  (ii) that its quality and suitability for particular purposes is not prejudiced, by the return of any substance from any premises to that main or pipe;

(b) for securing that water which is in any pipe connected with any such main or other pipe or which has been supplied to any premises by a water undertaker is not contaminated, and that its quality and suitability for particular purposes is not prejudiced, before it is used;

(c) for preventing the waste, undue consumption and misuse of any water at any time after it has left the pipes of a water undertaker for the purpose of being supplied by that undertaker to any premises; and

(d) for securing that water fittings installed and used by persons to whom water is or is to be supplied by a water undertaker are safe and do not cause or contribute to the erroneous measurement of any water or the reverberation of any pipes.

In other words, Section 74 outlines that the Water Regulations have been made to:

- make sure water isn't contaminated, and that its quality and suitability for a purpose isn't harmed before or after being supplied to a premise

- prevent waste, undue consumption and misuse of water supplied by the undertaker

- make sure that water fittings are safe and don't cause or lead to erroneous measurements or vibration and noise in pipes.

These show that the Regulations have been written to protect the water supply and to protect users against their own actions.

The Water Supply (Water Fittings) Regulations 1999

The Water Supply (Water Fittings) Regulations 1999 is made up of 14 Regulations. These are divided into three parts and are supported by three Schedules. The Schedules should be treated as part of the Regulations. A brief outline of the Regulations is given in Table 2.2.

Regulation	Content of Regulation
Part I	• Gives the date when the Regulations came into force and some interpretations to help understand the Regulations. • Makes statements as to how the Regulations should be applied.
Schedule 1	• Supports Part I. • Outlines the fluid risk categories that may occur within and downstream of a water-supply network. • Needed for the **backflow** requirements of Schedule 2.
Part II	• Defines what is expected of a person(s) installing water fittings. • Outlines how water fittings should be installed and used to prevent waste or contamination. • Puts conditions on materials and fittings that may be used. • Requires contractors to notify the water suppliers of certain installations, and encourages the introduction of **Approved Contractors** Schemes.
Schedule 2	• Supports Regulation 4(3) (Requirements for Water Fittings). • Has 31 separate requirements. • Looks at all aspects of water fittings. • Deals with the practical aspects of Part II Regulations.
Part III	• Deals with the enforcement of the Regulations. • Sets out penalties for breaking the Regulations. • Sets out disputes procedures.
Schedule 3	• Supports Regulation 14. • Lists Byelaws of various water undertakers which have been replaced with the Water Industry (Water Fittings) Regulations 1999.

Table 2.2 Summary of the main parts of the Water Supply (Water Fittings) Regulations 1999

WSR Part I, Reg 1

Important aspects of Part I

Part I helps with the interpretation of the Regulations by defining terms used in them and explaining what they apply to.

Regulation 1 definitions

Regulation 1 makes some important definitions:

- **Approved Contractor**
  - a person who has been approved by the water undertaker for the area where a fitting is installed or used
  - a person who has been certified as an approved contractor by an organisation specified in writing by the regulator.

- **The Regulator**
  - In England the Regulator is the Secretary of State. In Wales it is the National Assembly of Wales.

- **Material Change of Use** a change in how premises are being used, so that after they are changed they are used:
  - as a dwelling
  - as an institution
  - as a public building
  - for storage or use of substances that mix with water to make a category 4 or 5 fluid.

- **Supply Pipe** as much of any service pipe that is not vested in the water undertaker.

Find out

What are some examples of material change of use?

WSR Part I, Reg 2

The Water Supply (Water Fittings) Regulations 1999	
Do apply to:	**Do not apply to:**
Every water fitting installed or used where the water is supplied by the water undertaker.	Water fittings installed or used for any purpose not related to domestic or food production, so long as: • the water is metered • the supply does not exceed 1 month (3, with written consent) and • no water is returned to any pipe vested in a water undertaker. • Water fittings that are not connected to water supplied by a water undertaker. • Lawful installations used before 1 July 1999 (these do not have to be replaced).

Table 2.3 Summary of Regulation 2, which lists the water fittings that are covered by the Water Regulations

Important aspects of Part II

Part II of the Regulations contains information about the quality or standard of water fittings and their installation.

The aims of the Water Industry Act, Section 74(1), are given in detail in Regulation 3. Regulation 3 also states that any work on water fittings is to be carried out in a **workmanlike manner**.

Regulation 5 requires a person who proposes to install certain water fittings to notify the water undertaker, and not to commence installation without the undertaker's consent. The undertaker may withhold consent or grant it on certain conditions.

This requirement does not apply to some fittings that are installed by a contractor who is approved by the undertaker or certified by an organisation specified by the Regulator.

The installation of the following water fittings and systems requires notice to the water undertaker, except those items in bold italics, if carried out by an Approved Contractor:

- the erection of a building or other structure not being a swimming pool or pond
- the extension or alteration of any water system in a building other than a house
- a material change of use of any premises
- the installation of:
  - a bath having a capacity of more than 230 litres
  - *a bidet with ascending spray or flexible hose*
  - a single shower unit, not being a drench shower for Health and Safety reasons, approved by the Regulator
  - a pump or booster pump drawing more than 12 litres a minute
  - a unit that incorporates reverse osmosis
  - a water treatment unit that uses water for regeneration or cleaning
  - *an RPZ valve assembly or other mechanical device for backflow protection from fluid category 4 or 5*
  - a garden watering system, unless designed to be operated by hand
  - any water system laid outside a building less than 750 mm or more than 1350 mm underground.

Where an Approved Contractor installs, alters, connects or disconnects a water fitting, they must provide a certificate to the person who commissions the work stating that it complies with the Regulations.

WSR Part II, Reg 3

WSR Part II, Reg 5

Definition

Workmanlike manner – working in line with appropriate British and European Standards, to a specification approved by the regulator or the water undertaker

Remember

The aim is to prevent waste, misuse, undue consumption or contamination, or false measurement of the water supplied

WSR Part II, Reg 6

Important aspects of Part III

WSR Part III, Regs 7 to 14

A brief description of the Regulations in Part III is given in Table 2.4.

Regulation number	Content of Regulation
7 & 8	Provide for a fine not exceeding level 3 on the standard scale for contravening the Regulations. **It is a defence to show that the work on a water fitting was done by or under the direction of an Approved Contractor, and that the contractor certified that it complied with the Regulations.** This defence is extended to the offences of contaminating, wasting and misusing water under section 73 of the Water Industry Act 1991 (reg 8).
9	Enables water undertakers and local authorities to enter premises to carry out inspections, measurements and tests for the purposes of the Regulations.
10	Requires the water undertaker to enforce the Regulations (this is done by the Regulator or the Director General of Water Services).
11	Enables the Regulator to relax the requirements of the Regulations on the application of the water undertaker.
12	Requires the Regulator to consult water undertakers and organisations representing water users before giving an approval for the purpose of the Regulations, and to publicise approvals.
13	Provides for disputes arising under the Regulations between a water undertaker and a person who has installed or proposes to install a water fitting to be referred to arbitration.
14	Revokes the existing Water Byelaws made by water undertakers under section 17 of the Water Act 1945.

Table 2.4 Summary of regulations covered by Part III of the Water Industry Act 1991

Schedules

WSR Schedules

The Regulations contain three schedules:

- **Schedule 1** on fluid categories defines categories of fluids that may exist both within and downstream of water supply pipework.
- **Schedule 2** is about the requirements for water fittings and contains most of the detailed information plumbers need to know. We shall be covering these in greater detail later in the section.
- **Schedule 3** covers the Byelaws revoked and consists of a list of water companies whose Byelaws have been revoked. (Schedule 3 is not reproduced in this book.)

Approved (Guidance) Document

The Water Regulations contain little technical detail, so DEFRA has produced a Guidance Document which explains in more detail the expectations of the Regulations.

In addition to the Guidance Document, the Water Regulations Advisory Scheme (WRAS) have produced a Water Regulations Guide. This technical colour guide gives practical advice on compliance with the Regulations and how they can be achieved. It includes a copy of the Regulations, the DEFRA formal guidance and further commentary and guidance on the Regulations.

It's not compulsory to follow the WRAS Guide, but if you do you will satisfy the requirements of the Regulations. It is up to you to show proof that your installation complies with these requirements. This means that you must be competent in what you do, and that you need to have the qualifications, knowledge and practical experience to carry out your work in a workmanlike manner.

European legislation

Under the conditions of a Directive of the European Commission (EC), Britain must consult with the EC (and with other European countries) for approval before any technical Regulations are made.

This has led to changes in installation methods and the ways in which water is being used (for example, the use of pressure flushing valves for WCs and urinals).

Did you know?

You can defend yourself against a fine for breaking the Regulations if you can show that the work on a water fitting was done by or under the direction of an approved contractor, and that the contractor certified that the work complied with the Regulations. This defence can also be used for offences of contaminating, wasting and misusing water (see Section 73 of the Water Industry Act 1991)

Materials and substances in contact with water

At the end of this section you should be able to:

- list potential hazards to the water supply
- list the materials approved for pipes and fittings
- describe how to identify approved materials
- state which materials are specifically forbidden for use on wholesome water systems
- describe the problems caused by galvanic action
- state the factors to consider when selecting materials for an installation
- define galvanic action and describe its effects
- state the particular problems posed by lead when connected to copper.

WSR Schedule 2, para 2

Schedule 2 Paragraph 2 of the Regulations makes sure that materials or substances used (alone or in conjunction with any other material) will not cause any contamination that will affect the qualities of drinking water.

Materials that are used in water systems must not contain any substance that could be absorbed into the water, causing the water to become toxic or biologically unhealthy, affecting its colour, taste and smell or making it unfit to drink.

Materials include metallic, non-metallic and plastic substances that are used to make pipes, fittings and appliances, and materials used in jointing and for pipe coatings. Any material or substance in contact with water must not affect its quality.

The Requirements

Below are set out the requirements of Schedule 2, paragraph 2. Remember, this is an extract from the actual Regulations.

2(i) Subject to sub-paragraph (ii) below, no material or substance, either alone or in combination with any other materials or substances, or with the contents of a water fitting of which it forms a part, which causes or is likely to cause contamination of water, shall be used in the construction, installation, renewal, repair or replacement of any water fitting which conveys or receives, or may convey or receive, water supplied for domestic or food production purposes.

(ii) This requirement does not apply to a water fitting downstream of a terminal fitting (tap or valve) supplying wholesome water where:

(a) the use to which the water downstream is put does not require wholesome water; and

(b) a suitable arrangement or device to prevent backflow is installed.

Now take a look at the wording and get an understanding of the requirements.

Paragraph 2 concerns the supply of water for domestic and food production purposes, as well as water fittings that contain wholesome water. 'Water for domestic purposes' means wholesome water that is fit for drinking supplied by a water undertaker for general use. Paragraph 2 is also concerned with fittings containing wholesome water, such as a domestic tap, which could become contaminated, for instance by the attachment of a hosepipe, resulting in it no longer complying with the Regulations. Figure 2.1 shows this situation. If the end of the hose were to lie in a pool of contaminated water it could pose two hazards to the water supply:

- contamination
- backflow risk.

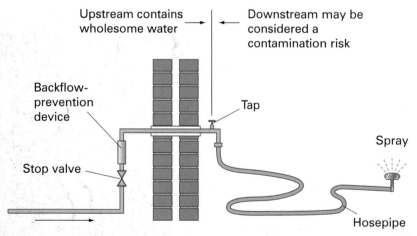

Figure 2.1 Preventing contamination risk

Backflow is the water 'downstream' in the hosepipe finding its way back through the tap into the water supply 'upstream', contaminating the supply. To prevent this, a backflow-prevention device is fitted.

Materials such as hosepipes, overflow pipes, flushing cisterns, feed and expansion cisterns and closed circuits may not be used to supply water for domestic purposes. Because of this, manufacturers don't have to be so careful when they make them. Therefore, where these pipes or components are connected to a pipe that supplies drinking water, the water supply will need to be protected from any possible contamination that the materials may cause.

Find out

Water supplied from a drinking tap inside a dwelling is wholesome water. What happens if you connect a washing machine to the tap? Would the changed situation comply with Regulations?

On the job: Connecting a hosepipe

It has been dry this summer and it looks as if a hosepipe ban is on the cards. Abe is worried about his perfect green lawn and decides it needs a good soaking to try to keep it as green as possible before a ban comes into force. He uses the sprinkler fitting on his hosepipe and attaches it to a cold-water tap.

1   How could this situation lead to contamination of the water?

2   How could this situation be a backflow risk?

3   How must Abe make sure he does not break Regulations with his hosepipe?

Identifying approved materials

Before buying or installing materials you should check to see if they comply with either a British or European Standard specification. However, you must note that British Standard specifications are guidance documents only and have no force of law, unless specifically referred to by the Regulations.

Figure 2.2 Manufacturer's product markings

Manufacturers will clearly mark the product or packaging to show that the product conforms to relevant recognised quality standards.

If you are unsure about materials you should seek advice from your local water undertaker or from the WRAS. Part of their work has involved producing the *Water Fittings and Materials Directory*, which lists all approved fittings. The directory is a very useful guide to those who aim to comply with or enforce Water Regulations.

Did you know?

Products that are certified to a BS or EN (European Number) standard have been put through numerous tests, which are themselves regularly checked under BSI or EC quality-control schemes

WBS teardrop symbol still shown on certain products tested before July 1999.

The WRAS approved product symbol shows that the product has been tested for approval and is listed in the Fittings and Materials Directory.

The CE mark indicates that the product has been tested to EN standards and may legally be placed on the market.

The BSI Kitemark, along with the BS number, shows that the product has been fully tested under the BSI quality testing scheme.

Figure 2.3 Common symbols for approved plumbing materials

Look for these symbols on materials. Remember: the symbol may be on the product or packaging.

Selecting materials

There are two materials referred to in the guidance documents as health hazards, and which are specifically forbidden for use on wholesome water systems. These are:

- lead and any material or substance containing lead
- any bituminous coatings produced from coal tar.

Lead can be a hazardous material having harmful effects, therefore where a defective length requires replacing, you should advise the customer of the need to remove as much of the pipe as possible, and preferably remove all of it.

All materials must be chosen with care, and consideration must be given to the purpose that they're intended for.

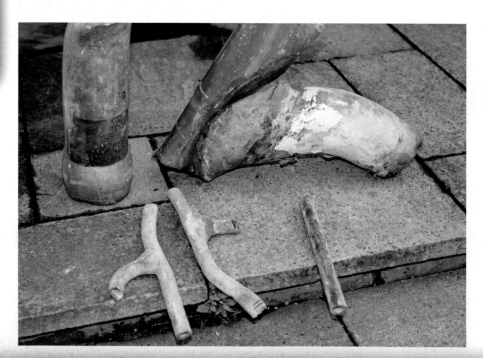

Old lead piping *must* be replaced

Find out

What do you think might be important factors that affect the selection of materials?

FAQ

What is galvanic action?

Galvanic action is the electrolytic corrosion of dissimilar metals in a damp or wet environment. It can cause severe corrosion of metal pipes and fittings.

For instance, where copper and lead pipes are connected, direct galvanic action takes place, causing the lead to be dissolved and taken into solution, contaminating the water.

Indirect galvanic action occurs where particles of one metal are taken into solution, and the resultant solution attacks the pipe or fitting of a different metal downstream of the pipework system.

The rate at which electrolytic corrosion takes place depends on the materials' position in the electro-chemical series (see Table 2.6).

Jointing materials

It is very important that you use materials that comply with a relevant British or European Standard specification, in particular those referred to in the Regulations or listed in the *Water Fittings and Materials Directory* or both (many materials will be covered in both ways).

Type	Used for	Standard
Soft solder	Capillary jointing of copper or copper alloy water fittings, may consist of: • Type 23 or 24 tin/copper alloy • Type 28 or 29 tin/silver alloy	BS EN 29453
Silver solder or silver brazing metal	Capillary jointing of copper or copper alloy, may consist of: • Type AG14 • Type AG20	BS 1845, Table 2
Copper phosphorus brazing filler metals	• Type CP1 to CP6 (all brazing alloys should be free from cadmium)	BS 1845, Table 3
Jointing compounds	Sealing screwed water fittings (other proprietary compounds and hemp aren't acceptable as they may promote microbiological growth)	BS 6956, Part 5
PTFE tape (un-sintered polytetrafluoroethylene tape)	Thread-sealing applications	BS 6974 and BS 6920, Part I

Table 2.5 Jointing materials

Safety tip

Solders used on wholesome water supply installations must indicate that they are 'lead free'.

Hemp or gaskin must not be used for any types of joints under any circumstances

Did you know?

Don't get confused over the terms taps, cocks and valves. They are all devices used for controlling the flow of liquids and gases. The term used for each fitting may vary on a regional basis according to the systems that are used within that area

The WRAS guidance document specifies the types of jointing materials that have been given individual approval for use.

The Regulations prohibit the use of solders containing lead for use on domestic hot and cold water pipework, although its use is permitted on central-heating installations where water isn't drawn off.

Water fittings

At the end of this section you should be able to:

- define what is meant by a water fitting
- list jointing materials for use on water fittings
- list the forms of stress supply pipes are subject to
- list appropriate frost-protection measures for a variety of fittings and locations
- state the requirements for underground pipes and fittings, cold surfaces, roof spaces and external situations
- list practical measures to prevent contamination of water-storage cisterns
- describe devices to be used when water-storage cisterns become flooded
- describe 'permeation' and its effects on water supplies
- state the specifications for the support of pipes and cisterns.

Water fittings in the Regulations are defined as 'meters, pipes (other than water mains), taps, cocks, valves, ferrules, cisterns, baths, water closets, soil pans and other similar apparatus used in connection with the supply and use of water'.

It's important to consider the way in which we use water fittings, so that the contamination or wasting of water will not occur as a result of the fittings installed. You are responsible for the fittings you use in your installation work.

Water meters can help reduce the wasting of water

WIA Schedule 2, paras 3 to 7

The Requirements

Paragraph 3 states that every water fitting shall:

(a) be immune to or protected from corrosion by galvanic action or by any other process which is likely to result in contamination or waste of water; and

(b) be constructed of materials of such strength and thickness as to resist damage from any external load, vibration, stress or settlement, pressure surges, or temperature fluctuation to which it is likely to be subjected.

Paragraph 4 states that every water fitting shall:

(a) be watertight;

(b) be so constructed and installed as to:

   (i) prevent ingress by contaminants, and

   (ii) inhibit damage by freezing or any other cause;

(c) be so installed as to minimise the risk of permeation by, or deterioration from, contact with any substance which may cause contamination; and

(d) be adequately supported.

Paragraph 5 states that:

Every water fitting shall be capable of withstanding an internal water pressure not less than 1½ times the maximum pressure to which that fitting is designed to be subjected in operation.

Paragraph 6 states:

No water fitting shall be installed, connected or used which is likely to have a detrimental effect on the quality or pressure of water in a water main or other pipe of a water undertaker.

Paragraph 7 states:

(1) No water fitting shall be embedded in any wall or solid floor.

WIA Part III, Reg 93(1)

(2) No fitting which is designed to be operated or maintained, whether manually or electronically, or which consists of a joint, shall be a concealed water fitting.

(3) Any concealed water fitting or mechanical backflow prevention device, not being a terminal fitting, shall be made of gunmetal or another material resistant to dezincification.

(4) Any water fitting laid below ground level shall have a depth of cover sufficient to prevent water freezing in the fitting.

(5) In this paragraph 'concealed water fitting' means a water fitting which:

   (a) is installed below ground;

   (b) passes through or under any wall, footing or foundation;

   (c) is enclosed in any chase or duct; or

   (d) is in any other position which is inaccessible or renders access difficult.

Water fittings need to be immune to, or protected from, galvanic action.

WSR Schedule 2, para 3(a)

The further apart metals are in the electro-chemical series, the more likely it is that corrosion will take place. If two dissimilar metals are placed in contact with each other, the metal at the lower base end of the scale will be the one to corrode.

A typical example of this corrosion can be seen in galvanised steel cisterns that are connected to a copper pipework system. From Table 2.6 you will see that copper and zinc are some distance apart, and zinc is the metal that will corrode.

Metal	Chemical symbol	Electrode potential (volts)	
Silver	Ag	0.80 +	cathode
Copper	Cu	0.35 +	noble end
Lead	Pb	0.12 –	anodic
Tin	Sn	0.14 –	base end
Nickel	Ni	0.23 –	
Iron	Fe	0.44 –	
Chromium	Cr	0.56 –	
Zinc	Zn	0.76 –	
Aluminium	Al	1.00 –	
Magnesium	Mg	2.00 –	
Sodium	Na	2.71 –	

Table 2.6 The electro-chemical series

Another example of galvanic corrosion occurs when connecting copper pipe directly into lead pipe. The lead, being at the lower base end of the scale, will corrode, resulting in it being taken into solution, contaminating the water. The lead will also be weakened by the corrosion, eventually resulting in leakage.

Sometimes **cathodic protection** can provide protection against galvanic action. A **sacrificial anode** can be put inside hot-water vessels, cisterns and tanks, and on pipelines. The anode will corrode instead of the fitting that it protects.

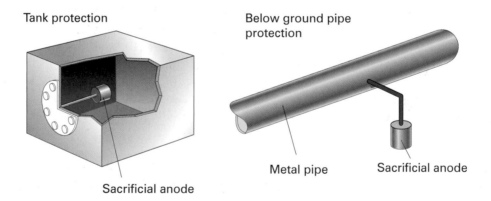

Tank protection

Below ground pipe protection

Metal pipe

Sacrificial anode

Sacrificial anode

Figure 2.4 Tank protection

WSR Schedule 2, para 3(b)

Water fittings are required to resist damage from external load, vibration, stress or settlement pressure surges or temperature fluctuations to which they're likely to be subjected.

Most water fittings, supply pipes, distributing pipes and discharge pipes are subject to the same forms of stress, whether by expansion and contraction, settlement and vibration or temperature change, e.g. cold and frost. All of these factors can put undue stress on materials, and when in service they must be capable of withstanding the stress, remaining watertight throughout. So, when selecting and installing water fittings you must ensure that any stress is kept to a minimum.

A typical example is where a pipe passes through a wall: it should be sleeved or ducted so as to protect it from stress caused by:

- expansion of the pipe
- movement of the wall.

Water fittings must be watertight and suitable for the working pressure and temperatures likely to be encountered within the installation. Basically, this means ensuring that water fittings are jointed using the proper techniques and remain watertight during service, carrying out all the correct commissioning checks, ensuring that all fittings are sound, and testing in accordance with the required standard.

Also, every water fitting must be capable of withstanding an internal water pressure not less than 1½ times the maximum pressure for which that fitting was designed.

Some of this requirement is the responsibility of the manufacturer of the water fittings, but you as the plumber have a responsibility to see that fittings are used and installed correctly and don't exceed the limits laid down in the manufacturer's instructions.

A typical example of the requirement is when installing or replacing copper storage vessels. The vessels are graded to suit various pressure conditions, e.g. a grade 3 copper cylinder to BS 1566 is suitable for a pressure of up to 10 metres head. If the vessel were to be fitted to an installation with a head greater than 10 metres, it would not be adequate for its purpose, invalidating the manufacturer's guarantee.

Ingress of contaminants

Every water fitting must be constructed and installed so as to prevent ingress by contaminants, such as chemicals, vermin, insects or other things. The most vulnerable fittings are those incorporating air gaps, where the water itself is open to the atmosphere. Examples of these include water-storage cisterns.

Water-storage cisterns are required to have rigid, close-fitting and securely fixed covers and to be fitted with screened overflows and vents to prevent insects and dust from entering the cistern.

Water-storage cisterns must not be installed in positions where they're likely to become flooded by rainwater, ground water or by the cistern overflowing. In instances where a cistern has to be installed in such a location, it must be installed in a watertight enclosure. This could be a concrete chamber or basement that's watertight. Installations of this type must additionally be fitted with an electric sump pump, to remove any unwanted water that collects at the base of the watertight enclosure. Audible or visual devices must also be fitted to show if the cistern reaches overflowing level, and to warn that water is accumulating at the base of the chamber.

Other components such as cisterns and cylinders should be installed so that they don't become contaminated by the build-up of sediment; and if necessary, filters should be fitted to prevent sediment passing through the pipework system.

Remember

It's very important when selecting water fittings that you ensure that they comply with the Regulations. The water fitting must be adequate for its purpose, complying with the appropriate standard, e.g. BS or EN specification.

WSR Schedule 2, para 4(a)

WSR Schedule 2, para 5

WSR Schedule 2, para 4(b)

WSR Schedule 2, para 16

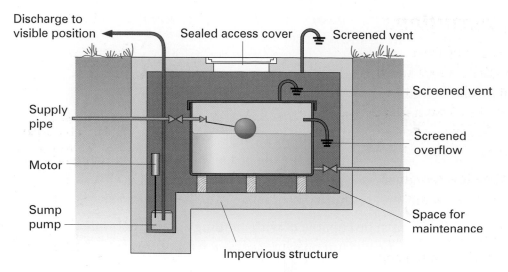

Figure 2.5 Design requirements for a watertight cistern enclosure

Regular cleansing and maintenance of the installation will help stop sediment accumulating: sediment can provide an ideal environment for bacteriological growth (e.g. legionella), particularly where the water could be warmed up.

Water fittings such as drain valves must not be installed in positions where they're likely to become submerged in any type of flooding – for example, below ground, below floors, or in sump chambers.

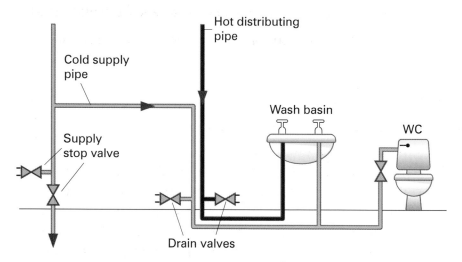

Figure 2.6 Drain valve positioning

This installation shows drain valves provided for a solid floor duct installation. The valves are sited so that they do not present a contamination risk (i.e. not in the duct). They are just above ground level, so they will not become submerged and are accessible for maintenance activities.

Frost precautions

When considering frost protection the obvious solution is thermal insulation, but this isn't always the answer, and sometimes in severe weather the insulation barrier can fail.

When putting in a new installation, you should consider whether to avoid installing pipes and fittings in areas that are known to be difficult to keep warm, such as:

- draughty areas within the building, e.g. doors, windows, ventilation openings, cold roof spaces and beneath suspended ventilated floors

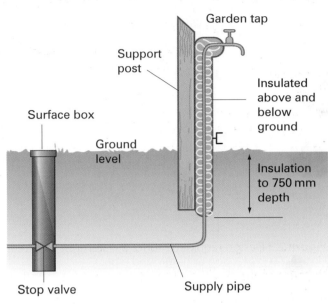

Figure 2.7 Frost precautions for external pipework installations

- any position external of buildings and above ground

- unheated parts of the building, e.g. cellars, roof spaces, garages and outhouses

- cold surfaces, e.g. where there is direct contact with external walls, ducts or chases within outside walls.

Underground pipes and fittings

You can't always avoid these areas, so you have to consider other protective measures. Any water fitting laid below ground level shall have a depth of cover sufficient to prevent water freezing in the fitting. The required depth of cover for frost protection should not be less than 750 mm. (This depth also provides protection from other causes of damage, e.g. ground movement, vibration due to traffic or garden digging.) In addition to this requirement the fittings should not be more than 1350 mm below ground: a depth greater than this makes accessibility awkward for maintenance or repair.

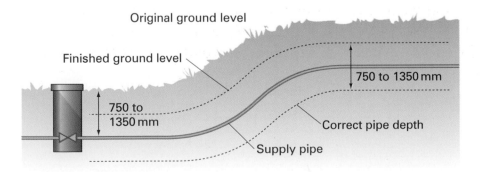

Figure 2.8 Pipe depth underground

WSR Schedule 2, para 4(b)

BS 6700

Definition

Thermal envelope – the area of a building that's enclosed within the walls, floor and roof and which is thermally insulated in accordance with the requirements of the Building Regulations. Examples of locations outside the thermal envelope are an unheated utility room with a washing machine, a cellar with a sink or a tap in a veranda or conservatory

Find out

What measures do you think could be taken to protect against freezing?

WSR Schedule 2, para 7(4)

WSR Reg 5

Where pipes are being laid in uneven ground, care must be taken to ensure that the minimum 750 mm depth of cover is maintained over the total length of pipe. Underground stop valves must not be brought up to a higher level merely for ease of access.

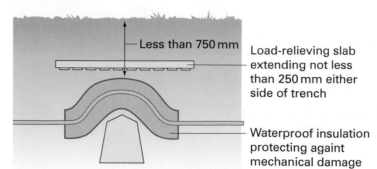

In situations where it is difficult or impossible to maintain the current depths of cover, you are required to notify the water undertaker of the work.

Figure 2.9 Protecting underground pipes

External above ground pipework

Where a pipe rises from below ground to fittings above ground, the pipe should be fitted with adequate waterproof insulation extending down into the ground to a depth of 750 mm.

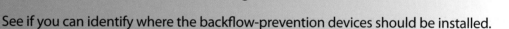

On the job: Backflow prevention

These illustrations show a fitting being fed from a service within a building and a fitting being fed from a service below ground.

1   See if you can identify where the backflow-prevention devices should be installed.

2   What can you say about the insulation requirements for both taps?

Figure 2.10

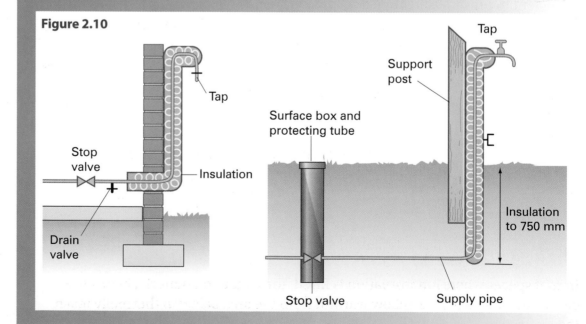

Supply pipes entering buildings

Where a supply pipe enters a building it should:

- be laid in a duct – this will permit access for renewal or repair, and provide some degree of insulation

- pass through or under the building foundations at a minimum depth of 750 mm

- rise vertically into the building at least 750 mm from the external wall. If less than 750 mm the pipe must be insulated.

Where the pipe rises vertically through a suspended ventilated floor, the pipe must be insulated from floor level to a depth of 750 mm below ground.

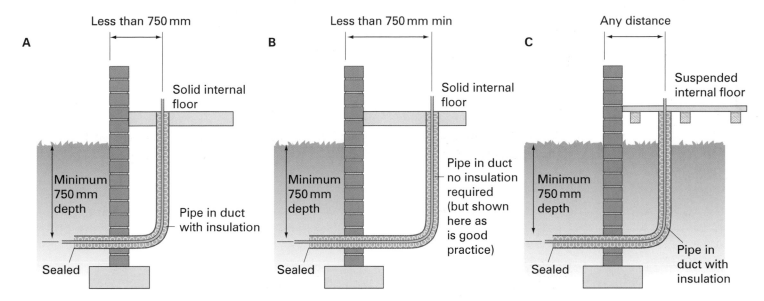

Figure 2.11 Examples of pipes in ducts

Cold surfaces and unheated spaces

Pipes and fittings must be spaced off external walls by using spacing clips. If saddle clips are used a baseboard must be fitted to the wall, to provide insulation protection from the cold surface.

Pipes and fittings must not be chased or ducted into external walls, or be installed in the cavity of an external wall. Where a pipe is installed within the internal leaf of an external wall, frost protection may be required. All pipes and water fittings within unheated spaces such as outbuildings and garages must be adequately insulated to protect from frost. If the pipe rises from below the floor, the insulation must be taken down to a depth of 750 mm.

Water fittings in roof spaces

In roof spaces where loft insulation is fitted, the insulation material below the cistern must be left out to allow warmth from the area below to thermally reach the cistern.

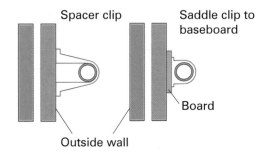

Figure 2.12 Fixing to cold surfaces

Remember

All water fittings located in roof spaces (this includes cisterns) must be insulated

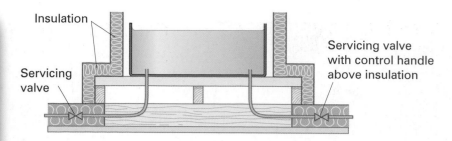

Figure 2.13 Installation details for water fittings in roof spaces

Ideally, and where possible, insulation pipework should be kept below the insulation material. If the cistern is sited in a location where a high temperature could occur from either heat gain or a warm roof space, there will be no requirement to insulate for frost protection. But insulation should be provided to the cistern and pipework, to stop the water from becoming unduly warm. This will restrict the growth of bacteria such as Legionella. Ideally, stored wholesome water should be kept at a temperature below 20°C.

External water fittings

External water fittings are extremely vulnerable to frost, wind conditions and wind chill.

Avoid installing water fittings and pipework externally to a building. Where external installations are unavoidable, insulation is essential, and the insulation must be suitable for external use, making provision for it to be waterproofed on completion.

Thermal insulation

The common understanding of insulation is that it prevents water pipes from freezing and is provided to keep the cold out. However, this is only part of the story: insulation is also provided to retain the heat energy in the water pipe. The British Standard provides several tables of calculated insulation thickness for pipework in various situations.

The document must be read with care, as some of the figures aren't economically viable. The guidance to Schedule 2 looks at the background criteria used for calculating insulation thickness and categorises them into two conditions:

- normal
- extreme.

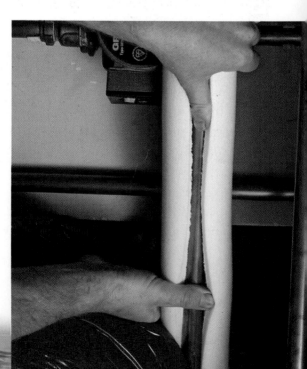

Insulation being fitted on to a pipe

	Normal conditions	Extreme conditions
Applies to:	Domestic accommodation and other types of premises where habitable rooms are normally heated for up to 12 hours each day (even though they are within the envelope of the heated accommodation, water fittings in unheated rooms need to be protected against freezing)	Water fittings: • installed externally to a building • inside any building or part of a building that is unheated, or only marginally heated, for fewer than 12 hours each day • inside a building but located outside the thermal envelope
Examples:	Water fittings in: • cloakrooms • store rooms • utility rooms • roof spaces located below the ceiling insulation etc.	Water fittings: • located under suspended ground floors • above the level of ceiling insulation in a roof space • in a communal staircase or corridor on flats • in domestic garages or other buildings • external above ground installations

Table 2.7 Conditions for calculating insulation thickness

The recommended minimum commercial and practical thickness of insulation for protection against freezing in unheated parts of a normally occupied building when the heating is turned off in the remainder of the building, such as overnight, is shown in Table 2.8.

Thermal insulation of insulation material at 0°C					
Watts per metre degrees Kelvin	0.02 W/(m.K)	0.025 W/(m.K)	0.03 W/(m.K)	0.035 W/(m.K)	0.04 W/(m.K)
15	20	30	25	25	32
22	15	15	19	19	25
28	15	15	13	19	22
25	15	15	9	9	13
42 & over	15	15	9	9	9

Table 2.8 Insulation thickness in mm

GD Figure G4:11

Safety tip

An absence of heating for more than 24 hours within a building is not considered normal

Drain valves

Drain valves to BS 2879 should be installed on all low points in the system to enable the entire system to be emptied, except for any position that may cause a contamination risk.

Ideally, when a dwelling is left empty for a considerable period of time, especially through the winter months, it should be completely drained down, eliminating the threat of frost damage.

The best form of frost protection is to maintain temperature levels in rooms that contain water fittings. This can be achieved by space heating or localised heating of the building, or heating the direct area where the water fittings are contained. Other acceptable methods are by the use of self-regulating **trace heating** to BS 6315, and this must also include a nominal thickness of insulation to the pipe, protecting against failure of the trace heating.

Permeation or deterioration

WSR Schedule 2, para 4(c)

Water fittings that minimise the risk of permeation, or deterioration, by contact with any substances that cause contamination should be installed. Some plastic pipes and fittings may be damaged when in contact with certain fluids or gases such as natural gas. Certain plastic pipes and fittings when exposed to petrol and oil may be penetrated by the substance, softening and weakening the material and resulting in contamination.

Gases and fumes have been known to permeate through the walls of plastic pipes, introducing a smell or taste to the water, although showing no apparent damage to the pipe.

Permeation can contaminate a water supply. A prime example is a water supply to a petrol filling station, which should not be of plastic material, unless protected from risk of petrol spillage; alternatively a suitable material should be used.

Where it is known that water fittings and pipework are at risk from contamination by oil, petrol (hydrocarbons) or other substances, alternative materials that are resistant to permeation must be used.

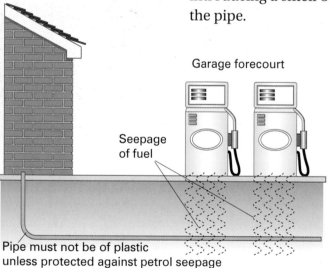

Garage forecourt

Seepage of fuel

Pipe must not be of plastic unless protected against petrol seepage

Figure 2.14 Permeation

Utility services

Figure 2.15 shows recommendations of the National Joint Utilities Group (NJUG) Report Number 6 for relative positions of **utility services** in a service trench. This should provide a 360° radius space between the gas and water services to prevent permeation of gas into the water supply. Additional protective measures are needed if these dimensions and depths can't be achieved.

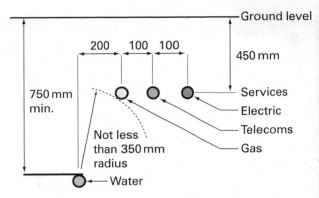

Figure 2.15 Recommended service locations in trenches

Providing adequate support

Every water fitting should be adequately supported. This means pipes, joints, valves, cisterns and cylinders (all components that carry water from the water undertaker's main).

Pipes must be securely fixed using appropriate clips or brackets and should be spaced at the recommended intervals set out in BS 6700.

Pipes that are to be insulated on completion of works should be secured on clips or brackets that allow adequate space between the pipe and supporting wall surface, for the insulation to be installed properly.

WSR Schedule 2, para 4(d)

Allowance must also be made for the expansion and contraction of pipes by forming expansion loops or introducing changes of direction in pipe systems, where there are long straight runs and few bends or offsets. This will be dependent on the pipe material and joints used.

Inadequate fixings to pipework may lead to damage, particularly from those manufactured from plastic materials, resulting in the pipe sagging. They may also encourage noise and vibration in the pipework system, causing nuisance and risk of damage to the pipes. Pipes connected to cisterns or cylinders need adequate support as their weight could cause stress on joints. This is very important where connections are made to plastic cisterns, and this requirement includes overflow and warning pipes. These are rarely supported adequately.

Safety tip

Pipes that are installed in ducts or roof spaces are often inadequately supported and should receive special attention

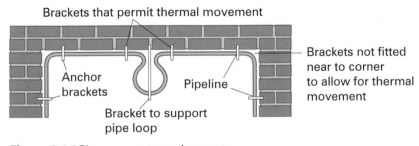

Figure 2.16 Pipe-support requirements

Pipe size		Copper (m)		Low Carbon Steel (LCS) (m)		Plastic (m)	
mm	in	horizontal	vertical	horizontal	vertical	horizontal	vertical
15	½	1.2	1.8	1.8	2.4	0.6	1.2
22	¾	1.8	2.4	2.4	3.0	0.7	1.4
28	1	1.8	2.4	2.4	3.0	0.8	1.5
35	1¼	2.4	3.0	2.7	3.0	0.8	1.7
42	1½	2.4	3.0	3.0	3.6	0.9	1.8
54	2	2.7	3.0	3.0	3.6	1.0	2.1

Table 2.9 Minimum clip-spacing requirements for steel and copper pipework

Water-storage cisterns

These carry a great deal of weight and require adequate support. When sited in roof spaces the weight must be spread across as many ceiling rafters as possible, or alternatively over a load-bearing wall.

The requirements for cisterns are:

Flexible cisterns	Rigid cisterns
• Continuous support is required over the whole base area of the cistern. • Cisterns must not be supported on chipboard or composite board that may be weakened by dampness. • No connections should be made to the base of plastic cisterns.	• The weight should be distributed over two or more timber joists (depending on the size of the cistern). • Continuous support of the base is not required for galvanised steel cisterns.

Table 2.10 The requirements for flexible and rigid cisterns

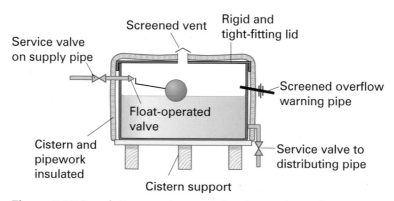

Figure 2.17 Regulation requirements for cisterns in roof spaces

Water quality

Paragraph 6 of Schedule 2 states that no water fittings shall be installed, connected or used, which are likely to have a detrimental effect on the quality or pressure of water. All water supplied by a water undertaker is required to be wholesome, and it's your responsibility as a plumber to make sure that anything you do will not allow water that has been used in any way to return back into the main.

WSR Schedule 2, para 6

Water pressure

Most distribution main systems operated by water suppliers experience pressure increases during periods of very low demand such as during the night. The amount of increase in pressure will vary depending on the size of the main, the number of consumers supplied and the sophistication of the plumbing and pressure-control systems. Pressure variations can also occur in pumped water-supply installations in buildings where booster pumps are running constantly.

Two important factors with booster-pump installations are:

- Water undertakers will insist on a break cistern being fitted, to separate the booster pump from the incoming supply. This will prevent excessive quantities of water from being drawn off, noticeably reducing the pressure in the water main.

- The plumber must notify the water undertaker of any pump or booster installation that draws off more than 12 litres per minute from the supply pipe.

Notifying the water undertaker of the installation gives them the opportunity to ensure that it will not have an adverse effect on the water main pressure.

Safety tip

All water fittings should be capable of withstanding an internal water pressure of not less than 1.5 times the maximum operating pressure

Concealed water fittings

Paragraph 7 of Schedule 2 states the requirements for concealed water fittings in sub-paragraphs 1 to 4:

WSR Schedule 2, para 7

(1) No water shall be embedded in any wall or solid floor.

This means that chasing pipes into walls or placing them in floor screeds to be encased isn't permitted. Pipes must be installed so that they're accessible and are protected from the corrosive effects of cement and plaster, making allowances for the expansion and contraction of the material.

(2) No fitting which is designed to be operated or maintained, whether manually or electronically, or which consists of a joint, shall be a concealed water fitting.

(3) Any concealed water fitting or mechanical backflow prevention device not being a terminal fitting shall be made of gunmetal, or another material resistant to dezincification, such as DZR brass.

(4) Any water fitting laid below ground level shall have a depth of cover sufficient to prevent water freezing in the fitting (at least 750 mm of cover).

Sub-paragraph (5) states what is meant by a concealed water fitting:

(a) a fitting installed below ground;

(b) a fitting passing through or under any wall, footing or foundation;

(c) a fitting enclosed in any chase or duct;

(d) a fitting in any position which is inaccessible or renders access difficult.

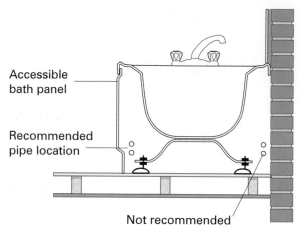

Accessible bath panel

Recommended pipe location

Not recommended

Figure 2.18 Regulation requirements for access to fittings behind a bath panel

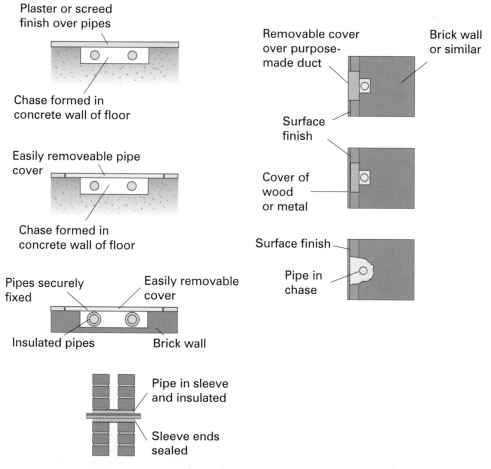

Plaster or screed finish over pipes

Chase formed in concrete wall of floor

Easily removeable pipe cover

Chase formed in concrete wall of floor

Pipes securely fixed

Easily removable cover

Insulated pipes

Brick wall

Removable cover over purpose-made duct

Brick wall or similar

Surface finish

Cover of wood or metal

Surface finish

Pipe in chase

Pipe in sleeve and insulated

Sleeve ends sealed

Figure 2.19 Requirements for pipes in walls and floors

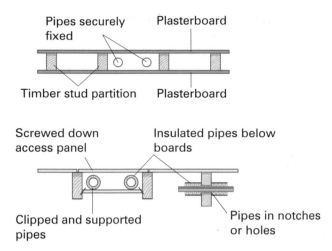

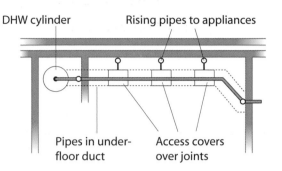

Figure 2.20 Requirements for pipes in suspended timber floors

Figure 2.21 Requirements for pipes in under-floor ducts

Design principles and system installation and commissioning

At the end of this section you should be able to:

- list the requirements for concealed water fittings
- make decisions affecting the choice of systems and components for domestic applications
- explain how to avoid damage when installing pipework and fittings
- explain the requirements for keeping water installations below 25°C
- state the requirements for installation fittings, taps, valves etc.
- describe commissioning procedures
- describe flushing, testing, disinfecting and disposal procedures.

Paragraphs 8 to 13 of Schedule 2 cover an assortment of topics relating to water systems design and installation. Paragraph 8 states: 'No water fitting shall be installed in such a position, or pass through such surroundings, that it is likely to cause contamination or damage to the material of the fitting or the contamination of water supplied by the undertaker.'

Basically, the requirement looks at preventing contamination of water fittings and the prevention of damage to pipes and fittings which could lead to the waste of water through leakage.

Paragraph 9 states: 'Any pipe supplying cold water for domestic purposes to any tap shall be installed that, so far as is reasonably practicable, the water is not warmed above 25°C.' This is to prevent contamination from microbiological growth.

Millions of litres of water are lost each year through leakage

Paragraph 10(1) states: 'Every supply pipe or distributing pipe providing water to separate premises shall be fitted with a stop valve, conveniently located to enable the supply to those premises to be shut off without shutting off the supply to any other premises.'

Paragraph 10(2) states: 'Where a supply pipe or distributing pipe provides water in common to two or more premises, it shall be fitted with a stop valve to which each occupier of those premises has access.'

Paragraph 11 states: 'Water supply systems shall be capable of being drained down and be fitted with an adequate number of servicing valves and drain taps so as to minimise the discharge of water when water fittings are maintained or replaced. A sufficient number of stop valves shall be installed for isolating parts of the pipework.'

Paragraph 12(1) states: 'The water system shall be capable of withstanding an internal water pressure not less than 1½ times the maximum pressure to which the installation or relevant part is designed to be subjected in operation ("the test pressure").'

Under paragraph 12(2) this requirement is deemed to be satisfied:

(a) in the case of a water system that does not include a pipe made of plastic, where –

   (i) the whole system is subjected to the test pressure by pumping, after which the test continues for one hour without further pumping;

   (ii) the pressure in the system is maintained for one hour; and

   (iii) there is no visible leakage throughout the test;

(b) in any other case, where either of the following tests is satisfied.

Test A

The whole system is subjected to the test pressure by pumping for 30 minutes, after which the pressure is noted and the test continues without further pumping.

 (i) the pressure is reduced to one third of the test pressure

 (ii) the pressure does not drop over the following 90 minutes

(iii) there is no visible leakage throughout the test.

Test B

The whole system is subjected to the test pressure by pumping for 30 minutes, after which the pressure is noted and the test continues without further pumping.

 (i) the drop in pressure is less than 0.6 Bar (60kPa) after a further 30 minutes, and less than 0.2 Bar after the next 120 minutes

 (ii) there is no visible leakage throughout the test.

Paragraph 13 states: 'Every water system shall be tested, flushed and where necessary disinfected before it is first used.'

Find out

See the Water Regulations guidance notes for a fuller description of these tests

Requirements of the Regulations

Pipes and fittings should be arranged to avoid any position where the water could become contaminated and damage to the material could occur. The guidance document discusses places considered to be contaminated, where water fittings should not be laid or pass through. These places include:

- foul soil

- refuse or a refuse chute

- an ash pit

- a sewer drain, cesspool, manhole or inspection chamber.

This prohibition is made regardless of any protection to the pipe or other fittings.

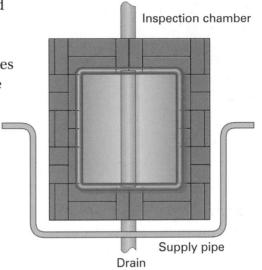

WSR Schedule 2, para 8

Figure 2.22 The correct method of laying pipes in the vicinity of inspection chambers

Installing to avoid damage

It is good practice to install pipework in a manner that will avoid the possibility of damage.

How can I avoid damaging installation pipework?

- Below ground pipework must be at a minimum depth of 750 mm to avoid damage from digging, vehicular traffic and frost.

- Pipes passing through walls should be sleeved and if below ground must be protected from ground movement or settlement.

- Pipes passing through walls or embedded in walls or floors should be wrapped or protected from any corrosive effects of cement, concrete and dampness.

- Pipes in duct or channels must be arranged so that they are accessible or can be withdrawn for repair.

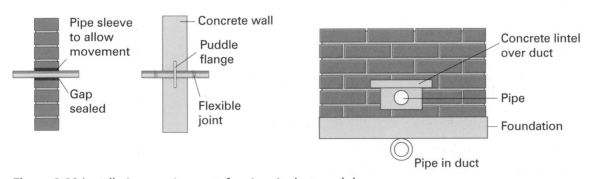

Figure 2.23 Installation requirements for pipes in ducts and sleeves

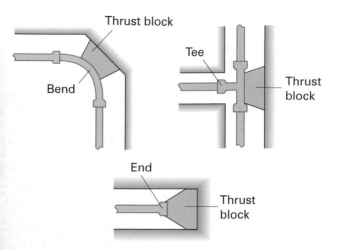

Figure 2.24 Examples of thrust blocks used to anchor pipes

- Pipes constructed from plastic materials should not be installed in areas where they are likely to come into contact with gases, petrol or oil, resulting in permeation and contamination of the water.

- Pipes and fittings, including appliances, must be adequately supported to prevent sagging, undue noise and vibration that may result in damage, ensuring the thermal movement of the fitting isn't restricted. Pipes installed below ground should be anchored at bends and tees, to prevent them pushing apart under pressure.

- Pipes must be arranged to avoid airlocks, allowing air to escape when filling. Pipes ideally should rise to vents and fall to outlets.

Keeping cold-water installations below 25°C

The requirement of paragraph 9 is quite straightforward and easy to understand: it requires any pipe supplying cold water for domestic purposes to be so installed that, so far as is reasonably practicable, the water isn't warmed above 25°C.

The temperature of the water in a main supplied from a water supplier can vary from approximately 4°C in winter to 25°C during warmer summer months. The wide variation in temperature will also depend on the type of water source – ground water sources average 10°C throughout the year. Mains water that's derived from surface water will have greater temperature variations.

The Regulation applies to any pipe, fitting and appliance that carries cold water, paying particular attention to pipes passing through airing cupboards and heated rooms. There will also be a special need to insulate pipes and cisterns in airing cupboards.

The Guidance Document states that the temperature of cold water should be kept below 20°C, not 25°C. This is also a recommendation of BS 6700. Both are good practice, but the Regulations are law and must be complied with.

On the job: Renewing a cold-water supply system

Martin has a job to renew a cold-water supply system. The system runs from the rising main which enters the dwelling in the kitchen, where it feeds the cold-water tap, central-heating boiler and washing machine, up to the cold-water supply cistern in the loft.

1   What should Martin do to reduce the risk of the cold-water supply pipes and fittings becoming unduly warmed?

2   What should he tell his customer when he finishes the work?

Stop valves to premises

WSR Schedule 2, para 10(1)

Paragraph 10(1)describes both **stop valves** and **service valves**. Every **supply pipe** or distributing pipe providing water to separate premises shall be fitted with a stop valve located to enable the supply to those premises to be shut off without shutting off the supply to any other premises. The Regulation deals generally with stop valves that are required to be fitted to control the whole supply of water to premises and those that are required to isolate individual sections of an installation.

Where a supply pipe or distributing pipe provides water in common to two or more premises, it shall be fitted with a stop valve to which each occupier of those premises has access. Every supply and distributing pipe providing water to premises must be of adequate size to suit the needs of the building and should be fitted with a stop valve located at the boundary of the premises, or elsewhere, to enable the supply to be shut off without shutting off the water supply to other premises. Stop valves to premises must be accessible so that the occupier, in the event of leakage or for other reasons, can isolate the supply.

> **Definition**
>
> **Stop valve** – a valve, other than a servicing valve, used for shutting off the flow of water in a pipe

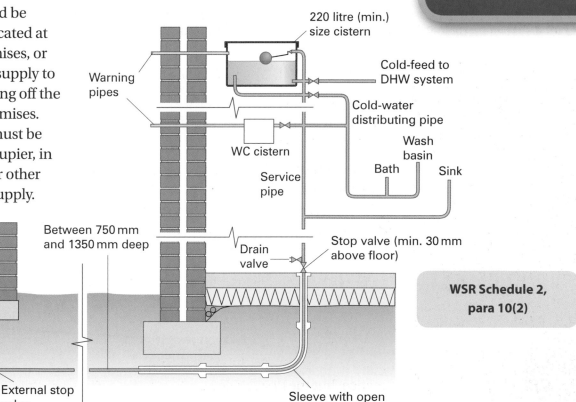

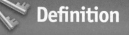

WSR Schedule 2, para 10(2)

Figure 2.25 Location of stop valves

The stop valve should ideally be positioned on the supply pipe, inside the premises, above floor level and as close as possible to the point of entry.

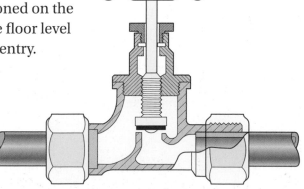

Figure 2.26 Internal workings of a stop valve

> **Definition**
>
> **Service valve** – a valve for shutting off, for the purpose of maintenance or service, the flow of water in a pipe connected to a water fitting

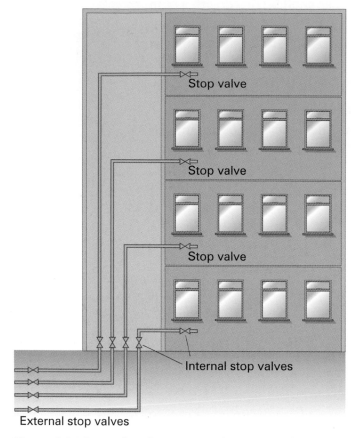

Figure 2.27 Stop valves for communal properties fed by separate supply pipes

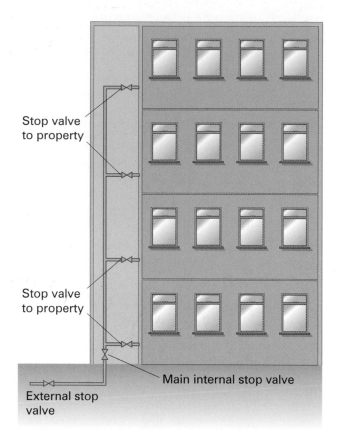

Figure 2.28 Recommended locations of stop valves for a block of flats fed from a common supply pipe

Figure 2.27 shows the recommended locations of stop valves in a block of flats where individual flats are fed by separate supply pipes. Separate internal stop valves are provided on entry to the property to ensure that the individual supply to the property can be isolated easily without having to gain access to the property.

Where a block of flats is fed from a common supply pipe, the recommended positions of the stop valves are as shown in Figure 2.28.

Where a block of flats is fed from a common distributing pipe from a cistern, the recommended positions of stop valves are as shown in Figure 2.29.

Other examples are existing terraced houses fed by a common supply pipe, shown in Figure 2.30. This type of installation would today be allowed only under exceptional circumstances.

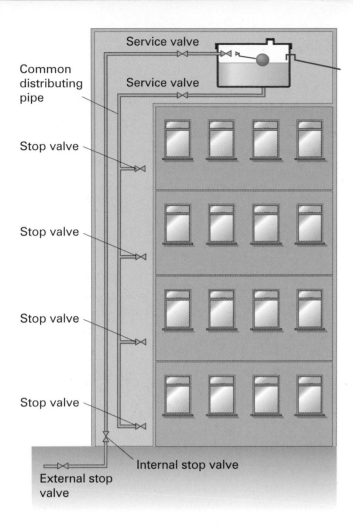

Service valve

Common distributing pipe

Service valve

Stop valve

Stop valve

Stop valve

Stop valve

Internal stop valve

External stop valve

Figure 2.29 Recommended locations of stop valves for a block of flats fed from a common distributing pipe

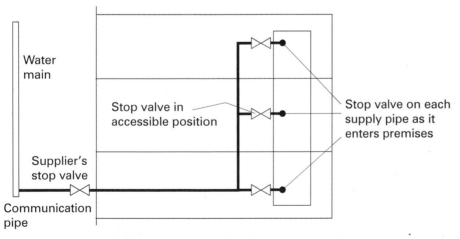

Water main

Stop valve in accessible position

Stop valve on each supply pipe as it enters premises

Supplier's stop valve

Communication pipe

Figure 2.30 Stop-valve locations for existing terraced houses (multi premises) fed by a common supply pipe

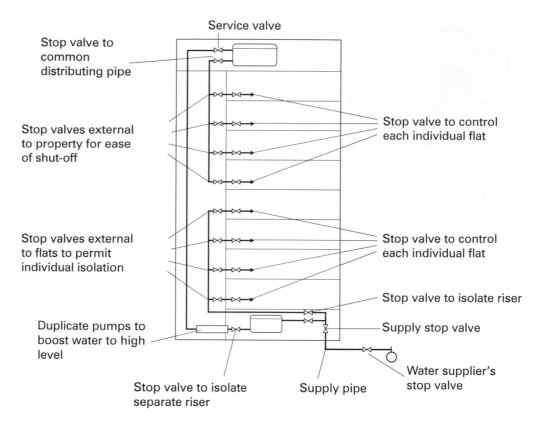

Service valve

Stop valve to common distributing pipe

Stop valves external to property for ease of shut-off

Stop valves external to flats to permit individual isolation

Duplicate pumps to boost water to high level

Stop valve to control each individual flat

Stop valve to control each individual flat

Stop valve to isolate riser

Supply stop valve

Water supplier's stop valve

Stop valve to isolate separate riser

Supply pipe

Figure 2.31 Requirements for tall buildings

Where distributing pipes supply separately chargeable premises from a common storage cistern all separate premises are fitted with stop valves in similar positions to those supplied by a common supply pipe. These will usually be tall buildings that have fittings above the limit of the mains supply, working in conjunction with booster pumps.

WSR Schedule 2, para 11

Water-supply systems shall be capable of being drained down, and shall be fitted with an adequate number of servicing valves and drain taps so as to minimise the discharge of water when water fittings are maintained or replaced. A sufficient number of stop valves should also be installed for isolating parts of the pipework. Complying with this requirement will give full control over the installation, allowing sections of pipework or individual appliances to be isolated and drained down, without isolating the supply to other parts of the building.

The provision of service valves also applies to mechanical backflow prevention devices, where the backflow risk is being protected against. Servicing valves should be fitted as close as reasonably practicable to float-operated valves or other inlet devices of an appliance, and they should be readily accessible.

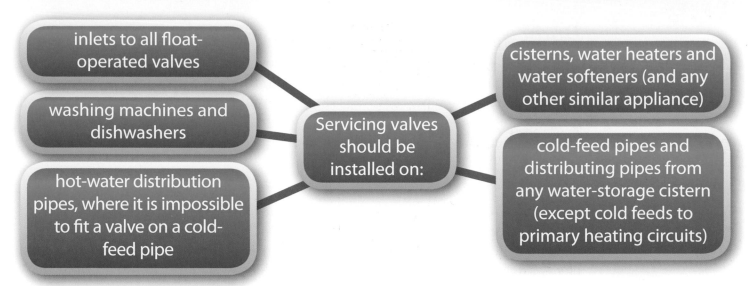

Figure 2.32 Where to install servicing valves

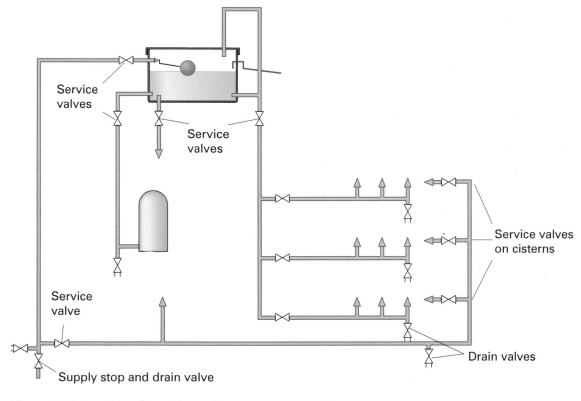

Figure 2.33 Location of servicing valves

Stop valves must be installed to isolate parts of pipework for maintenance and for isolating sections of the supply should leakages occur. On larger installations servicing valves or stop valves should be fitted to:

- isolate pipework on different floors
- isolate various parts of an installation
- isolate branch pipes to a range of appliances.

BS 2879

Drain valves

Sufficient drain valves must be provided to all low points on the system. They must be installed in a position that makes them readily accessible and allows the connection of a hosepipe. Drain taps should be of the screw-down type and located in a frost-free location. They should not be located below ground or in such a position where they're likely to become submerged or cause contamination.

Stop and service valves

There are several types of valves that are suitable for use as stop valves and service valves. The valves should comply with the relevant British Standard or EN standard, and any valve used below ground or in an inaccessible position should be manufactured from gunmetal or a material that's resistant to dezincification. Stop valves installed below ground should be of the crutch type or square cap type, permitting shut-off by a key.

WC cisterns should be fitted with spherical plug valves (slot type) to prevent unauthorised interference.

Approved types of valves for above or below ground above 50 mm diameter are:

- flanged gate valve to BS 5163 (key operated).

Approved types of valves for above or below ground up to 50 mm diameter:

- screw-down stop valve to BS 5433
- plug valve to BS 2580.

> **Remember**
>
> Drain valves are also a precautionary measure against frost

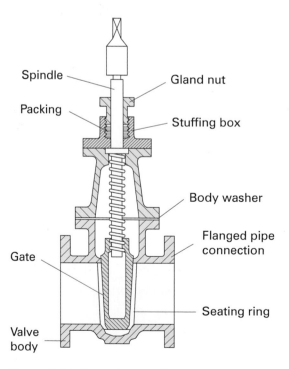

Figure 2.34 Flanged gate valve to BS 5163

Spindle • Gland nut • Packing • Stuffing box • Body washer • Flanged pipe connection • Gate • Seating ring • Valve body

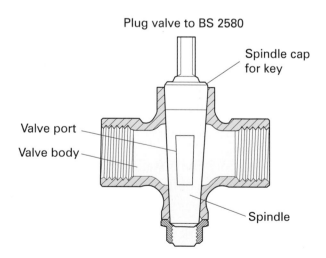

Plug valve to BS 2580

Spindle cap for key • Valve port • Valve body • Spindle

Figure 2.35 Plug valve to BS 2580

Approved types of valves for above ground only:

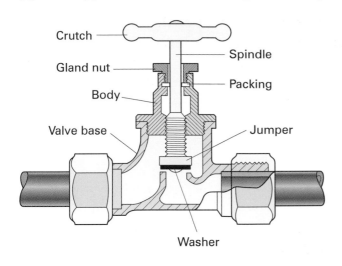

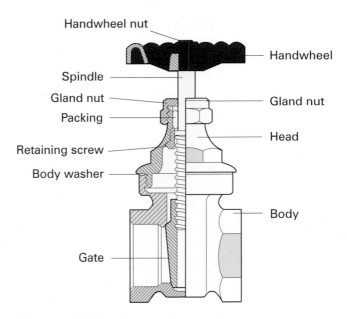

Figure 2.36 Screw-down stop valve to BS 1010

Figure 2.37 Wheel-operated gate valve to BS 5154

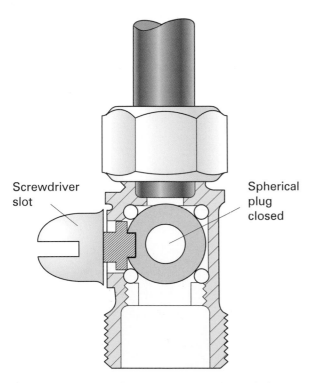

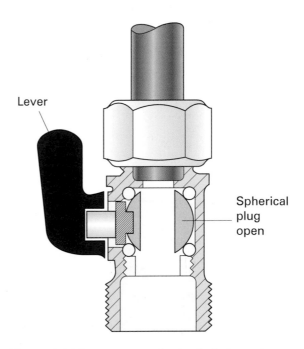

Figure 2.38 Screwdriver-operated spherical plug valve to BS 6675

Figure 2.39 Lever-operated spherical plug valve to BS 6675

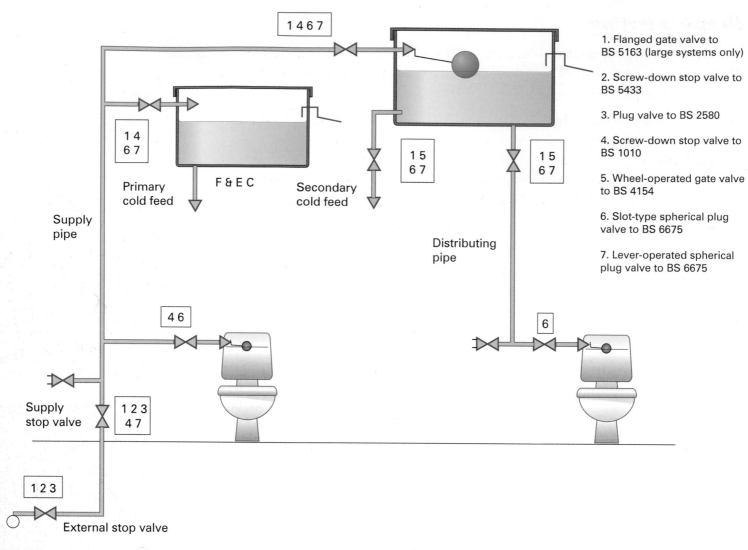

1. Flanged gate valve to
BS 5163 (large systems only)

2. Screw-down stop valve to
BS 5433

3. Plug valve to BS 2580

4. Screw-down stop valve to
BS 1010

5. Wheel-operated gate valve
to BS 4154

6. Slot-type spherical plug
valve to BS 6675

7. Lever-operated spherical
plug valve to BS 6675

Figure 2.40 Positioning and types of valves suitable for installation

**WSR Schedule 2, paras
12 & 13**

Commissioning

Commissioning a water installation includes:

- making a visual inspection of the installation
- soundness testing (testing for leaks)
- flushing and disinfection
- performance testing
- final checks/hand over.

All the above procedures are considered to be good practice, but unfortunately they aren't always carried out correctly and are sometimes completely missed.

Definition

Commissioning
– completing an installation, checking for faults, putting the system in use, and ensuring that it operates safely and efficiently and is to the customer's satisfaction

Pressure testing

The **test pressure** applies to all tests and all installations. The requirement does not distinguish between installation sizes, new or replacement work or location, e.g. above or below ground.

WSR Schedule 2, para

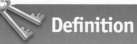

BS 6700

Additional guidance on testing, flushing and disinfection of water installations can be seen in BS 6700, but you must remember the British Standard specifications give guidance only and are not law. However, if you carry out testing, flushing and disinfection in a workmanlike manner 'Regulation A(5) following the recommendations of BS 6700, then you're doing what's reasonably expected, to satisfy the law'.

Paragraph 12(2) sets out the 'test criteria', giving separate criteria for:

WSR Schedule 2, para 12(2)

- systems that contain no plastics, e.g. copper, steel, cast iron

- systems that do contain plastic pipes and fittings.

Testing non-plastic systems

For non-plastic systems there are three test requirements:

WSR Schedule 2, para 12(2)(9)

- The installations shall be pumped up to test pressure or the maximum operating pressure plus an allowance for any expected surge pressure (whichever is the greatest).

- The test shall be for one hour.

- During this there should be no visible leaks and no loss of pressure.

Water-soundness testing should be carried out on all completed installations, including supply pipes, distributing pipes, fittings and components.

Test gauge

Control valves

Water supply with backflow protection via pump from mains

Connection point

Main under test

Figure 2.41 Testing equipment and requirements for a water main

Definition

Test pressure – an internal water pressure of not less than 1½ times the maximum pressure to which the installation or relevant part of it is designed to be subjected in operation (WSR Schedule 2, 12(1))

Safety tip

It's very important that the manufacturer's recommendations and advice are taken into consideration when using chemical agents for flushing and disinfection of installations and equipment

Remember

The final test of an installation is a crucial part of the commissioning procedure, and any buried pipework or concealed pipework must be successfully tested before backfilling or encasing takes place

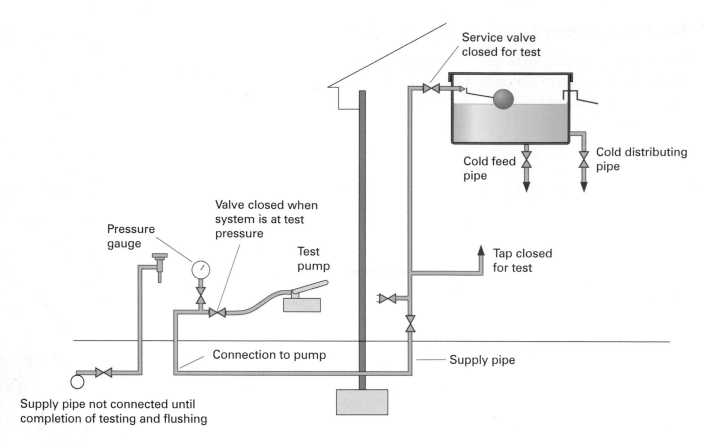

Service valve
closed for test

Cold distributing
pipe

Cold feed
pipe

Valve closed when
system is at test
pressure

Pressure
gauge

Test
pump

Tap closed
for test

Connection to pump

Supply pipe

Supply pipe not connected until
completion of testing and flushing

Figure 2.42 Testing a supply pipe and the soundness testing of a hot-water distributing system

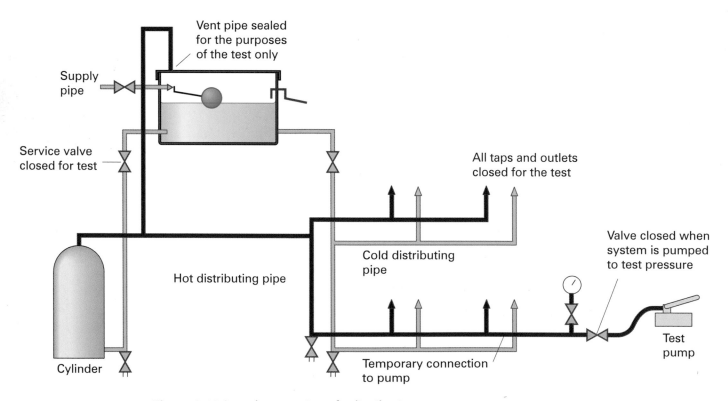

Vent pipe sealed
for the purposes
of the test only

Supply
pipe

Service valve
closed for test

All taps and outlets
closed for the test

Valve closed when
system is pumped
to test pressure

Cold distributing
pipe

Hot distributing pipe

Cylinder

Temporary connection
to pump

Test
pump

Figure 2.43 Soundness testing of a distributing system

Testing systems containing plastics

There are two tests for installations containing plastics. The Guidance Document systems containing plastic pipes are tested in accordance with the recommendations of BS 6700.

WSR Schedule 2,
para 12(2)(b)

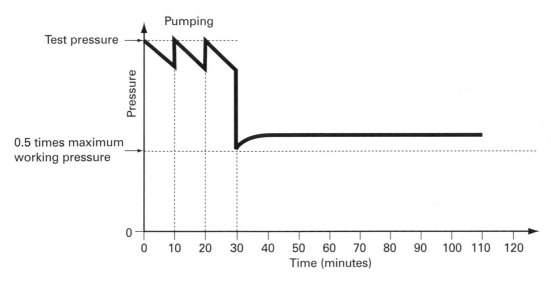

Figure 2.44 Test A procedure for plastic pipes

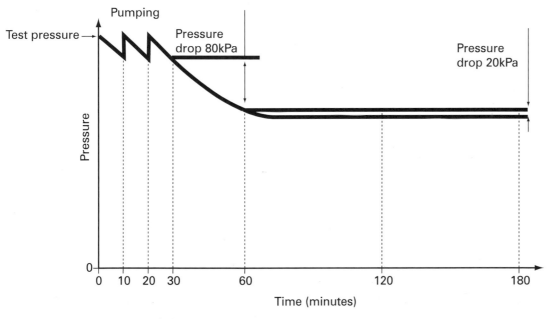

Figure 2.45 Test B procedure for plastic pipes

Did you know?

The testing procedure varies slightly between BS 6700 and the Water Regulations. The variation takes into consideration the fact that some plastic materials, when subjected to a test pressure, suffer stress that can be retained in the pipe material once the test is over. This can result in pipe failure at a later date, so for this reason pipes of an **elastomeric** material are allowed to be subjected to a less severe test

Remember

Acceptable test results are shown on page 44

System flushing

Every water system shall be flushed out before it is first used. This applies to new installations, alterations, extensions or any maintenance to existing installations. This makes sure that any debris, including excessive **flux** that may have collected in the pipework during installation, is removed from the system.

WSR Schedule 2, para 13

System disinfecting

Every water system shall where necessary be disinfected before it is first used. The Guidance Document recommends that, after system flushing, the system should be disinfected. This applies to:

WSR Schedule 2, para 13

- new installations, except private dwellings occupied by a single family
- major extensions or alterations except private dwellings occupied by a single family
- underground pipework, except localised repairs or insertion of junctions
- where contamination may be suspected, e.g. fouling by sewage, drainage, animals, insects and vermin
- after physical entry by personnel for interior inspection, painting or repairs
- where a system has not been in regular use and not regularly flushed. Regular use means periods of up to 30 days without use depending on the characteristics of the water.

Find out

Why does the pipework need to be free of debris?

Disinfection recommendations

The Guidance Document refers to the recommendations of BS 6700 for flushing and disinfection of systems, and following these procedures will satisfy the requirements of the Regulation.

BS 6700

BS 6700 sets the following guidance on the sterilisation of pipework fittings:

- Where pipework is under mains pressure, or has a backflow device fitted downstream, the water undertaker should be notified.
- If water used for disinfection is to be discharged to a sewer, drain or water course the authority responsible for the sewer, drain or course should be notified.
- Chemicals used for disinfection of drinking-water installations must be chosen from a list of substances compiled by the drinking-water inspectorate, which is listed in the *Water Fittings and Materials Directory* published by the WRAS.

Did you know?

Private dwellings include a normal house containing a single family. In instances where a larger house has been divided into separate units, e.g. bed-sits or student flats, disinfection will be required

Unless a specific chemical disinfectant is specified, sodium hypochlorite (diluted chlorine) can be used. Other disinfectants are available in tablet form, and these are easily obtainable and safe to use.

Safety factors

- Prior to carrying out disinfection, the system must be taken out of use, marking all outlets 'Do Not Use DISINFECTION IN PROGRESS'.

- All operatives carrying out the disinfection procedure must receive appropriate Health and Safety training under the COSHH Regulations.

- No other chemicals, e.g. sanitary cleaners etc. should be added to the water during disinfection (as this could generate toxic fumes).

- All occupants within the premises must be notified that disinfection is taking place.

- Extreme care must be taken when using disinfectants: some can be hazardous, and operatives must wear safety goggles and protective clothing and refrain from smoking.

The procedure for disinfection

This can be used for both hot-water and cold-water installations and is the general procedure for the disinfection of a system whether chlorine or any other approved disinfectant is used.

- Thoroughly flush the system prior to disinfection.

- Introduce a disinfection agent at specified concentrations into the system, filling systematically to ensure a total saturation. If using chlorine use initial concentrations of 50 mg per litre (50 ppm).

- Leave the system for a contract period of one hour. If using chlorine, check the free residual chlorine levels at the end of the contract period. If this is less than 30 mg per litre, then the procedure requires repeating.

- Immediately following successful disinfection, the system should be drained and thoroughly flushed with clean water until the residual chlorine level is at the same level as the drinking water supplied.

After flushing, samples should be drawn off and taken for bacteriological analysis. If the test result proves unsatisfactory, the disinfection and sampling test procedures should be repeated.

When disinfecting supply pipes or any hot or cold storage vessels that are connected directly to mains pressure, the chlorine solution should be injected into the lower end of the pipe near its point of connection to the communication pipe, using a suitable injector point. The procedures for disinfecting should then be followed.

Water should be introduced into the system by systematically opening individual taps, working away from the point of connection, until the whole of the system is filled with water of the specified concentration (50 ppm) (ppm = parts per million).

Safety tip

The correct sequence for system disinfection should follow the flow of water into the premises, i.e. first the water mains, then the supply pipe and cisterns and finally the distribution systems

Did you know?

When disinfecting supply pipes or any hot or cold storage vessels that are connected directly to mains pressure, the chlorine solution should be injected into the lower end of the pipe near its point of connection to the communication pipe using a suitable injector point

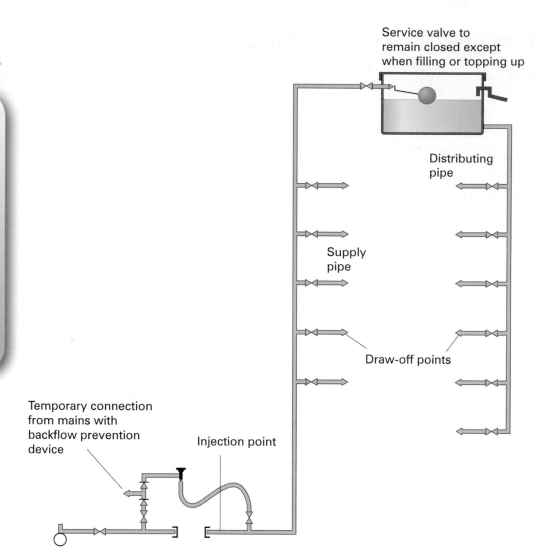

Figure 2.46 Disinfection connections for a whole system

Disposal of disinfection fluid

Generally, water and other fluids that have been used for disinfecting water systems can be safely discharged into a public sewer. However, particularly in rural areas, the discharge may have an adverse effect on sewage treatment (e.g. cesspools) and the sewage undertaker should be consulted if in doubt.

Prevention of cross connection to unwholesome water

At the end of this section you should be able to:

- state the requirements of the Water Industry Act 1991, and your duties under it
- distinguish between wholesome and unwholesome water by means of colour-coding of pipes
- describe the means of identifying water suitable for drinking purposes at the point of draw-off.

The Water Industry Act 1991 places duties and responsibilities on suppliers of water that it should be clean, free from impurities and fit for drinking. It is the duty of you the installer not to contaminate the water supplied. This also applies to water users under the Water Supply (Water Fittings) Regulations 1999.

The Water Regulations Advisory Committee has made recommendations for requirements suggesting that the use of recycled water could make significant contributions to water conservation, and considers that systems making use of recycled water in future years could become more common. This pre-empts the fact that as the use of these waters increase, so will the risk of cross connection and backflow.

Identification

Pipes can be identified by colour-coded pigmentation incorporated in plastic pipes, permanent marks or labels, or colour painting of the pipes themselves.

> **WSR Schedule 2, para 14(1)**

Pipes located above ground within buildings should be colour-coded to BS 1710 to distinguish them from others. Generally, this applies only to industrial and commercial buildings, but in cases where a house or small building uses water other than wholesome water supplied by the undertaker, then colour-coding of the pipes will be required.

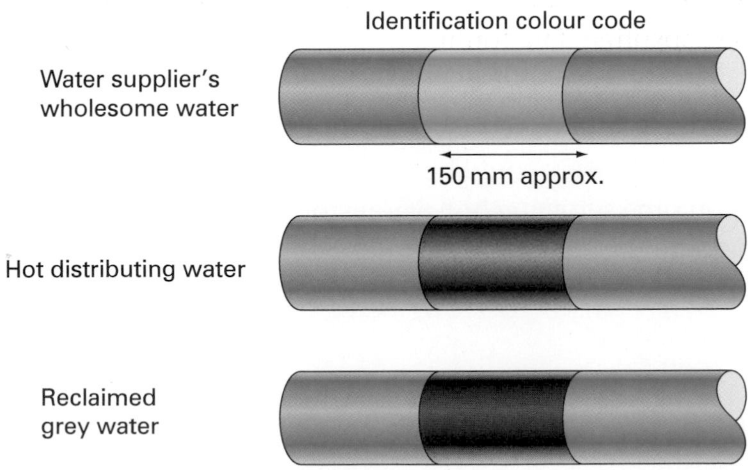

Identification colour code

Water supplier's wholesome water

150 mm approx.

Hot distributing water

Reclaimed grey water

Figure 2.47 Colour-coding of pipes for **non-potable** water should follow this scheme

Find out

Find out the colour-coding for a cold distributing pipe and a pipe carrying water for fire-fighting purposes

Remember

The identification requirement also applies to all water fittings, including cisterns and valves. It's particularly important that taps should be labelled, identifying those that are suitable for drinking purposes and those that are not

WSR Schedule 2, para 14(2)

Colour identification should be fitted on pipes at junctions, inlets and outlets of valves, and service appliances where a pipe passes through walls (both sides).

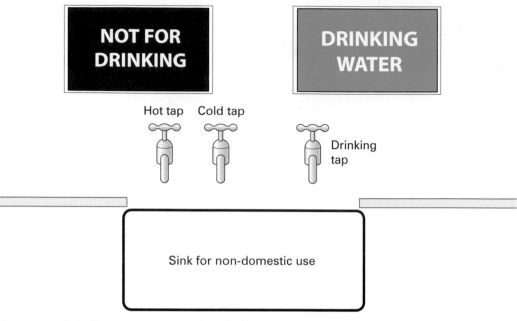

Figure 2.48 Labelling requirements for wholesome and unwholesome water

Ideally, on larger installations, accurate pipe-layout drawings should be handed over to the customer, identifying the locations of pipes, within and below ground level, feeding the building.

Any pipe carrying fluid that isn't wholesome water must not be connected to a pipe that's carrying wholesome water, unless a suitable backflow prevention device is fitted.

The requirement would be satisfied if wholesome water (fluid category 2) feeding into a cistern containing grey water was fed into the cistern via a backflow prevention device or an arrangement suitable for protection against a fluid 5 category risk (examples are Type AA, AB and AD air gaps).

It's very important that you remember that water derived from a supply pipe is considered to be wholesome water, but water derived from a distribution pipe may not be. This depends on the quality of the water contained in the cistern supplying the distribution pipe and on what the distribution pipe is serving.

A distributing pipe from a cistern containing wholesome water, servicing taps over sinks, baths, wash basins, and showers could be considered to be servicing wholesome water. A distribution pipe servicing hot-water storage vessels, or hot-water distribution pipes, should not be considered as supplying wholesome water and should not have any connections made into it for drawing wholesome water.

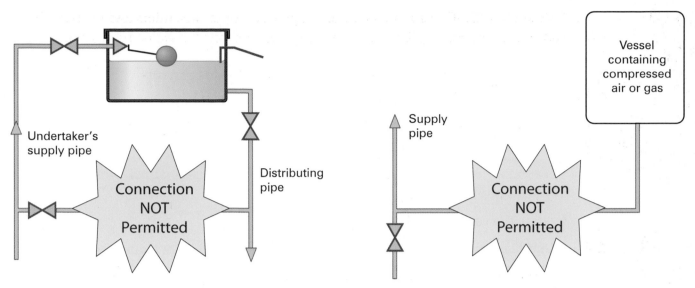

Figure 2.49 Unlawful connection between a supply pipe and distributing pipe (left) and an unlawful connection between a supply pipe and a vessel containing compressed air or gas (right)

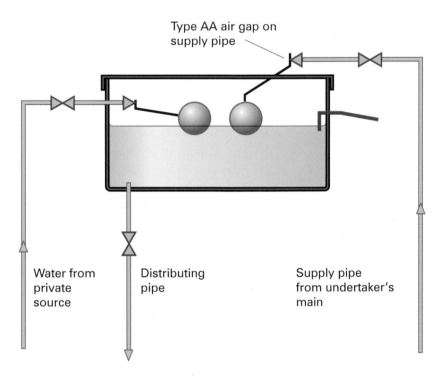

Figure 2.50 Correct connection for a cistern containing reserve water for fire-fighting or other industrial use

Backflow prevention and fluid categories

At the end of this section you should be able to:

- describe the purpose of backflow-prevention devices
- state the main requirements of the Regulations as they apply to backflow prevention and fluid categories
- explain what causes backflow, and the differences between back pressure and back siphonage
- state the various fluid risk categories
- name the various backflow-prevention devices including mechanical and non-mechanical
- determine what backflow-prevention methods can be used for various fluid risk categories.

Backflow prevention is probably the most important element of the Water Regulation requirements.

When water is supplied by the water undertaker to premises etc. it is essential that this water isn't allowed to return into the undertaker's main, because of the possibility that it may become contaminated after being supplied.

> **Definition**
>
> **Backflow** – water flowing in a direction contrary to the direction intended

The Requirements

Paragraph 15 of Schedule 2 states that:

(1) Subject to the following provisions in this paragraph, every water system shall contain an adequate device or devices for preventing backflow of fluid from any appliance, fitting or process from occurring.

(2) Paragraph (1) does not apply to:

    (a) a water heater where the expended water is permitted to flow back into a supply pipe, or

    (b) a vented water storage vessel supplied from a storage cistern, where the temperature of the water in the supply pipe or the cistern does not exceed 25°C.

(3) The device used to prevent backflow shall be appropriate to the highest applicable fluid category to which the fitting is subject downstream before the next such device.

(4) Backflow prevention shall be provided on any supply or distributing pipe:

    (a) where it is necessary to prevent backflow between separately occupied premises, or

(b) where the water undertaker has given notice for the purposes of this Schedule that such prevention is needed for the whole or part of any premises.

Because of the importance of backflow prevention more detailed requirements are given in the Regulator's Specification on the prevention of backflow (from 1 May 2000). The specification describes a range of backflow-prevention devices, giving essential information on the application and installation of them.

Backflow can be caused by either back pressure or back siphonage.

An example of back pressure is the expansion of water from an unvented hot-water vessel or heater, allowed to flow back into the supply pipe, and this happens when the water in the vessel or heater is heated. The expanded water must not be permitted to reach points where the water can be drawn off.

Back siphonage backflow is caused by siphonage of water from a cistern or appliance back into the pipe which feeds it.

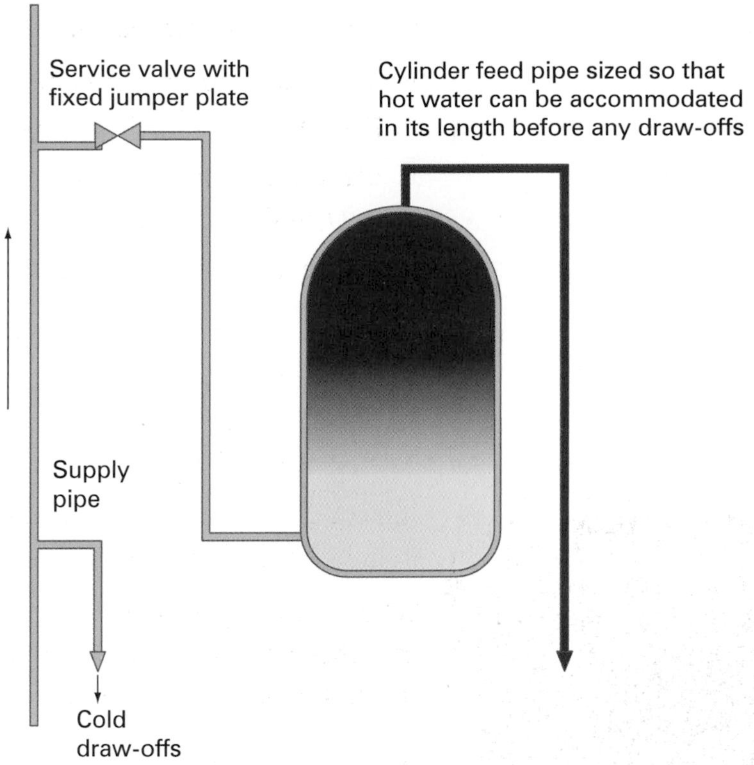

A typical example of back siphonage is where a hosepipe is used to fill a storage vessel (this being higher than the water mains). If the main is interrupted or a break in the main occurs, this causes the water within the storage vessel to be siphoned back along the supply pipe towards the main.

Service valve with fixed jumper plate

Cylinder feed pipe sized so that hot water can be accommodated in its length before any draw-offs

Supply pipe

Cold draw-offs

Figure 2.51 Back pressure

Backflow-prevention devices

A backflow-prevention device stops contamination of drinking water by backflow. Devices can be mechanical, non-mechanical or an arrangement in the pipework system preventing backflow (e.g. check valves, double check valves, air-gap arrangements).

Fluid risk categories

Schedule 1 of the Water Regulations recognises and implements a five fluid risk category list, based on that developed by the Union of Water Supply Associations of Europe (EUREAU 12), and which is also currently used in North America and Australia.

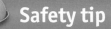

Safety tip

All safety devices have been removed from Figure 2.51 for simplicity

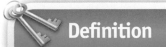

Definition

Back pressure – the reversal of flow in a pipe caused by an increase in pressure in the system

WSR Schedule 1

Fluid category	Description	Application
1	Wholesome water supplied by a water undertaker complying with the Requirements of the Regulations made under Schedule 67 of the Water Industry Act 1991	
2	Water that would be classed as fluid category 1 except for odour, appearance or temperature. These changes in water quality are aesthetic changes only and the water is not considered a hazard to human health	(a) water heated in a hot-water secondary system (b) mixtures of fluids from categories 1 and 2 discharged from combination taps or showers (c) water that has been softened by a domestic common salt regeneration process

3	These fluids represent a slight health hazard and aren't suitable for drinking or other domestic purposes	(a) in houses or other single-occupancy dwellings
		  (i) water in primary circuits and heating systems, whether additives to the system have been used or not
		  (ii) water in wash basins, baths or shower trays
		  (iii) washing machines and dishwashers
		  (iv) home dialysing machines
		  (v) hand-held garden hoses with flow-control spray or shut-off control
		  (vi) hand-held garden fertiliser sprays
		(b) in premises other than a single-occupancy dwelling
		(c) where domestic fittings such as wash basins, baths or showers are installed in premises other than a single-occupancy dwelling (that is, commercial, industrial or other premises) these appliances may still be regarded as a fluid category 3, unless there's a potentially higher risk. Typical premises that justify a higher fluid risk category include hospitals and other medical establishments
		(d) house, garden or commercial irrigation systems without insecticide or fertiliser additives, with fixed sprinkler heads not less than 150 mm above ground level
		(e) fluids that represent a slight health hazard because of the concentrations of substances of low toxicity include any fluid that contains:
		  – ethylene glycol, copper sulphate solution or similar chemical additives or
		  – sodium hypochlorite (chloro's and common disinfectants)

continued from previous page

4	These fluids represent a significant health hazard and aren't suitable for drinking or other domestic purposes. They contain concentrations of toxic substances, and include:	(a) water containing chemical carcinogenic substances or pesticides (b) water containing environmental organisms of potential health significance (micro-organisms, bacteria, viruses and parasites of significance for human health which can occur and survive in the general environment) (c) water in primary circuits and heating systems other than in a house, irrespective of whether additives have been used or not (d) water treatment or softeners using other than salt (e) water used in washing machines and dishwashing machines for other than domestic use (f) water used in mini-irrigation systems in a house garden without fertiliser or insecticide applications such as pop-up sprinklers, permeable hoses or fixed or rotating sprinkler heads fixed less than 150 mm above ground level
5	Fluids representing a serious health risk because of the concentration of **pathogenic organisms**, radioactive or very toxic substances, including any fluid which contains: (a) faecal material or other human waste, or (b) butchery or other animal waste, or (c) pathogens from any other source	(a) sinks, urinals, WC pans and bidets in any location (b) permeable pipes or hoses in other than domestic gardens, laid below ground or at ground level with or without chemical additives (c) grey-water recycling systems (d) washing machines and dishwashers in high-risk premises (e) appliances and supplies in medical establishments

Table 2.11 The fluid risk categories which describe water, based on how drinkable it is and how dangerous to health it may be, depending on impurities

Definition

Pathogenic organisms – micro-organisms such as bacteria, viruses or parasites which are capable of causing illness, especially in humans, e.g. salmonella, vibrio cholera, campylobacter. These generally form in living creatures and can be released into the environment, for example in faecal matter, animal wastes or body fluids

What type of backflow device should be fitted?

Before installing a backflow-prevention device, you will need to know the correct type suitable for the particular installation, and your selection will also depend on the fluid risk category. Guidance for these two important factors is given in the regulator's specification on the prevention of backflow where several tables are provided (look at DEFRA's website, www.defra.gov.uk/environment/water/industry/wsregs99).

Table A.1 in Appendix A is a list of non-mechanical backflow-prevention devices acceptable under the WSR. Table A.2 lists mechanical backflow-prevention devices.

On the job: Preventing backflow

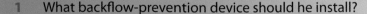

Martin is about to install a commercial washing machine (for clothes) in a laundry. This represents a fluid category 4 risk.

1   What backflow-prevention device should he install?

On determining the fluid category and the appropriate backflow-prevention device, you will need to look at any installation requirements that may apply.

The prevention of backflow can be achieved by using one or more of the devices listed in Appendix A – remember, the type of device will depend on the severity of the risk.

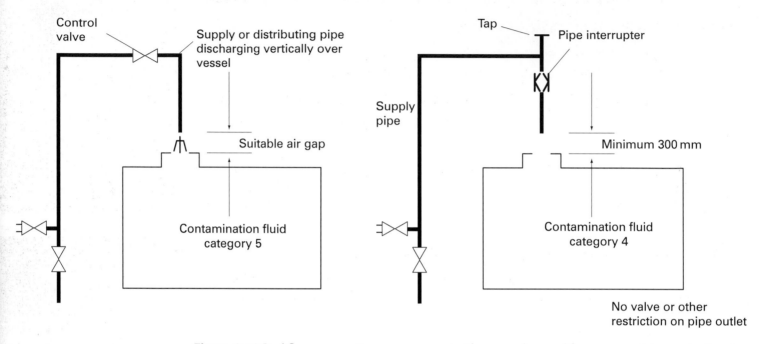

Figure 2.52 Backflow-prevention arrangements. These can be used for commercial or agricultural applications, e.g. farms or chemical plants

Positioning of backflow-prevention devices

GD G15.2 and G15.5-7

Backflow-prevention devices are normally point-of-use devices, which means that they're to protect against individual risk at or near the point of supply or at the point where the risk is likely to occur with an individual appliance.

Further advice on the installation of backflow-prevention devices is given in the Regulator's Specification on the Prevention of Backflow and states:

- G15.2 Backflow prevention can be achieved by good system design and the provision of suitable backflow-prevention arrangements and devices.

- G15.5 Where practicable, systems should be protected against backflow without relying on mechanical backflow protection; preferred protection is by the use of tap gap or air gap at the point of use.

- G15.6 Permanently vented distributing pipes will provide good 'secondary' protection (whole-site or zone protection) in many cistern-fed installations.

- G15.7 Mechanical backflow-prevention devices, which, depending on the type of device, may be suitable for protection against back pressure or back siphonage or both, should be installed.

    (a) They should be readily accessible for inspection, operational maintenance and renewal, and

    (b) except for types HA and HUK1 (backflow-prevention devices for protection against fluid categories 2 and 3) they should not be located outside premises;

    (c) they should not be buried in the ground;

    (d) vented or verifiable devices or devices with relief outlets are not to be installed in chambers below ground or where liable to flooding;

    (e) if used for category 4 devices, they should have line strainers fitted upstream (before the backflow-prevention device) and a servicing valve upstream of the strainer; and

    (f) the lowest point of the relief outlet from any reduced pressure zone valve assembly or similar device should terminate with a type AA air gap located not less than 300 mm above the ground or floor level.

Non-mechanical prevention devices

These types of backflow-prevention devices are called air gaps. The WSR refers to seven different types of air gap, each of which is fluid category or risk 3–5. These are outlined in detail in Appendix A.

Mechanical prevention devices

Back-pressure devices are devices or arrangements for prevention of back pressure, and are those where the outlet control valves or taps are positioned downstream of the backflow-prevention device; such devices are B, C and E families. Typical examples are **reduced pressure zone (RPZ) valves**, non-verifiable disconnectors and pressurised inlet valves.

Back-siphonage devices are devices or arrangements for prevention against back siphonage and are those where the control valve is located prior to the device; such devices are type A, D or H families. A typical example is an air gap or tap gap, a pipe interrupter or anti-vacuum valve.

Unlike with non-mechanical backflow-prevention devices, it's very important that mechanical devices are periodically inspected and tested to maintain correct operation. This is a particular requirement when RPZ valves are fitted. The RPZ valve is fairly new to the United Kingdom and is ideal for protection against fluid risk 4 applications.

The installation of an RPZ valve requires a contractor's certificate, and the water undertaker must be notified. Approved contractors are not required to notify the water undertaker.

Remember

- The advantage of air gaps is that because they're non-mechanical they require virtually no maintenance – but they must be correctly installed

- Remember that some devices are suitable for back pressure and for back siphonage, but they may not be suitable for the same fluid category

Did you know?

The RPZ valve must only be installed by a competent person and the device must be tested every year. To install, commission and maintain this type of valve you'll need to go on a special course. Furthermore, special test equipment needs to be used to commission this type of valve. When installing, altering, or disconnecting an RPZ, approved contractors must send a signed certificate of competence to the customer, with a copy to the water undertaker

WSR Part II, Reg 5

The use of backflow-prevention devices

Backflow can be prevented by good system design or by using suitable backflow-prevention devices or arrangements.

WCs and urinals

WC pans and urinals are considered to be a fluid category 5 risk – a serious health hazard irrespective of whether they are installed in a domestic dwelling or industrial or commercial premises.

There are two suitable backflow-prevention devices to protect against the risk. These are:

1   An interposed cistern type AUK 1. This means a siphonic or non-siphonic flushing cistern that may be used in premises of any type.

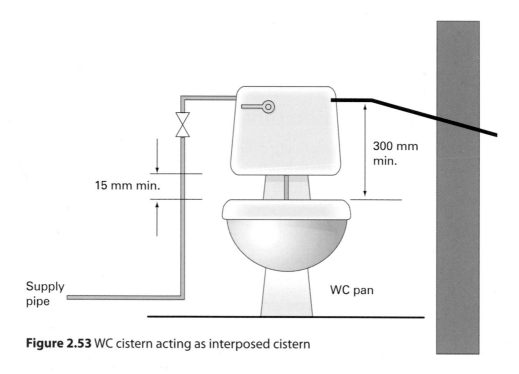

Figure 2.53 WC cistern acting as interposed cistern

2   A pipe interruptor with a permanent atmospheric vent type DC installed to the outlet of a manually operated pressure-flushing valve. It may be connected to a supply pipe or distributing pipe (excluding domestic dwellings).

There should be no other obstruction between the outlet of the pipe interruptor and the flush-pipe connection to the appliance.

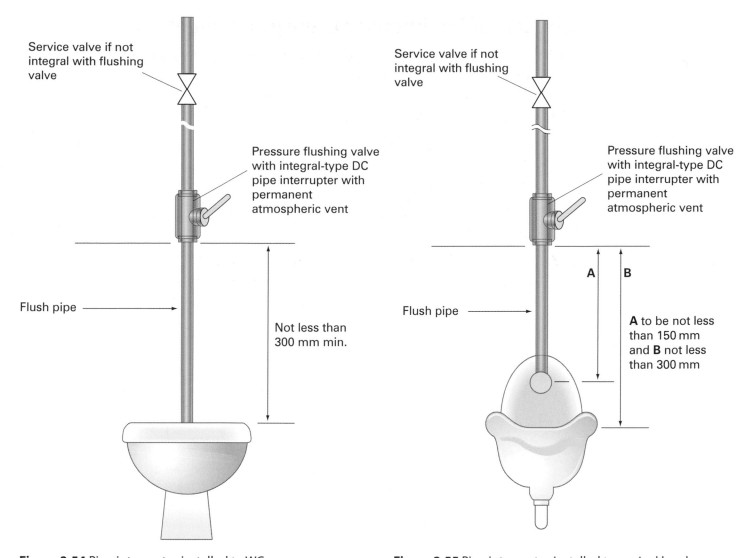

Figure 2.54 Pipe interruptor installed to WC

Figure 2.55 Pipe interruptor installed to a urinal bowl

Bidets

Other types of sanitary appliance with a high fluid risk category are bidets. This includes WCs adapted as bidets with flexible hose and spray handset fittings or with submerged water inlets. Bidets can be divided into two groups, each of which requires different protection. These are:

1 over-rim types – supplied from a tap at the back edge of the appliance

2 ascending-spray or submerged-inlet types – have a water spray jet situated below the rim of the spill-over level.

Included in the second group are bidets that are supplied from a tap that uses flexible spray attachments.

Over-rim type bidet

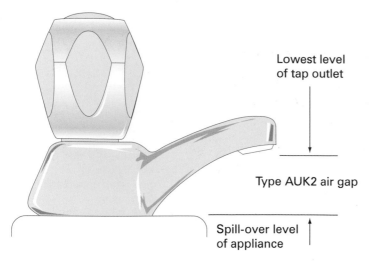

Lowest level
of tap outlet

Type AUK2 air gap

Spill-over level
of appliance

Over-rim bidets

Bidets installed in domestic dwellings that are of the over-rim type, having no ascending spray or flexible hose spray, can be supplied with cold and hot water through individual or combination tap assemblies, from either a supply or distribution pipe, providing that a type AUK 2 air gap is maintained between the outlet of the water fitting and the spill-over level of the bidet.

Figure 2.56 Air-gap requirements for over-rim bidets with no spray attachments

Ascending-spray bidets

When making connections to appliances of this type you must remember that bidets with an ascending spray aren't permitted to be connected directly from the supply pipe. They must be supplied from a storage cistern. Also, the hot and cold connections must be taken from independent dedicated distributing pipes that do not supply other appliances.

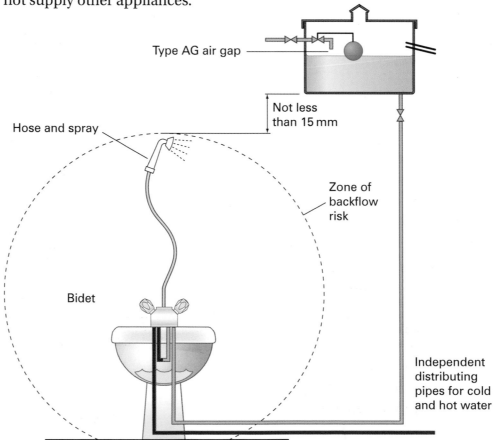

Type AG air gap

Hose and spray

Not less
than 15 mm

Zone of
backflow
risk

Bidet

Independent
distributing
pipes for cold
and hot water

Figure 2.57 Installation requirements for ascending-spray bidet

Exceptions to this can be made for the following cases:

- the common distributing pipe serves only the bidet and a WC urinal flushing cistern
- the bidet is the lowest appliance served from the pipe, with no likelihood of any other fittings being connected at a later date. The connection to the distribution pipe isn't less than 300 mm above the spill-over level of the bowl
- an over-rim bidet with a flexible spray connection, the connection should not be less than 300 mm above the spill-over level of any appliance that the spray outlet may reach.

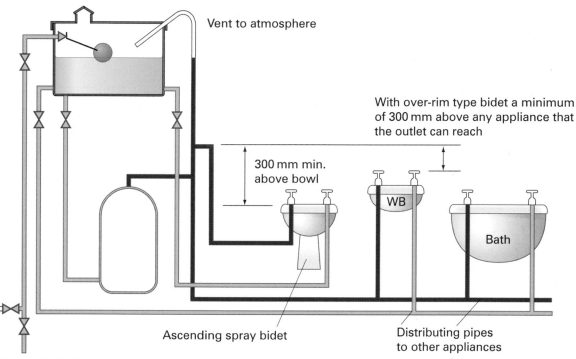

Figure 2.58 Pipework layout for bidet

Showerheads and tap inlets

The Regulations require (except where suitable additional backflow protection is provided) that all single tap outlets, combination tap outlets, fixed showerheads terminating over wash basins, baths, or bidets in domestic situations should discharge above the spill-over level of the appliance with a tap gap type AUK 2.

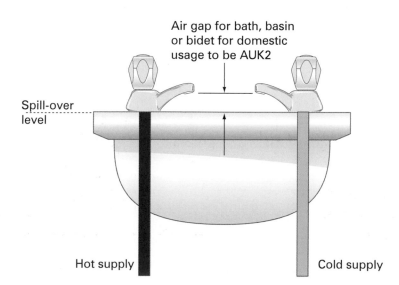

Figure 2.59 Air gap requirements for baths, basins or bidets for domestic usage

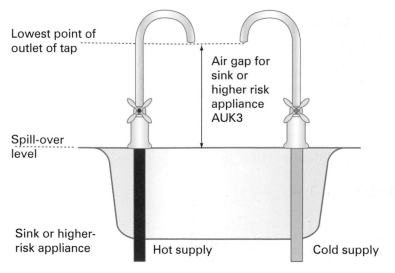

Lowest point of outlet of tap

Air gap for sink or higher risk appliance AUK3

Spill-over level

Sink or higher-risk appliance

Hot supply

Cold supply

Sinks in domestic and non-domestic situations are considered to be a fluid category 5 backflow risk and the minimum protection should be a type AUK 3 air gap/tap gap. Generally with sinks this isn't a problem, as sinks require additional space for access to work and for the filling of buckets etc.

Figure 2.60 Air gap requirements for sinks

Baths and wash basins

Baths and wash basins fitted in domestic dwellings that have submerged tap outlets are considered to give a fluid category 3 risk and should be supplied with water from a supply or distributing pipe through double check valves.

Submerged tap outlets to baths or wash basins in non-domestic situations are considered to be a fluid category 5 risk, and appropriate backflow protection is required for this higher risk level.

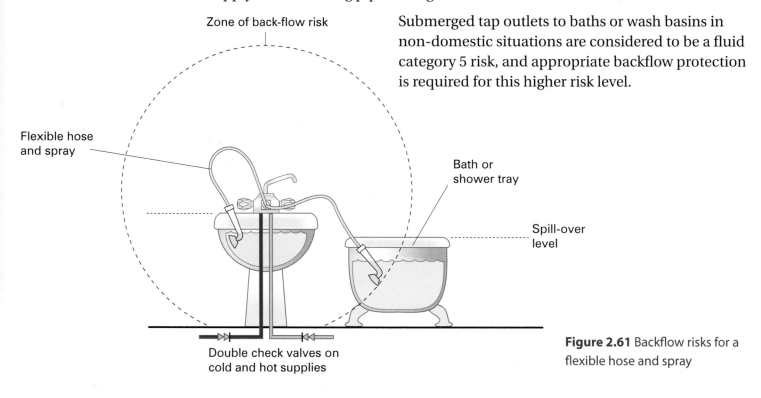

Zone of back-flow risk

Flexible hose and spray

Bath or shower tray

Spill-over level

Double check valves on cold and hot supplies

Figure 2.61 Backflow risks for a flexible hose and spray

Where installations consist of a spray or jet served from a tap or combination tap assembly or mixer fitting located over a wash basin, bath or shower tray, the zone of backflow risk must be ascertained. If the spray or jet on the end of the hose is capable of entering a wash basin, bath or shower tray located within the zone of backflow risk, then a fluid category 3 prevention device such as a double check valve must be fitted on each inlet pipe to the appliance types EC or ED.

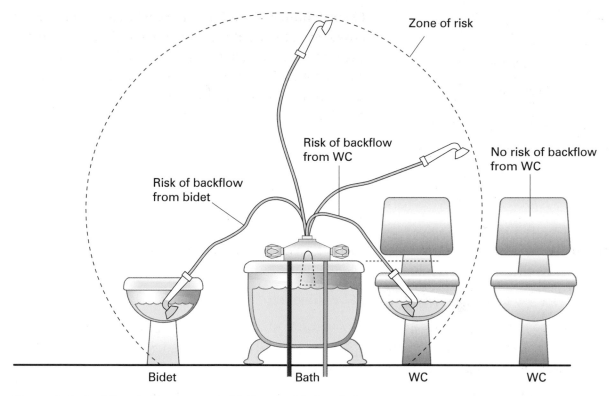

Figure 2.62 Backflow fluid category 5 risk for a bath mixer fitting

Washing machines

Domestic household washing machines, including washer-dryers and dishwashers, are manufactured to satisfy a fluid category 3 risk. Prior to installing the appliance, the Water Fittings and Materials Directory should be consulted. If the hoses are approved, they will be listed under the WRAS. In instances where the hose isn't approved, an appropriate fluid category 3 risk check valve must be fitted.

Commercial machines such as those used in hotels, restaurants, launderettes etc. are a fluid category 4 risk. Machines that are used in healthcare premises and hospitals are classed as a fluid category 5 risk and higher protection will be required. Most machines now incorporate a type AD device to guard against any type of risk.

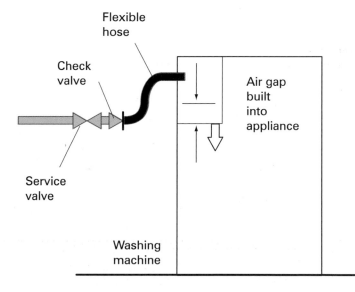

Figure 2.63 Connection to washing machine. A single check valve is shown because non-approved hoses are used, which may result in the possibility of tainting the water supply, leading to a category 2 risk

Drinking fountain

Drinking fountains

Drinking-water fountains have their own requirements. These should be designed so that there's a minimum 25 mm air gap between the water delivery jet nozzle and the spill-over level of the bowl. The nozzle should be provided with a screen or hood to protect it from contamination.

Domestic garden installations

Hand-held hosepipes for garden/other use must be fitted with a self-closing mechanism at the hose outlet. This will reduce the risk of backflow into the supply pipe if the end is dropped on the ground, and additionally promotes water conservation.

Any garden tap that enables a hose connection to be made to it must:

- be fitted with a double check valve
- be positioned where it will not be subject to frost damage.

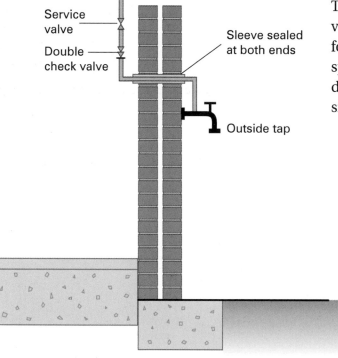

The installation of a double check valve is also adequate protection for hand-held hosepipes used for spraying fertilisers or domestic detergents in domestic garden situations.

Figure 2.64 Installation requirements for an outside tap

Irrigation and porous-hose systems

Irrigation and porous-hose systems, laid above or below ground, are a serious potential backflow risk, and the pipe supplying such systems must be protected against a fluid category 4 or 5 risk. Mini-systems and smaller applications are category 4, whereas commercial systems are category 5.

Irrigation systems that consist of fixed sprinkler heads located no less than 150 mm above ground level, which are not intended to be used in conjunction with insecticides, fertilisers or additives, are considered as a fluid category 3 risk.

The following diagrams show how pop-up sprinklers or porous hoses should be installed in a domestic situation (category 4 risk).

Irrigation in action

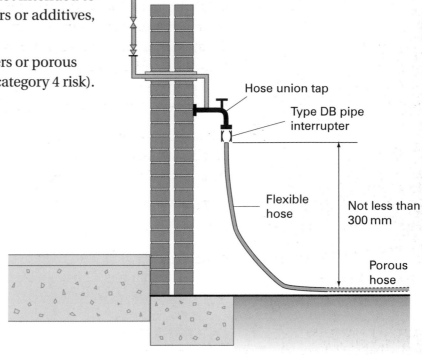

Figure 2.65 Installation details for a pop-up sprinkler

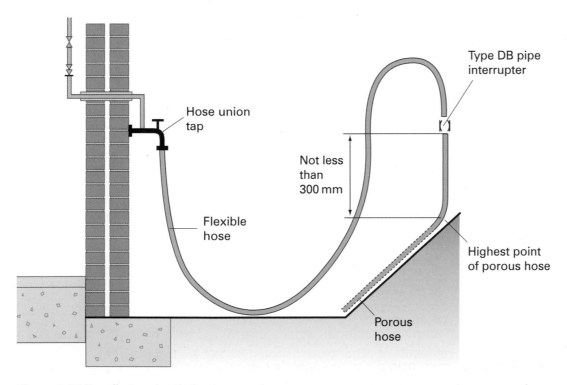

Figure 2.66 Installation details for a porous hose

Existing garden-tap installations

While the Water Regulations aren't retrospective, appropriate measures must be taken against any known situation where there's a potential backflow risk from hoses. Theoretically, if a tap was installed legally under previous Byelaws (before 1 July 1999) and has a hose connected to it, it remains legal. However, as soon as the hose is disconnected and reconnected, the installation becomes illegal unless appropriate steps are taken.

Where an external tap is being replaced the following factors apply:

GD G15.21

- if practicable, a double check valve should be provided on the supply to the tap type EC or ED. This should be installed inside the building, or where it isn't practicable to locate a double check valve within a building, the tap could be replaced with:

  - a hose union tap that incorporates a double check valve type HUK1, or

  - a tap that has a hose union backflow preventer type HA or a double check valve type EC or ED, fitted and permanently secured to the outlet of the tap.

External taps and systems in commercial situations

Taps and fittings used for non-domestic applications such as commercial, horticultural, agricultural or industrial purposes (this may include small catering establishments and hotels) must be supplied with backflow-protection devices appropriate to the downstream fluid category and, where appropriate, an additional zone-protection system.

Animal drinking troughs or bowls

The water inlet to an animal or poultry drinking trough should be provided with a float-operated valve or other similar effective device.

The inlet device should be of type AA or AB air gap installed to prevent backflow from a fluid category 5 risk and contamination of the supply pipe. The inlet device and backflow arrangements must be protected from damage. The installation arrangements of the trough will be accepted as being satisfied providing they comply with the requirements of BS 3445: fixed agricultural water troughs and water fittings.

A service valve should be provided on the inlet pipe to every drinking appliance for animals or poultry.

Where a number of animal drinking troughs are supplied with water from a single trough, the spill-over levels of other troughs must be at a higher level than the initial drinking trough where the water inlet device is located.

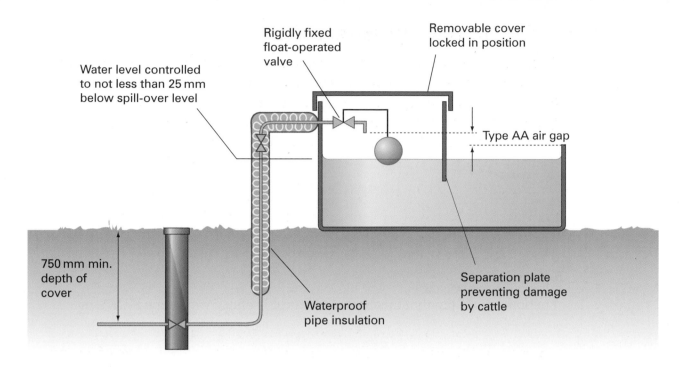

Water level controlled to not less than 25 mm below spill-over level

Rigidly fixed float-operated valve

Removable cover locked in position

Type AA air gap

750 mm min. depth of cover

Waterproof pipe insulation

Separation plate preventing damage by cattle

Figure 2.67 A cattle trough installation with a Type AA air gap

Animal drinking trough

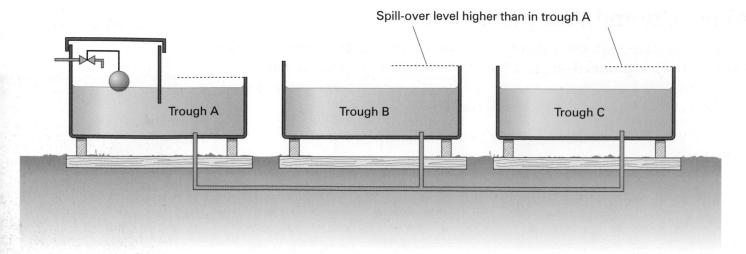

Figure 2.68 A combination of inter-connected cattle troughs

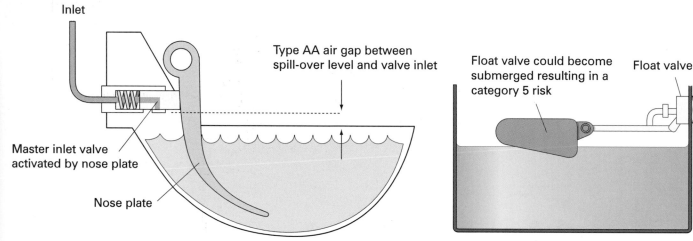

Figure 2.69 Animal drinking bowl with air gap built into appliance

Figure 2.70 Animal drinking bowl with submerged water outlet or inadequate air gap at the appliance

Where animal drinking bowls are to be installed, the source of the water supply will depend on the type of bowls being installed.

The type of bowl shown in Figure 2.69 has a spring-return or float-valve device which operates when depressed by the animal's mouth. These may be connected directly from a supply pipe or distributing pipe, providing the type AA air gap is maintained and that the animal's mouth can't come into direct contact with the outer nozzle.

Where the outlet nozzle is below the spill-over level or is likely to be contaminated by animals' mouths, the bowl must be supplied from a dedicated distributing pipe that only supplies similar appliances.

Whole-site and zone protection

The WSR Guidance Document states that whole-site or zone backflow-prevention devices should be provided on the supply pipe, such as a single check valve, or double check valve, or other no less effective backflow-prevention device, according to the level of risk as deemed by the water undertaker, where:

- a supply or distributing pipe conveys water to two or more separately occupied premises

- a supply pipe conveys water to premises that are required to provide sufficient water storage for 24 hours of ordinary use.

It requires that whole-site or **zone protection** should be provided in addition to individual requirements at points of use and within the system. The use of zone protection is extremely important in premises where industrial, medical or chemical processes are undertaken alongside the supply of water for domestic purposes such as drinking.

GD G15.24

Did you know?

Whole-site protection is used to protect one building from another. This used to be termed secondary protection

GD G15.25

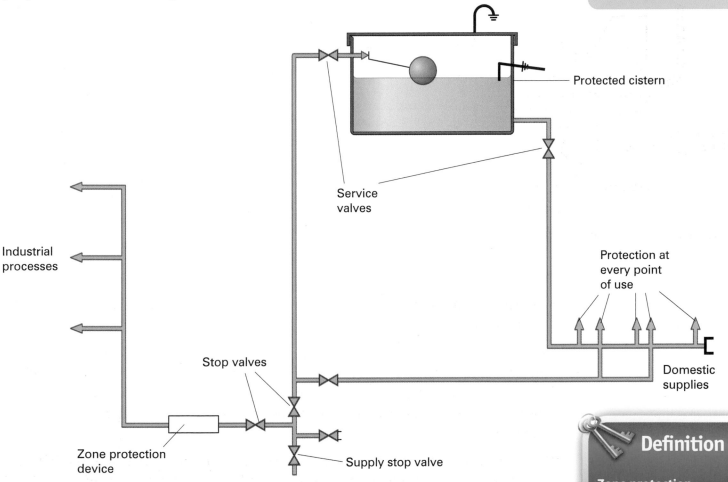

Figure 2.71 Example of zone protection

Definition

Zone protection – used to protect one part (zone) of a building from another part

Backflow protection to fire systems

GD G15.27–29

Fire-protection systems require backflow protection to suit the level of risk. Wet sprinkler systems (that contain no additives), fire-hose reels and hydrant landing valves are considered to be a fluid category 2 risk and will require the minimum protection of a single check valve.

Wet sprinkler systems in exposed situations often have additives in the water to prevent freezing at low ambient temperatures, and these are considered to be a fluid category 4 risk. Also included in this risk category are systems that contain hydro-pneumatic pressure vessels; therefore the system requires either a verifiable backflow preventer (RPZ or type BA) or it must be fitted with a suitable air gap (type AA, AB, AD or AUK1).

Sprinkler head

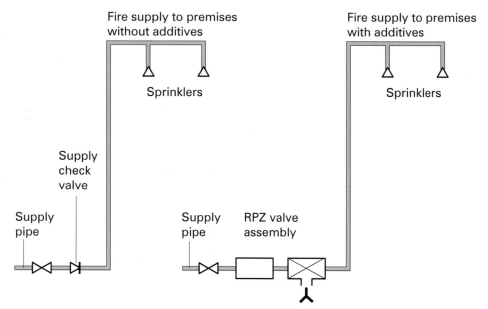

Figure 2.72 A sprinkler system (no additives used) with a single check valve installed, and a system (additives used) with a type BA (RPZ) valve

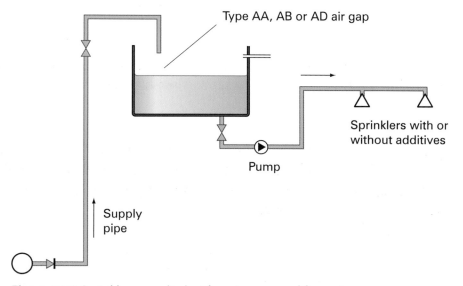

Figure 2.73 Sprinklers supplied with water pumped from storage

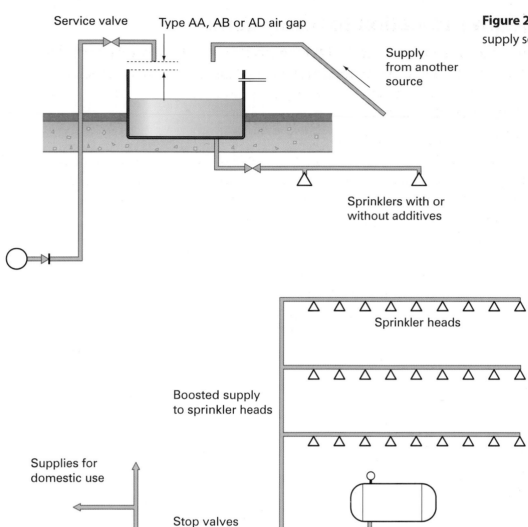

Figure 2.74 Storage cistern with supplementary supply servicing water for sprinkler use only

Figure 2.75 A boosted fire-protection system using a hydro-pneumatic pressure vessel. Used when the mains supply is not sufficient and water storage cannot be provided

In situations where fire-protection systems and drinking-water systems are served from a common domestic supply, the connection to the fire system should be taken from the supply pipe directly on entry to the building, and an appropriate backflow-protection device must be installed.

Safety tip

Some water suppliers may insist on independent service pipes for domestic supplies and fire protection

Cold-water services

At the end of this section you should be able to:

- name the valves used for shutting off the inflow of water into a storage cistern
- name the types of cisterns covered by the Regulations
- state the position in which servicing valves should be located
- describe suitable methods for alerting consumers to a cistern overflowing
- explain why and how cisterns should be covered.

WSR Schedule 2, para 16

In this section you will look at cold-water services and the requirements concerned with cold-water storage cisterns. This includes the control of incoming water, overflow pipes and warning pipes, and preventing waste and contamination in cisterns.

Paragraph 16 of Schedule 2 ensures that water supplied by the water undertaker for domestic purposes remains wholesome. It also looks at the provision of servicing valves on inlet and outlet pipes to cisterns, and the requirements of thermal insulation to minimise freezing and undue warming.

Paragraph 16 states that:

(1) Every pipe supplying water connected to a storage cistern shall be fitted with an effective adjustable valve capable of shutting off the inflow of water at a suitable level below the overflowing level of the cistern.

(2) Every inlet to a storage cistern, combined feed and expansion cistern, WC flushing cistern or urinal flushing cistern shall be fitted with a servicing valve on the inlet pipe adjacent to the cistern.

(3) Every storage cistern, except one supplying water to the primary circuit of a heating system, shall be fitted with a servicing valve on the outlet pipe.

(4) Every storage cistern shall be fitted with:

    (a) an overflow pipe, with a suitable means of warning of an impending overflow, which excludes insects;

    (b) a cover positioned so as to exclude light and insects; and

    (c) thermal insulation to minimise freezing or undue warming.

(5) Every storage cistern shall be so installed as to minimise the risk of contamination of stored water. The cistern shall be of an appropriate size, and the pipe connections to the cistern shall be so positioned as to allow free circulation and to prevent areas of stagnant water from developing.

Storage cisterns should be:

- fitted with an effective inlet-control device to maintain the correct water level
- fitted with servicing valves on inlet and outlet pipes
- fitted with a screened warning/overflow pipe to warn against impending overflow

- supported to avoid damage or distortion that might cause them to leak
- installed so that any risk of contamination is minimised, and arranged so that water can circulate and stagnation will not take place
- covered to exclude light or insects and insulated to prevent heat losses and undue warming
- corrosion-resistant and watertight and must not deform unduly, shatter or fragment when in use
- have a minimum unobstructed space above them of not less than 350 mm.

Remember

The entire base of all cold-water storage cisterns must be adequately supported to avoid distortion or damage. They must be installed in a position where the inside may be readily inspected

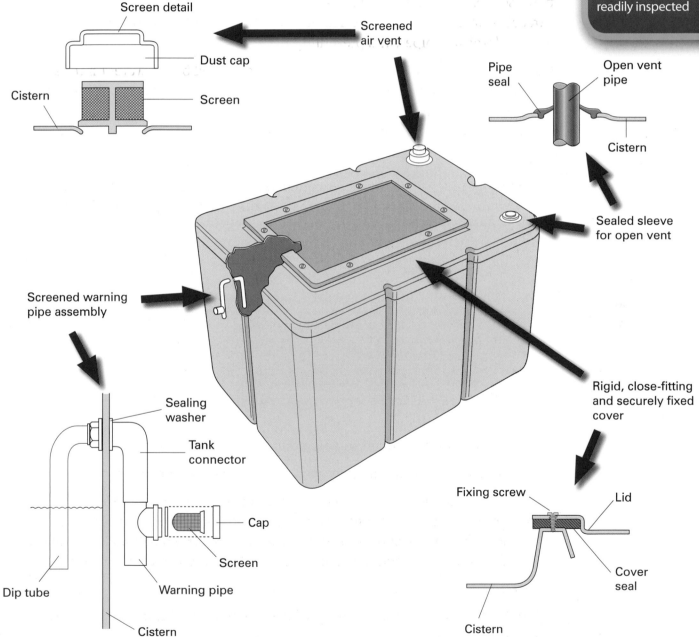

Figure 2.76 Requirements for protected cisterns used to store drinking water conforming to fluid category 1

WSR Schedule 2, para 16(1)

BS 1212

In situations where two or more cisterns are used to provide the required storage capacity, the cisterns should be connected in parallel, and to avoid stagnation, the float-operated valves should be adjusted so that they all operate to the same maximum water level. The cisterns must be connected in such a manner that there's an equal flow of water through each cistern.

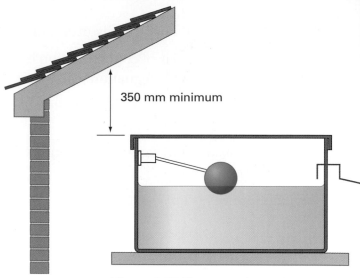

350 mm minimum

Figure 2.77 Clearance above cistern

Cistern inlet controls

Every pipe supplying water to a storage cistern shall be fitted with an effective adjustable shut-off device which will close when the water reaches its normal full level below the overflowing level of the cistern. Generally, the device will be a float-operated valve, although larger cisterns may be fitted with a float switch, connected to an electrically operated valve or pump.

Where float-operated valves are used, they should comply with one of the following standards (which cover valves up to 50 mm in diameter):

- BS 1212 – Part I – Portsmouth type
- BS 1212 – Part II – diaphragm valve (brass)
- BS 1212 – Part III – diaphragm valve (plastic)
- BS 1212 – Part IV – compact-type float-operated valve.

Float-operated valves used in WC cisterns should comply with these or with Part IV, which are specially designed for use in WC cisterns.

There are many float valves available that do not comply with the requirements of BS 1212. Look in the WRAS Water Fittings and Materials Directory for other acceptable types.

If installing valves above 50 mm in diameter you will need to ensure that they meet Water Regulations standards. This can be done by checking in the Water Fittings and Materials Directory, asking the Water Regulations Advisory Scheme for advice or contacting the local water undertaker.

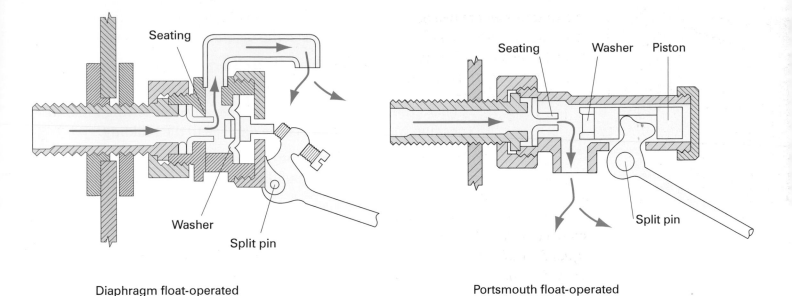

Diaphragm float-operated
valve BS 1212 Pts. 2/3

Portsmouth float-operated
valve BS 1212 Pt. 1

Figure 2.78 Internal workings of cistern inlet controls

Cistern control valves

Every inlet to a storage cistern, combined feed and expansion cistern shall be provided with a servicing (isolation) valve on the inlet pipe adjacent to the cistern. This requirement also applies to WC and urinal cisterns.

> **WSR Schedule 2, para 16(2)**

The servicing valve should be fitted as close as is reasonably practical to the float-operated valve or other device. This does not apply to a pipe connecting two or more cisterns with the same overflowing levels.

The requirements of paragraph 16.3 state that every cistern (except one supplying water to a primary circuit of a heating system) shall be provided with a servicing (isolation) valve on the outlet(s). The valve should be fitted as close as is reasonably practical to the cistern.

Warning and overflow pipes

Every storage cistern shall be fitted with an overflow pipe, with a suitable means of warning of an impending overflow. This requirement excludes urinal-flushing cisterns.

> **WSR Schedule 2, reg 16(4)(a)**

The requirement for overflow and warning pipes will vary depending on the water-storage cistern capacity. Cisterns up to 1000 litres or less actual capacity require only a single warning/overflow pipe. Where a cistern has a greater actual capacity than 1000 litres it is recommended that a warning pipe and an overflow pipe should be provided.

All warning pipes must discharge in a conspicuous position, and the water overflow pipe should discharge in a suitable position elsewhere.

Remember

The warning pipe must be installed so that it discharges immediately when the water in the cistern reaches the defined overflowing

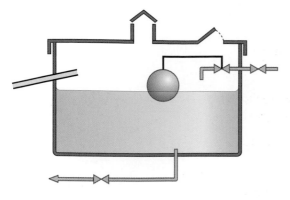

Figure 2.79 Overflow and warning pipes

To ensure that you have the correct understanding of an overflow pipe and warning pipe, look at the distinctions between the two:

- An overflow pipe is a pipe from a cistern in which water flows only when the water level in the cistern reaches a predetermined level.

- The overflow pipe is used to discharge any overflowing water to a position where it will not cause damage to the building.

- A warning pipe is a pipe from a cistern that gives warning to the owners or occupiers of a building that a cistern is overflowing and requires attention.

Either pipe must not have an internal diameter of less than 19 mm, and the diameter of the pipe installed must be capable of taking the possible flow in the pipe arising from any failure of the inlet valve.

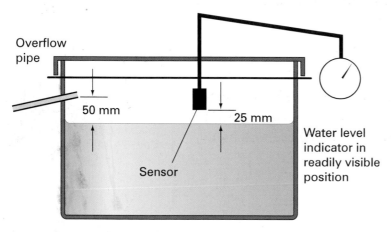

Figure 2.80 Cistern water level indicator

Larger cisterns

Cisterns that have an actual capacity greater than 5000 litres should be provided with an overflow that operates when the water level is 50 mm above the set shut-off level. It is acceptable to omit the warning pipe, but a level indicator should be provided, and the installation must include an audible or visible alarm which operates when the water reaches 25 mm below the opening of the overflow.

Alternative methods are a float-operated **water level indicator** and overflow pipe, as shown in Figure 2.80.

In situations where the cistern is larger than 10,000 litres capacity, the cistern must be fitted with either:

- a warning pipe and an overflow pipe (same criteria as for medium and large cisterns)

- an audible or visual alarm (electrically operated) which clearly indicates a rise in water level to within 50 mm of the cistern overflowing level

- a hydraulic audible or visual alarm that clearly indicates when the water level rises to within 50 mm of the overflowing level.

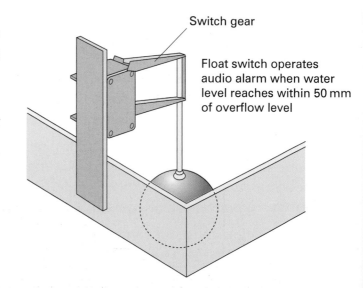

Figure 2.81 Large-capacity cistern warning device

Installation factors

There are several important installation factors to consider when installing overflow/warning pipes. These are:

- The overflow/warning pipe must be capable of removing the excess water from the cistern without the inlet device becoming submerged in the event of an overflow.

- Warning pipes are to discharge in a conspicuous position, preferably in an external location.

- Warning/overflow pipes must fall continuously from the cistern to the point of discharge.

- Feed and expansion cisterns must have separate warning pipes from those serving cold-water cisterns.

- Warning pipes and overflow pipes must be fitted with some means of preventing the ingress of insects etc. (usually in the form of screens or filters).

- When the installation consists of two or more cisterns, the warning or overflow pipe must be arranged so that the cistern can't discharge one into the other.

Paragraph 16.4(b) requires that cisterns are to be fitted with a cover and be positioned so as to exclude light and insects, and 16.4(c) requires that insulation shall be fitted to minimise freezing or undue warming – this includes insulation to the overflow and warning pipe.

WSR Schedule 2, para 16(4)

Contamination of stored water

Paragraph 16.5 requires that cisterns are to be installed to minimise the risk of contamination of stored water. This can be achieved to a certain extent by installing a 'protected cistern', especially in cases where water is supplied for domestic purposes.

WSR Schedule 2, para 16(5)

Further requirements of the paragraph are that the cistern must be of an appropriate size and connections positioned to allow circulation and prevent areas of stagnation from developing. To reduce the potential risk of contamination in cisterns, the following factors should be considered:

- Cistern outlet connections should be connected as low as possible; this will allow sediment to pass through the taps rather than settle in the base of the cistern.

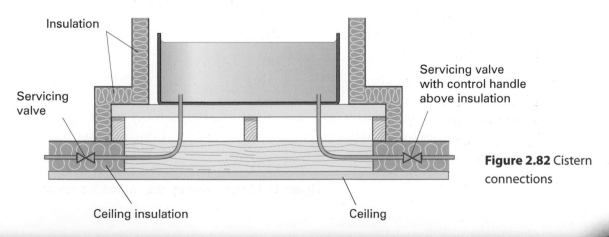

Figure 2.82 Cistern connections

Remember

Make sure the connections are made into the bottom of the cistern and not the side wall

- Cisterns must be adequately sized and not oversized, thus reducing the risk of legionella and ensuring that there's a speedy replenishment of fresh water when stored water is being drawn off.
- Cistern outlet connections should be installed so as to allow movement of water throughout the entirety of the cistern. This can be achieved by connecting at least one outlet pipe to the appropriate end of the inlet connection.

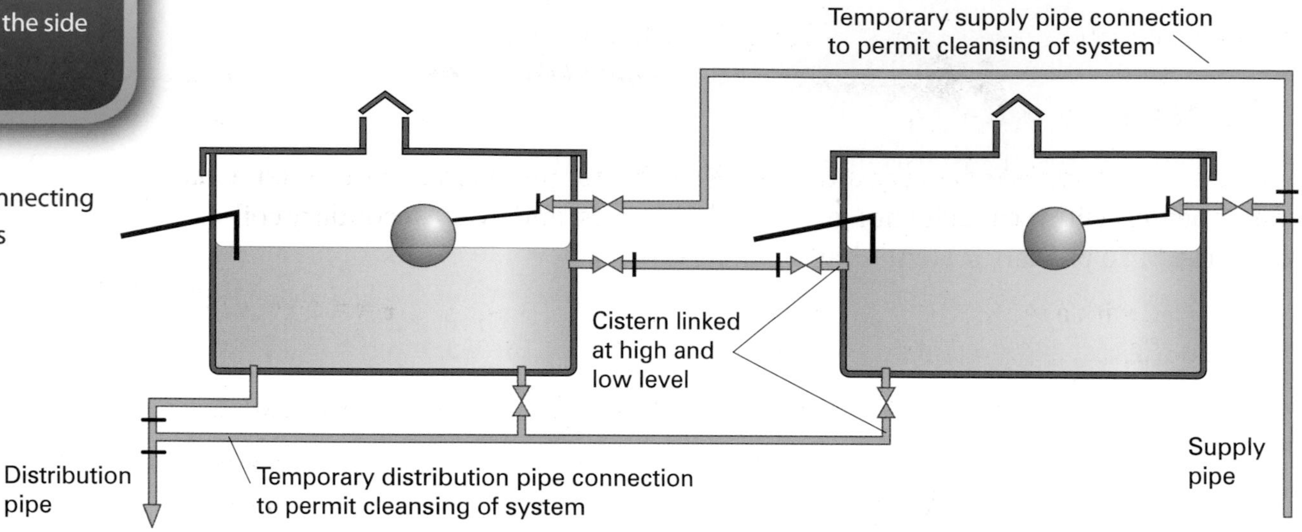
Temporary supply pipe connection to permit cleansing of system
Cistern linked at high and low level
Supply pipe
Distribution pipe
Temporary distribution pipe connection to permit cleansing of system

Figure 2.83 Connecting cisterns in series

- In instances where storage cisterns are linked together they should be installed in such a manner that they can be drained and cleaned easily. Cisterns that are connected in series should also be installed in such a manner as to allow a good throughput of water to reduce the risk of stagnation.

Hot-water services

At the end of this section you should be able to:

- state the difference between vented and unvented hot-water systems
- understand the sizing requirements for vent pipes
- state the methods of preventing tampering with control valves
- state the purposes of thermal insulation
- state the reasons for preventing stagnation
- explain the difference between an overflow pipe and a warning pipe.

WSR Schedule 2, para 17–24

The Regulation requirements for hot-water services are generally concerned with the prevention of wasted water and the overall safety of the building and its occupants where services are installed.

Paragraphs 17 to 24 of Schedule 2 mostly regulate the installation of the hot-water service within the building, and look at the following issues:

- expansion in hot-water systems
- the measures required to accommodate expansion in vented and unvented systems
- control of water temperature and safety devices
- the discharge from temperature-relief valves and **expansion valves**
- backflow prevention to closed circuits (filling loops etc.).

The Requirements

Paragraph 17.1 states that every unvented water heater (not being an instantaneous water heater) with a capacity not greater than 15 litres, and every secondary coil contained in a primary system, shall:

- be fitted with a temperature-control device and either a temperature-relief valve or a combined temperature- and pressure-relief valve, or
- be capable of accommodating expansion within the secondary hot-water system.

Paragraph 17.2 states that an expansion valve shall be fitted to ensure that the water is discharged in a correct manner in the event of a malfunction of the expansion vessel or system.

Paragraph 18 states that appropriate vent pipes, temperature-control devices and combined temperature- and pressure-relief valves shall be provided to prevent the temperature of the water within a secondary hot-water system from exceeding 100°C.

Paragraph 19 states that discharges from temperature-relief valves and expansion valves shall be made in a safe and conspicuous manner.

Paragraph 20.1 states that no vent pipe from a primary circuit shall terminate over a storage cistern containing wholesome water for domestic supply or for supplying water to a secondary system.

Paragraph 20.2 states that no vent pipe from a secondary circuit shall terminate over any combined feed and expansion cistern connecting to a primary circuit.

Paragraph 22.1 states that every expansion valve, temperature-relief valve, or combined temperature- and pressure-relief valve connected to any fitting or appliance shall close automatically after a discharge of water.

Paragraph 22.2 states that every expansion valve shall:

- be fitted on the supply pipe close to the hot-water vessel and without any intervening valves
- only discharge water when subjected to a water pressure of not less than 0.5 bar (50 kPa) above the pressure to which the hot-water vessel is, or is likely to be, subjected in normal operation.

Paragraph 23.1 states that a temperature-relief or combined temperature- and pressure-relief valve shall be provided on every unvented hot-water storage vessel with a capacity greater than 15 litres.

Paragraph 23.2 states that the valve shall:

- be located directly on the storage vessel in an appropriate location, and have a sufficient discharge capacity to ensure that the temperature of the stored water does not exceed 100°C

- only discharge water at below its operating temperature when subjected to a pressure of not less than 0.5 bar (50 kPa) in excess of the greater of the following:

  - the maximum working pressure in the vessel in which it is fitted

  - the operating pressure of the expansion valve.

Paragraph 23.3 states that 'unvented hot-water storage vessel' means a hot-water storage vessel that does not have a vent pipe to the atmosphere.

Paragraph 24 states that no supply pipe or secondary circuit shall be permanently connected to a closed circuit for filling a heating system unless it incorporates a backflow-prevention device in accordance with a specification approved by the regulator for the purpose of this Schedule.

Vent pipes

WSR Schedule 2, para 18

Paragraph 18 requires open-vented systems to use a 'vent' to dissipate excessive heat from the system, should the storage vessel overheat or boil. This is normally in the form of a vent pipe (sometimes termed the open safety vent). The vent pipe will provide a safe path for boiling water and steam to exit from the top of the storage vessel, dissipating over into the feed cistern.

WSR Schedule 2, para 20

Paragraph 20 of the Regulations looks at the termination of vent pipes. Paragraph 20.1 states: no vent pipe from a primary circuit shall terminate over any storage cistern containing wholesome water for domestic supply or for supplying water to a secondary system. The reason behind the requirement is that any water discharged from the primary circuit could contaminate the stored water supply of the secondary circuit.

WSR Schedule 2, para 20(2)

Paragraph 20.2 states that no vent pipes from a secondary circuit shall terminate over any combined feed-and-expansion cistern connected to a primary circuit.

Figure 2.84 shows a double-feed indirect hot-water system with open venting arrangements to primary and secondary circuits.

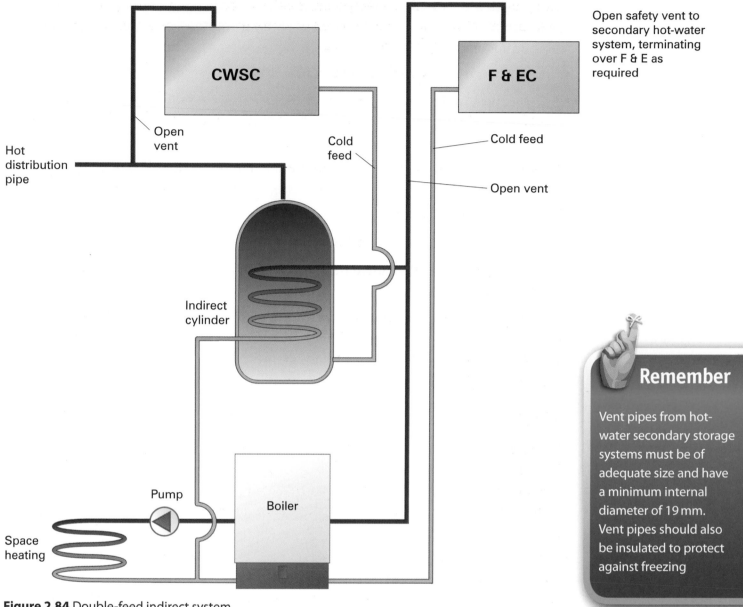

Figure 2.84 Double-feed indirect system

In the diagram:
- CWSC
- F & EC
- Open safety vent to secondary hot-water system, terminating over F & E as required
- Open vent
- Hot distribution pipe
- Cold feed
- Cold feed
- Open vent
- Indirect cylinder
- Pump
- Boiler
- Space heating

Expansion in vented systems

Expansion in vented systems

Paragraph 21 is concerned with the control of water expansion in vented primary circuits, as previously stated in unvented systems. Unless expansion is allowed to take place within the system there's a potential danger of the hot-water vessel being damaged or even exploding.

When water in a vented hot storage vessel is heated it expands, and cooler water at the bottom of the vessel is displaced, returning back into the cold-feed pipe that supplies the vessel.

WSR Schedule 2, para 21

> **Remember**
>
> Vent pipes from hot-water secondary storage systems must be of adequate size and have a minimum internal diameter of 19 mm. Vent pipes should also be insulated to protect against freezing

> **Find out**
>
> If you know the approximate percentage volume expansion of water when heated from 4°C to 100°C, what volume has to be allowed for expansion in a vessel containing 100 litres?

Every feed-and-expansion cistern (this includes cold-water combined feed-and-expansion cisterns connected to a primary or heating circuit) must be capable of accommodating any expansion water from the circuit, installed in a manner so that the water level isn't less than 25 mm below the overflowing level of the warning pipe, when the primary or heating circuit is in use.

Vent pipe to terminate not less than twice the diameter of the vent pipe above the top of the float-operated valve A or the top of the overflow pipe B, whichever is greater

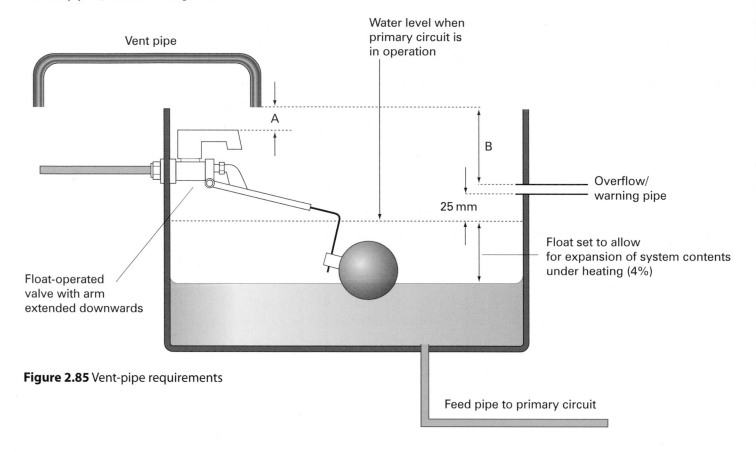

Figure 2.85 Vent-pipe requirements

WSR Schedule 2, para 24

Paragraph 24 (the final paragraph) looks at connections to primary circuits. The requirement states: no supply pipe or distributing pipe shall be permanently connected to a closed circuit for filling a heating system unless it incorporates a backflow device in accordance with a specification approved by the regulator for the purpose of this Schedule.

Primary and other closed circuits have to be filled with water initially and may require additional 'top-ups' at intervals during use. Primary circuits may contain additives and the water can be heavily contaminated, therefore they are not to be permanently connected to any supply pipe without an adequate backflow-prevention device.

Where a connection is made to a supply pipe or a distributing pipe, in some instances for supplying water for filling or replenishing water in a closed circuit, such as a hot-water primary circuit and/or a space heating system, it is essential that:

- there is no backflow of water, at any time, from the primary circuit into the water supply, and

- the water supply is disconnected, or vented to the atmosphere, during the periods between filling and subsequent replenishing of the water in the primary circuit.

Under normal operating conditions, the pressure in the primary heating circuit is less than that of the pipe supplying water to the circuit. However, in the event of a malfunction of an expansion valve or pressure-relief valve in the primary circuit, pressure may rise above the pressure in the supply pipe.

In such an instance, a mechanical backflow-prevention device could become damaged and cease to function. If there is no discontinuity or venting to atmosphere and (as frequently happens) the valve controlling the water supply has been left in the open position, fluid from the primary circuit may return into the supply pipe.

It is therefore essential that when the filling or replenishing of the primary circuit is complete, there should be a discontinuity at the point of connection, or the type of backflow-prevention device installed shall be of a type that allows any fluid resulting from excess pressure in the primary circuit to discharge to waste.

The type of backflow-prevention device required should be suitable for a category 3 risk in the case of a house or for fluid category 4 risks with installations in premises other than a house. It is, however, essential that there is a discontinuity in the connecting pipework, or a backflow-prevention arrangement is used, in which any fluid resulting from backflow from the primary circuit is discharged to waste.

To avoid control valves being tampered with or left in the open position it is recommended that all control valves used in connection with filling loops should be lockshield-type valves with a loose key.

A satisfactory method of filling or replenishing a primary circuit in a house is shown in the first illustration in Figure 2.86, where the temporary connecting pipe is completely disconnected after filling or replenishment.

Another method that is considered acceptable for fluid category 3 risk in a house is the installation of a Type CA 'Non-verifiable disconnector with different pressure zones backflow-prevention device', which is shown in the second illustration in Figure 2.86 .

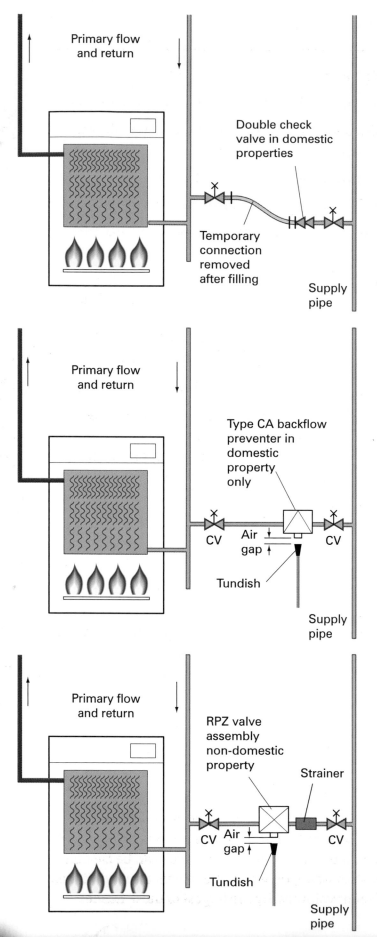

In buildings other than a house, where backflow protection against fluid category 4 risk is required, a Type BA 'Verifiable backflow preventer with reduced pressure zone (RPZ) valve' or a backflow preventer with a strainer on the inlet could be used, which is shown in the third illustration in Figure 2.86.

Figure 2.86 Filling a primary circuit

Other appliances

At the end of this section you should be able to:

- state the requirements for water conservation with respect to WCs and urinals
- state the different devices permitted to be used for WCs and urinals in domestic and non-domestic premises
- describe the variety of permissible warning devices to alert users of an overflow
- state the requirements for plugs to contain water within a basin or sink
- describe the water conservation measures for dishwashers, washing machines and washer-driers.

This section will cover a wide scope of appliances and look at the requirements of the related paragraphs 25 to 29. Paragraph 25 covers WCs, flushing devices and urinals. The concerns of the paragraph are primarily with the 'conservation of water', and deal in particular with the flushing arrangements of WCs and urinals.

WCs and urinals can use large quantities of water, especially WCs – their use can account for approximately 25 per cent of the water consumed within a dwelling.

The Regulations following previous Byelaws made new provisions and reduced the standard flushing volume for wash-down WCs from 7.5 litres to a maximum of 6 litres. The requirement applies to WC suites installed after 1 January 2001. Further conservation measures have been implemented, with manufacturers being encouraged to produce dual-flush cisterns. These cisterns have a maximum full flush of 6 litres and a reduced flush of 4 litres.

The aim of the new requirements of the Regulations for flushing cisterns and urinals is to bring the UK in line with European practices, encouraging new and innovative flushing arrangements, with the result that a number of changes have been made to previous practices.

Also since 1 January 2001, WC pans and urinals in non-domestic premises have been permitted to be supplied by a flushing cistern, or by pressure-flushing valve, provided that suitable backflow-prevention devices are installed. Siphonic and non-siphonic devices are also permitted.

The Requirements

Paragraph 25 of Schedule 2 states:

(1) Subject to the following provisions of this paragraph:

    (a) every water-closet pan shall be supplied with water from a flushing cistern, pressure-flushing cistern or pressure-flushing valve, and shall be so made and installed that after normal use its contents can be cleared effectively by a single flush of water, or, where the installation is designed to receive flushes of different volumes, by the largest of those flushes;

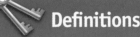

Definitions

Pressure-flushing cistern – a WC flushing device that utilises the pressure of water within the cistern supply pipe to compress air and increase the pressure of water available for flushing a WC pan

Pressure-flushing valve – a self-closing valve supplied with water directly from a supply pipe or a distributing pipe which when activated will discharge a predetermined flush volume

Trap – a pipe fitting, or part of a sanitary appliance, that retains liquid to prevent the passage of foul air

Warning pipe – an overflow pipe whose outlet is located in a position where the discharge of water can be readily seen

(b) no pressure-flushing valve shall be installed:

    (i) in a house, or

    (ii) in any building not being a house where a minimum flow rate of 1.2 litres per second cannot be achieved at the appliance;

(c) where a pressure-flushing valve is connected to a supply pipe or distributing pipe, the flushing arrangement shall incorporate a backflow-prevention device, consisting of a permanently vented pipe interruptor located not less than 300 mm above the spill-over level of the WC pan or urinal;

(d) no flushing device installed for use with a WC pan shall give a single flush exceeding 6 litres;

(e) no flushing device designed to give flushes of different volumes shall have a lesser flush exceeding two-thirds of the largest flush volume;

(f) every flushing cistern, other than a pressure-flushing cistern, shall be clearly marked internally with an indelible line to show the intended volume of flush, together with an indication of that volume;

(g) a flushing device designed to give flushes of different volumes:

    (i) shall have a readily discernible method of actuating the flush at different volumes; and

    (ii) shall have instructions, clearly and permanently marked on the cistern or displayed nearby, for operating it to obtain the different volumes of flush;

(h) every flushing cistern, not being a pressure-flushing cistern or a urinal cistern, shall be fitted with a warning pipe or with a no less effective device;

(i) every urinal that is cleared by water after use shall be supplied with water from a flushing device which:

    (i) in the case of a flushing cistern, is filled at a rate suitable for the installation;

    (ii) in all cases, is designed or adapted to supply no more water than is necessary for effective flow over the internal surface of the urinal and for replacement of the fluid in the trap; and

(j) except in the case of a urinal which is flushed manually, or which is flushed automatically by electronic means after use, every pipe which supplies water to a flushing cistern or trough used for flushing a urinal shall be fitted with an isolating valve controlled by a time switch and a lockable isolating valve, or with some other equally effective automatic device for regulating the periods during which the cistern may fill.

(2) Every water closet, and every flushing device designed for use with a water closet, shall comply with a specification approved by the regulator for the purposes of this Schedule.

(3) The requirements of sub-paragraphs (1) and (2) do not apply where faeces or urine are disposed of through an appliance that does not solely use fluid to remove the contents.

(4) The requirement in sub-paragraph (1)(i) shall be deemed to be satisfied:

    (a) in the case of an automatically operated flushing cistern servicing urinals, which is filled with water at a rate not exceeding:

        (i) 10 litres per hour for a cistern servicing a single urinal;

        (ii) 7.5 litres per hour per urinal bowl or stall, or, as the case may be, for each 700 mm width of urinal slab, for a cistern servicing two or more urinals;

    (b) in the case of a manually or automatically operated pressure-flushing valve used for flushing urinals, which delivers not more than 1.5 litres per bowl or position each time the device is operated.

(5) Until 1 January 2001 paragraphs (1)(a) and (d) shall have effect as if they provided as follows:

    (a) every water-closet pan shall be supplied with water from a flushing cistern or trough of the valveless type which incorporates siphonic apparatus;

    (b) no flushing device installed for use with a WC pan shall give a single flush exceeding 7.5 litres.

(6) Notwithstanding sub-paragraph (1)(d) a flushing cistern installed before 1 July 1999 may be replaced by a cistern which delivers a similar volume and which may be either single flush or dual flush; but a single-flush cistern may not be so replaced by a dual-flush cistern.

Flushing methods for WC pans

Paragraph 25 (a) states that WCs may be flushed using a flushing cistern, or a pressure-flushing valve. Once flushed, the WC flushing device must clear the contents of the bowl effectively using a single flush of water.

WSR Schedule 2, para 25(a)

A single-flush WC

The illustration below shows a typical siphon-type single-flush cistern.

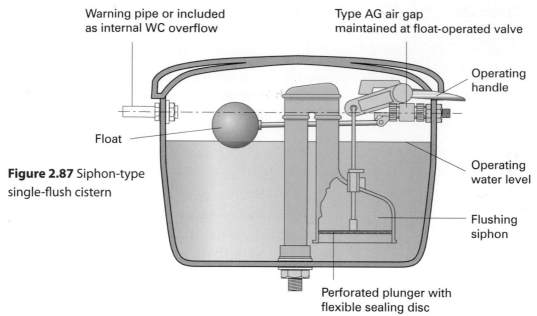

Figure 2.87 Siphon-type single-flush cistern

Cisterns can be arranged to provide for single- or dual-flush action, but where dual-flush devices are installed, the single lower flush volume should be adequate to clear urine and paper. The larger flush volume should clear faeces and paper.

Figure 2.88 shows a valve-type WC cistern that may be installed in houses and which tends to be the most common type that is now installed.

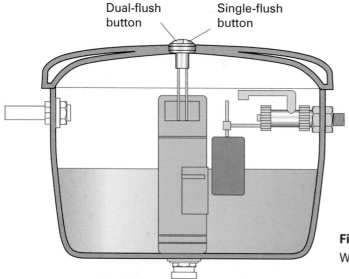

Figure 2.88 Valve-type WC cistern

Flushing valves

WSR Schedule 2, para 25(b)

Paragraph 25(b)(i) states that WCs may be flushed using manually operated flushing valves but only:

- in premises other than a house
- where a minimum flow rate of 1.2 litres per second can be achieved at the appliance – Paragraph 25(b)(ii).

WSR Schedule 2, para 25(c)

Paragraph 25(c) states that pressure-flushing valves may be supplied from a supply pipe or from a distributing pipe. The requirement also asks for backflow protection for WCs supplied through pressure-flushing valves. The flushing valves are required to be fitted with a permanently vented pipe interruptor to protect against a fluid category 5 backflow risk.

Pipe interrupters should be positioned:

- not less than 300 mm above the spill-over level of the pan
- at the outlet of the flushing valve.

There should be no valve, tap or restriction to the flow at its outlet.

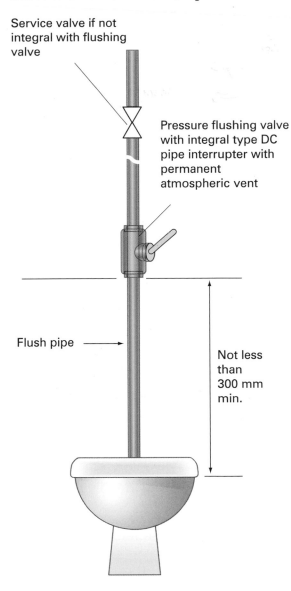

Service valve if not integral with flushing valve

Pressure flushing valve with integral type DC pipe interrupter with permanent atmospheric vent

Flush pipe

Not less than 300 mm min.

Figure 2.89 Pressure-flushing valve

Paragraph 25(d) states that no flushing device installed for use with a WC pan shall give a single flush exceeding 6 litres, and paragraph 25(e) goes on to say no flushing device designed to give a flush of different volumes shall have a lesser flush exceeding two thirds of the largest flush volume. So theoretically the permitted flushing volumes for WCs are:

- single flush: 6 litres
- dual flush: 6 litres full flush, 4 litres lesser flush.

> **WSR Schedule 2, para 25(d–e)**

Paragraph 25(g)(ii) states: flushing devices designed to give flushes of different volumes shall have instructions clearly and permanently marked on the cistern or displayed nearby for operating it to obtain the different volumes of flush.

> **WSR Schedule 2, para 25(g)**

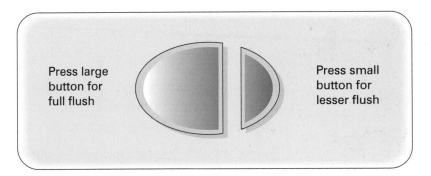

Press large button for full flush

Press small button for lesser flush

Figure 2.90 Operating directions

Exemptions to the Requirements

Paragraph 25(5) exempts flushing cisterns installed before 1 January 2001. This is the date that the Regulator's Specification for WC suite performance came into force. This was written to compliment the Byelaws and requires manufacturers of WC suites to meet stringent standards to give more efficient and effective flushing.

> **WSR Schedule 2, para 25(5)**

Permitting innovative flushing mechanisms brings us more in line with current European standards. However, it should be noted that although the dual-flush cistern was discontinued back in 1986 their installation has been re-permitted since 1 July 1999.

WSR Schedule 2, para 25(6)

Paragraph 25(6) exempts flushing cisterns installed before 1 July 1999, and cisterns that were installed under the current byelaw can be replaced by a cistern that delivers a similar volume to the original one.

Throughout your career you will encounter many different types of WC suite installations, with flushing cisterns ranging in capacity from the new Regulation 6 litres through to 7.5 litres, 9 litres and possibly larger. The types that you come across will include both single- and dual-flush mechanisms.

WC cistern warning pipes

WSR Schedule 2, para 25(1)(h)

The requirement of paragraph 25(1)(h) is that all flushing cisterns or other urinal flushing or pressure-flushing cisterns are to be fitted with a warning pipe or some no less effective device. The warning pipe must:

- have a minimum diameter of 19 mm or at least one size greater than the inlet pipe (whichever is the greater)
- terminate externally to the building where it will be readily noticed.

The water supply industry considers that WCs that have an internal overflow discharging into the WC pan shall be deemed to meet the requirements of the Regulations, in that the internal overflow will be regarded as a 'no less effective device' (in place of a warning pipe); many WCs are now supplied with an integral overflow as opposed to having a separate overflow pipe.

A warning pipe may also discharge directly into the flush pipe (without a tundish), as this can be considered equivalent to an internal overflow. These two interpretations are conditional upon measures to reduce the likelihood of the internal overflow being used.

For compact inlet valves (those manufactured to BS 1212: part 4) the provision of a gauze strainer, incorporated in or fitted upstream of the float-operated inlet valve to trap debris (swarf etc.) which might cause premature failure of the valve, is required.

Manufacturers are encouraged by the water supply industry to supply, as a unit, the cistern complete with a valve and a strainer. The water-supply industry considers the following to be a no less effective device:

- a visible warning, for example a tundish sight glass, mechanical signal or an electrically operated device such as an indicator lamp
- an audible signal
- a mechanical device that enables the flush, thereby indicating to the user when there is a fault in the WC flushing system
- a device that detects when the water level rises above the maximum operating level and closes the water supply to the float-operated valve.

Ranges of WCs

Individual warning pipes from a range of WCs are permitted to discharge into a common warning pipe, provided that:

- the common pipe serves WCs only and that they are at the same level
- the individual warning pipes discharge into a tundish which is visible and must incorporate a Type AA air gap.

Other arrangements

Internal overflow pipes are permitted to discharge into the WC pan, on condition that a Type AG air gap is provided between the lowest level of the outlet of the float valve and the discharge from the overflow pipe.

The installations below, although legal, will tend to become obsolete due to the changes in the Regulations, and with the introduction of the 6-litre flush cistern that will incorporate some type of internal overflow arrangement.

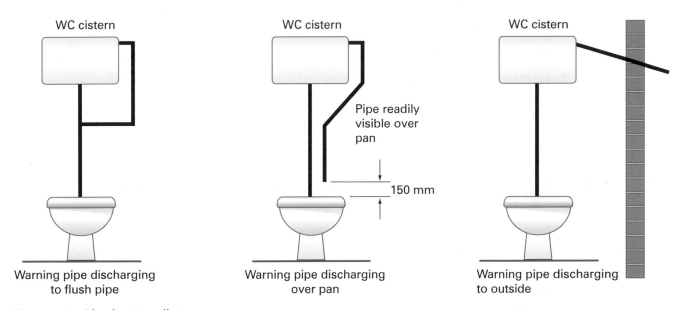

WC cistern

WC cistern

Pipe readily visible over pan

150 mm

WC cistern

Warning pipe discharging to flush pipe

Warning pipe discharging over pan

Warning pipe discharging to outside

Figure 2.91 Obsolete installations

Flushing urinals

Flushing devices for urinals should be designed so that they do not supply more water than is necessary to effectively clear the urinal and replace the trap seal. The acceptable flushing methods to meet the Regulations are:

- by flushing cistern, operated manually or automatically
- by flushing valve, operated manually or automatically.

Paragraph 25(4) sets out the maximum volumes of water permitted for urinal flushing.

WSR Schedule 2, para 25(4)

Manually operated cistern with chain

Automatically operated flushing cisterns

These should supply no more water than:

- 10 litres per hour for a single urinal bowl or stall

- 7.5 litres per hour, per urinal position, for a cistern servicing two or more urinal bowls, stalls or per 700 mm slab position.

Manually operated (chain-pull or push-button) cisterns to a single urinal bowl

These are required to flush no more than 1.5 litres each time the cistern is operated.

Pressure-flushing valves operated manually or automatically

These should not flush more than 1.5 litres each time the valve is operated.

Pressure-flushing valves may be fed with water from either a supply pipe or a distributing pipe. The outlet of the pressure-flushing valve should incorporate (or be provided with a pipe interruptor with a permanent atmospheric vent) the flushing valve being installed, so that the level of the lowest vent aperture is not less than 150 mm above the **sparge outlet** and not less than 300 mm above the spill-over level of the urinal.

Unless a servicing valve is integral with the pressure-flushing valve, it is recommended that a separate servicing valve be provided on the branch pipe to each pressure-flushing valve.

Service valve if not integral with pressure-flushing valve

Pressure-flushing valve with integral-type DC pipe interrupter with permanent atmospheric vent

Flush pipe

A

B

A to be not less than 150 mm and B not less than 300 mm

Urinal

Water-saving valves

Paragraph 25(1)(j) of the Regulations focuses on water-saving controls for urinals, and states that any urinals supplied either manually or electronically from a flushing cistern must have a time switch (and a lockable isolating valve) fitted to its incoming supply, or some other equally effective automatic means of regulating the periods during which the cistern may fill.

> **WSR Schedule 2, para 25(1)**

Figure 2.92 Pressure-flushing valve to urinal

The prevention of water-flow to urinal cisterns during periods when the building is not occupied can be achieved in several ways:

- by incorporating a time-operated switch controlling a solenoid valve which cuts off the water supply when other appliances are used

- by an 'impulse'-initiated automatic system that allows water to pass to a urinal cistern only when other appliances are used

- by proximity or sensor devices (infra-red).

Figure 2.93 shows an example of a system containing a timing device and an automatic isolation valve.

Figure 2.94 shows the urinal operation controlled by a hydraulic valve.

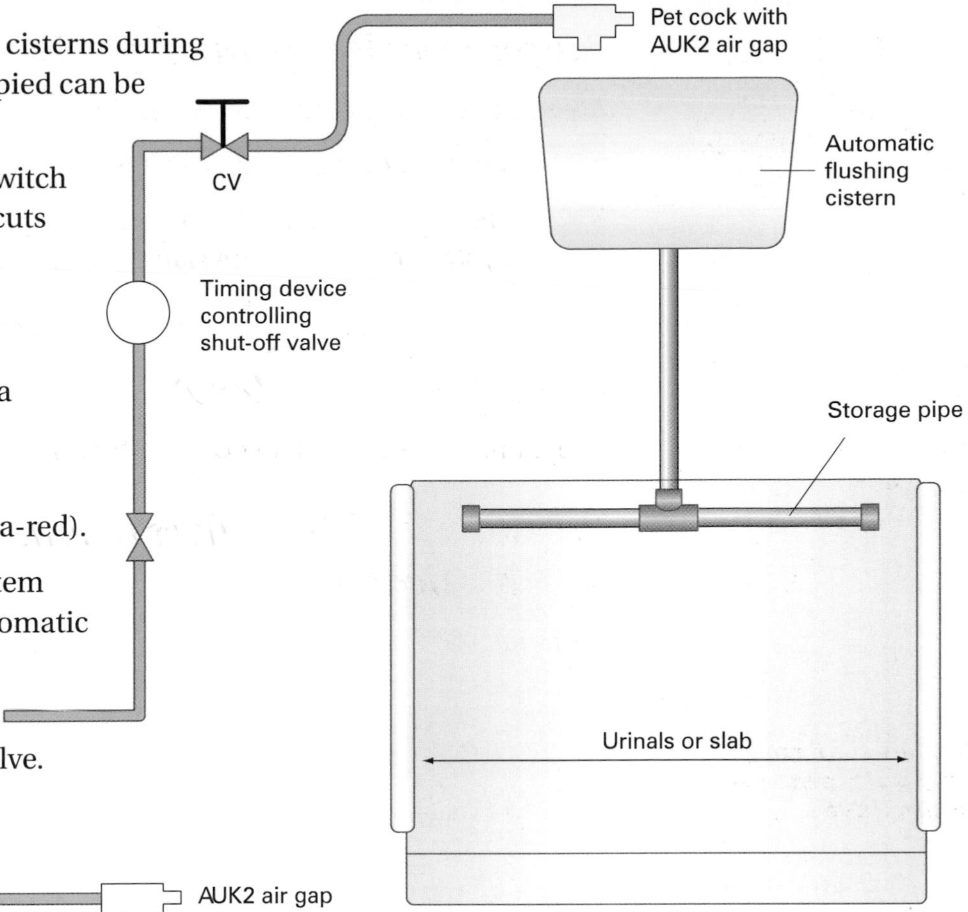

Pet cock with AUK2 air gap

CV

Timing device controlling shut-off valve

Automatic flushing cistern

Storage pipe

Urinals or slab

Figure 2.93 System timing device

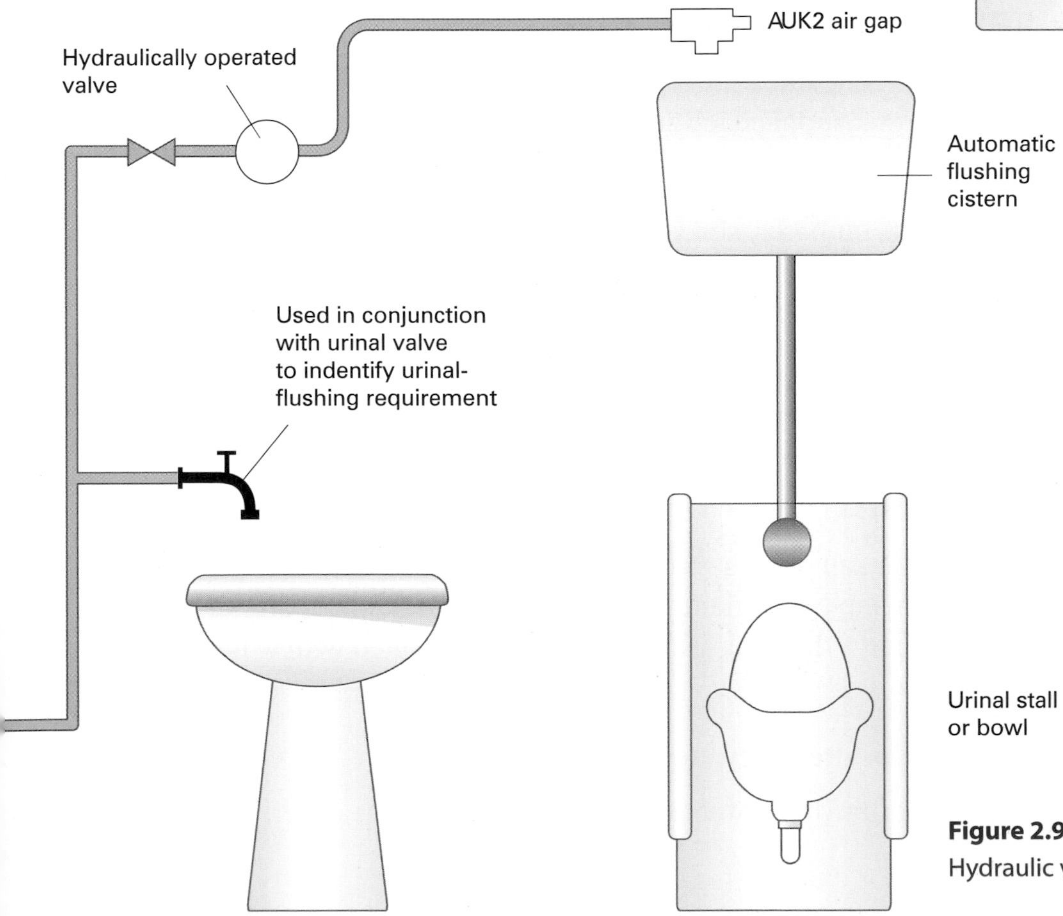

Hydraulically operated valve

AUK2 air gap

Used in conjunction with urinal valve to indentify urinal-flushing requirement

Automatic flushing cistern

Urinal stall or bowl

Figure 2.94 Hydraulic valve

Below is a table giving volumes and flushing intervals for urinals.

| Number of bowls, stalls or per 700 mm of slab | Volume of automatic flushing cistern | | | | Maximum fill rate in litres per hour |
| | 4.5 litres | 9 litres | 13.5 litres | 18 litres | |
	Shortest period between flushes in seconds				
1	27	54	81	108	10
2	18	37	54	72	15
3	12	24	36	48	22.5
4	9	18	27	36	30
5	7.2	14.4	21.6	28.8	37.5
6	6	12	18	24	45

Table 2.12 Volumes and flushing intervals for urinals

Baths, sinks, showers and taps

WSR Schedule 2, para 26–29

Paragraphs 26 to 29 of Schedule 2 cover the prevention of contamination and wastage of water relating to water supplied to various sanitary appliances, covering:

- the supply of drinking water for domestic uses
- plugs, waste outlets for baths, sinks, showers and taps
- the economical use of water in washing machines, washer-driers and dishwashers.

The concerns of the requirements are basically quite straightforward, with a conservation measure being implemented in paragraph 29: this introduces water-economy measures for washing machines and dishwashers.

The requirements

Paragraph 26 states that all premises supplied with water for domestic purposes shall have at least one tap conveniently situated for the drawing of drinking water. Paragraph 27 states that a drinking-water tap shall be supplied with water from:

- a supply pipe
- a pump delivery pipe drawing water from a supply pipe, or
- a distributing pipe drawing water exclusively from a storage cistern supplying wholesome water.

Paragraph 28(1) states that, subject to paragraph (2), every bath, wash basin, sink or similar appliance shall be provided with a watertight and readily accessible plug, or other device capable of closing the waste outlet.

This requirement does not apply to:

- an appliance where the only taps provided are spray taps

- a washing trough or wash basin whose waste outlet is incapable of accepting a plug and to which water is delivered at a rate not exceeding 0.06 litres per second exclusively from a fitting designed or adapted for that purpose

- a wash basin or washing trough fitted with self-closing taps

- a shower bath or shower tray

- a drinking-water fountain or similar facility

- an appliance that is used in medical, dental, or veterinary premises and is designed or adapted for use with an unplugged outlet.

Paragraph 29(1) states that, subject to paragraph (2), washing machines, washer-driers and dishwashers shall be economical in the use of water.

The requirements of this paragraph shall be deemed to be satisfied in the case of machines having a water consumption per cycle of not greater than:

- for domestic horizontal-axis washing machines, 27 litres per kilogram of wash load for a standard 60°C cotton cycle

- for domestic washer-driers 48 litres per kilogram of wash load for a standard 60°C cotton cycle

- for domestic dishwashers 4.5 litres per place setting.

Dishwashers must be economical in their use of water

Drinking-water supplies and points

Paragraph 26 concerns the supply of water for domestic purposes through at least one tap, conveniently situated for supplying drinking water (wholesome water). In a house, a drinking-water tap should be situated over the kitchen sink, connected to the incoming supply pipe.

In premises where a water softener is used, an un-softened 'drinking-water' tap must be provided.

The Water Industry Act 1991 refers to water for domestic purposes as water used for:

- drinking
- washing
- sanitary purposes
- cooking
- central heating.

The watering of gardens and washing of vehicles are included in domestic purposes if not done using a hosepipe.

Where it is not possible to provide the drinking-water tap with water from the supply pipe, the tap should be supplied from a cistern containing water of drinking quality. This appropriately takes us to the requirements of paragraph 27, which states that a drinking-water supply shall be supplied with water from:

Supply of drinking water is a basic human need

- a supply pipe
- a pump delivery pipe drawing water from a supply pipe, or
- a distributing pipe drawing water exclusively from a storage cistern supplying wholesome water.

In instances where it is not possible to supply water directly off the supply pipe due to there being insufficient water pressure available, it may be necessary to install pumps or a booster system.

If the amount of water required is less than 0.2 litres per second it is permissible to pump direct from the supply pipe. In cases where a greater flow capacity is required to serve the premises, then written consent from the water supplier will be required for the direct or indirect pumping (via a cistern or closed vessel) from the supply pipe. If an indirect pumping system is installed, it must be of a type that minimises the possibility of the water quality deteriorating.

The following shows an installation in which water is boosted from a break cistern. A pneumatic pressure vessel is included so that the pump is not continually shutting on and off; the pump is controlled by a low- and high-pressure switch sited in the pressure vessel to activate the pump on and off. The principles of this system are used in domestic properties fed from wells or bore-holes, so if you understand them, you should be able to transfer them to domestic installations.

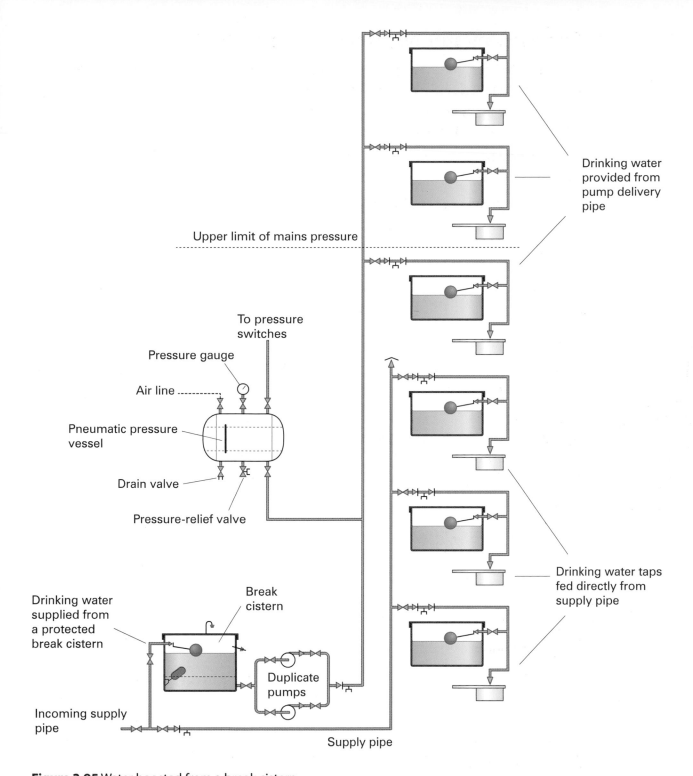

Figure 2.95 Water boosted from a break cistern

When drinking water is supplied direct from a storage cistern, it is recommended that:

- the interior of the cistern is kept clean
- the quantity of stored water is restricted to the minimum essential amount required, so that the through-put of water is maximised
- the stored water temperature is kept below 20°C, taking into consideration that cisterns sited in roof spaces or voids can be subjected to varying temperatures

- the cistern is insulated and ventilated, and fitted with a screened warning/overflow pipe
- the cistern is regularly inspected and cleaned internally.

The following illustration shows an alternative method of providing drinking water using a large header pipe as an alternative to the pneumatic pressure vessel.

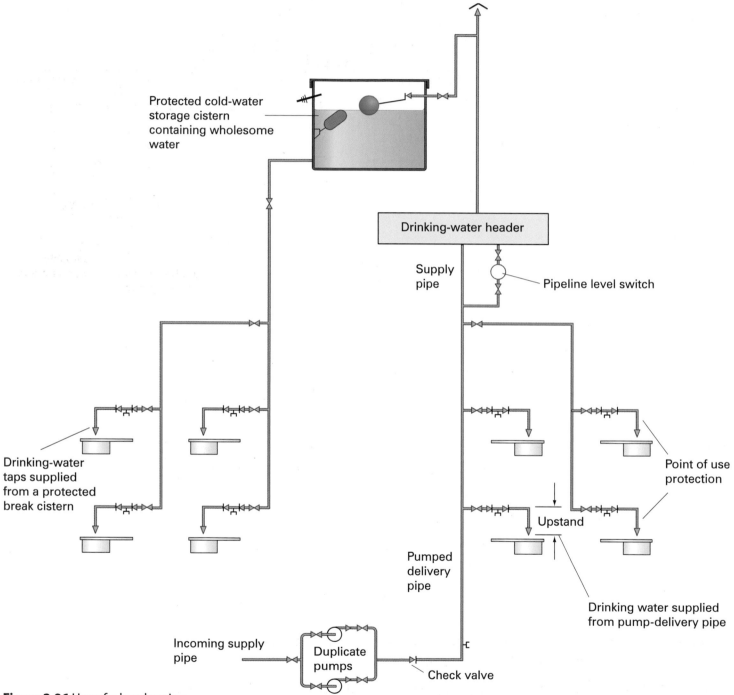

Figure 2.96 Use of a header pipe

When considering pumped supplies to domestic showers, invariably these are going to be supplied from storage. There are two different types of pump that can be used:

- single impeller – installed between the shower mixer outlet and the shower head (these tend not to be so common as they are only really suitable for piping totally concealed showers)

- double impeller, sited before the shower mixer and pumping both hot and cold supplies together.

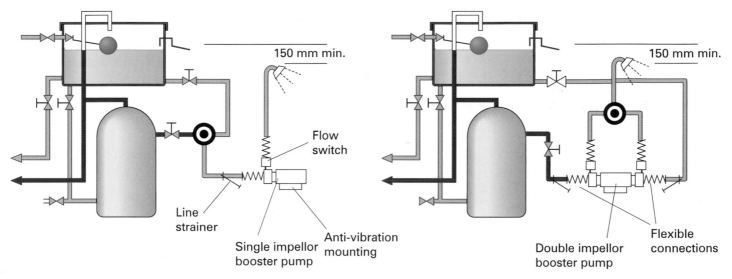

Figure 2.97 Pumped supply to shower

The key features of a pumped shower system are that:

- the pump must be properly sized to do the job

- the storage cistern must be properly sized so that there is an adequate supply of stored hot and cold water

- the pump should be installed in a position not likely to cause a noise nuisance and should be mounted on anti-vibration mounts with pipework connected using flexible connections

- the shower mixer should be of the thermostatic type

- the electrical connections must be made in accordance with the wiring Regulations and the pipework properly earthed

- pump-inlet connections require strainers to be fitted to prevent damage to the pump.

Different manufacturers have different recommended piping arrangements, so refer to their installation instructions before fitting.

Figure 2.97 shows the operation of the pump by a flow switch. This usually requires a minimum head of water in the storage cistern above the shower head: in the example shown this is 150 mm. An alternative to this is to control the pump by pressure switch: in this case the cistern can be sited below the shower head, possibly on a lower floor.

Find out

Using a manufacturer's catalogue, identify a suitable commissioning procedure for a pumped shower in a domestic property

In domestic dwellings it is the occupier's responsibility to maintain the quality of water used; in commercial and industrial buildings the responsibility lies with the operators or owners. It is therefore important that all taps within the building that supply drinking water are prominently distinguished from those that do not, by labelling the types accordingly.

The requirement of paragraph 28(1) is very simple: it states that every bath, wash basin, sink or similar appliance must have an accessible watertight plug fitted to its waste outlet. The usual method is by a plastic/rubber push-in plug attached by a chain to the appliance, but more advanced methods are by a lever plug operated through a combination tap arrangement.

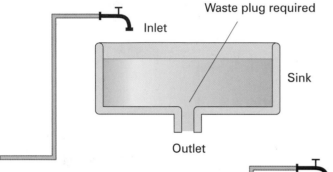

The aim of the requirements is to prevent undue water consumption and conserve water, so it is important that no water is allowed to run to waste during the use of the appliance.

Paragraph 28(2) permits several exceptions to the previous requirements, these all being appliances that are fitted with low-flow taps or inlet valves.

Figure 2.98 Waste-plug requirements

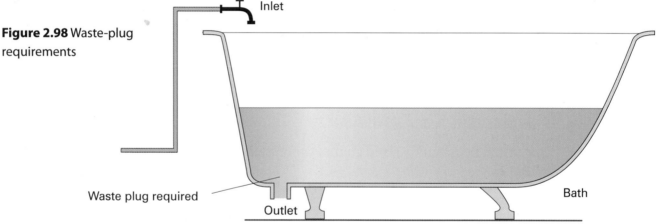

Outlet plugs or other closing outlet devices are not required for:

- appliances where the only taps provided are spray taps, e.g. wash basins

- wash basins or washing troughs fitted with self-closing taps

- wash basins or washing troughs where the waste outlet is incapable of accepting a plug, and where water is delivered at a rate not exceeding 0.06 litres per second (3.6 litres per minute)

- shower trays or shower baths

- drinking-water fountains or similar appliances

- appliances that are used in medical, dental or veterinary premises and are designed or adapted for use with unplugged outlets.

Water for outside use

At the end of this section you should be able to:

- explain how to prevent contamination of the water supply in animal drinking troughs
- state what devices are used to control the flow of water to animal drinking vessels
- state the permitted construction details of ponds and pools and methods for replenishing ponds, pools and fountains.

Paragraph 30 is about animal drinking troughs and bowls, and states requirements for the provision of float-operated valves, stop valves, servicing valves and the prevention of contamination of the supply pipe.

Paragraph 31 deals with the wastage of water via leaks or seepage of water from ponds, pools or fountains irrespective of their use.

The Requirements

Paragraph 30 of Schedule 2 states that every pipe that conveys water to a drinking vessel for animals or poultry shall be fitted with:

- a float-operated valve or some other no less effective device to control the inflow of water, which:
  - is protected from damage and contamination
  - prevents contamination of the water supply
- a stop valve or servicing valve, as appropriate.

Paragraph 31 of Schedule 2 states that every pond, fountain or pool shall have an impervious lining or membrane to prevent the leakage or seepage of water.

Many of the requirements of paragraph 30 have already been covered in previous sessions, but we will take another look at the requirements in relation to this subject.

The water supply to animal drinking troughs must be fitted with a float-operated valve or some other equally effective inlet device. The inlet device should have a Type AA or Type AB air gap installed, so as to prevent backflow from a fluid category 5 risk, and prevent contamination of the supply pipe.

The inlet and backflow devices should be protected from damage; the installation arrangements of the trough will be deemed as acceptable if the watering trough complies with BS 3445.

The requirements ask for a servicing valve to be fitted on the inlet pipe adjacent to the drinking appliance. Figure 2.99 shows the location of a servicing valve that is suitable for underground use.

WSR Schedule 2, para 30–31

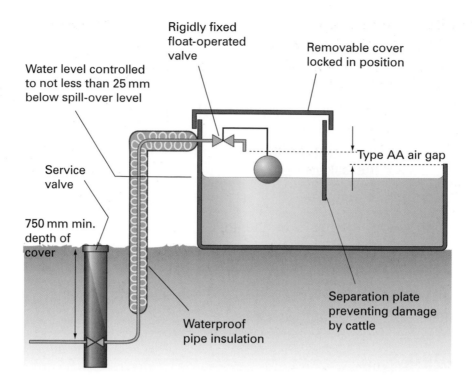

Figure 2.99 Watering trough

In instances where animal drinking troughs are supplied with water from a single trough, the spill-over levels of the additional troughs must be at a higher level than the initial feeder trough in which the inlet device is located.

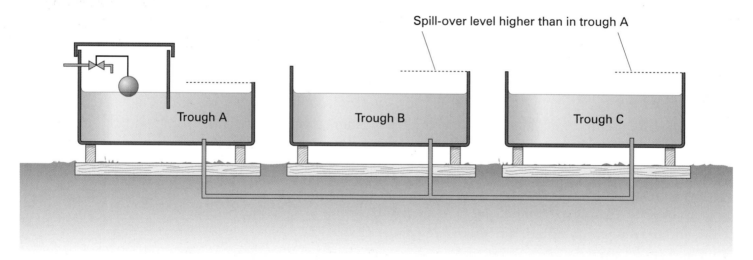

Figure 2.100 Multiple troughs

For individual animal drinking bowls that are supplied with water, the source of the water supply will depend on the type of bowl being installed.

Drinking-bowl types that have inlet valves that could become submerged during use must only be supplied with water from a dedicated storage cistern that only supplies similar appliances.

Drinking-bowl types that have an air gap incorporated within the inlet device may be fed with water from a supply pipe, providing that the air gap is equivalent to a Type AA.

The final requirement of the Regulations covers ponds, fountains or pools, and states quite simply that they shall have an impervious lining or membrane fitted to prevent the leakage or seepage of water. The requirement applies to all ponds, fountains or pools, large or small, irrespective of their use.

Ponds and pools that are constructed of concrete are acceptable if they have been designed, constructed and tested in accordance with BS 8007 1987 – Design of Concrete Structures for Retaining Aqueous Liquids.

Two important factors to remember are that:

- ponds, fountains or pools should not be permanently or directly connected to a supply pipe or distribution pipe
- ponds, fountains or pools can be replenished by automatic means provided that a method of backflow prevention suitable for a fluid category 5 risk is provided.

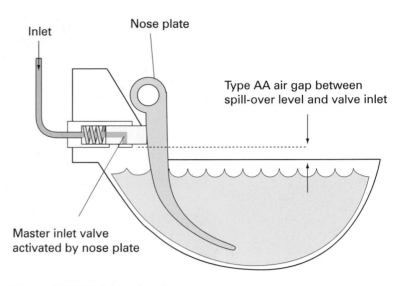

Figure 2.101 Drinking bowl

Domestic swimming pools are subject to BS 8007

FAQ

Do I really need to know all these Regulations?

It is vital that you understand the Regulations. This will help ensure you do a good job and avoid being prosecuted for not complying with them! Don't try to remember them all in one go – start with the everyday ones and build on your knowledge.

Knowledge check

1   Who has right of entry to premises under the Water Regulations?

2   Why is water passing through a hosepipe classified as unfit for human consumption?

3   A customer wishes you to install an outside tap for watering the garden and topping up the ornamental pond. What should you incorporate into the pipework?

4   A washing machine has been installed in a utility room; the hot and cold copper pipes pass in a shallow chase underneath the external door, covered by a wooden board. What conditions should apply?

5   If a commercial building collects rainwater for toilet flushing, how would you expect to distinguish the pipes supplying it from other pipes?

6   A customer wishes to install an original Victorian bath, with globe taps. What would the fluid category risk be, and what device must be fitted to make these taps compliant?

7   What fluid category would water heated in a hot-water secondary system be classified as?

8   What are the two main types of bidet?

9   What is the minimum unobstructed space above a cistern required for maintenance access?

10 Under what circumstances is it permissible to discharge individual warning pipes from a range of WCs into a common warning pipe?

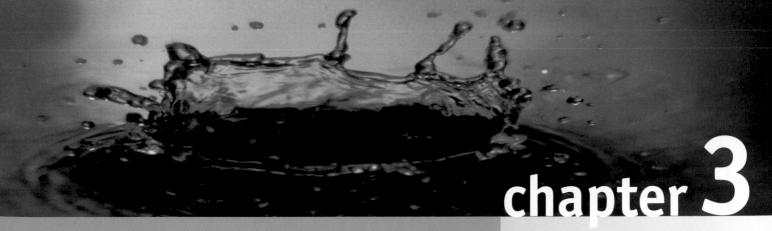

chapter 3

Unvented hot-water systems

OVERVIEW

This section leads to a final industry unvented hot-water systems assessment, which will be completed in the final stage of your advanced apprenticeship programme.

The installation of unvented hot-water systems must be carried out by properly qualified plumbers as identified in Approved Document G of the Building Regulations. The systems provide potentially high water-flow rates to plumbing systems and components such as very powerful showers. They also minimise the risk of freezing by not having components such as cold-water storage cisterns that are usually sited in exposed places.

What you will learn in this unit:

- Introduction to unvented hot-water systems
- Unvented system components and their operation
- Unvented hot-water systems legislation
- Installation and commissioning of unvented systems
- Service and maintenance of unvented systems.

Introduction to unvented hot-water systems

At the end of this section you should be able to:

- recognise the key differences between vented and unvented systems
- identify why safety is an important feature of unvented system design
- identify the advantages and disadvantages of unvented systems
- state the various types of unvented systems and appliances.

The difference between vented and unvented hot-water systems

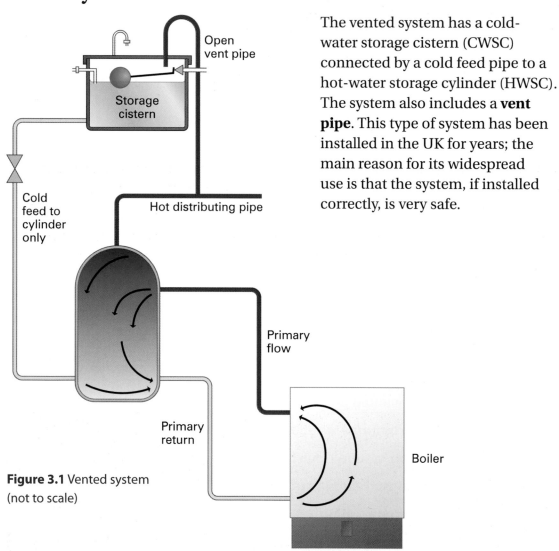

The vented system has a cold-water storage cistern (CWSC) connected by a cold feed pipe to a hot-water storage cylinder (HWSC). The system also includes a **vent pipe**. This type of system has been installed in the UK for years; the main reason for its widespread use is that the system, if installed correctly, is very safe.

Open vent pipe

Storage cistern

Cold feed to cylinder only

Hot distributing pipe

Primary flow

Primary return

Boiler

Figure 3.1 Vented system (not to scale)

Why is safety an issue with hot-water systems?

When water is heated it expands and its volume increases. Figure 3.2 represents the expansion of water when it is heated.

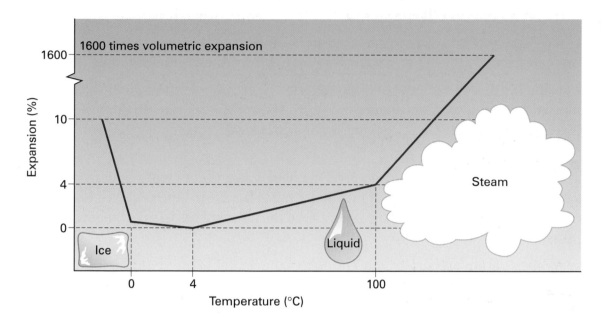

Figure 3.2 Water expansion

Below 0°C water expands as it cools; this means that water in an enclosed space (such as a pipe) increases in volume when it freezes. With nowhere for the frozen water to go, the pipe ruptures.

Similarly, when water is heated from 4°C to 100°C it expands. This expansion has to be allowed for in any hot-water system, to stop pressure build-up and prevent damage to the components.

A typical increase in volume for water heated in a 120-litre hot-water cylinder is:

120 x 0.04 (4 per cent) = 4.8 litre increase in volume

Water expands by up to 4 per cent under normal system heating conditions. This must be catered for somewhere in the system!

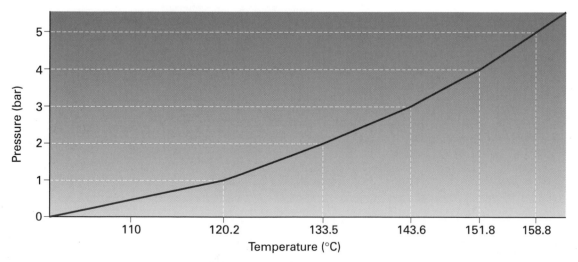

Figure 3.3 Effects on the boiling point of water at pressures greater than atmospheric pressure

Remember

It is extremely important that both the temperature and pressure are controlled in hot-water systems

The boiling point of water increases as the pressure in a system increases. For example, water at 2 bar pressure has a boiling point of 133.5°C. If the temperature in a closed vessel is increased, then the pressure in that vessel is increased.

The vented system

The vented system is open to the atmosphere and is not designed to work at pressures above atmospheric, so the system has safety built into its design.

As water is heated in the system it expands from the cylinder through the cold feed pipe and into the cold-water storage cistern. The volume of water in the system increases as the water heats, leading to the water level rising in the cold-water storage cistern.

The vent pipe provides a dual feature. First, it acts as a vent to remove air from the system. But more importantly, it provides a safety back-up if the cold feed becomes blocked or does not work – there is a route to the atmosphere to safely relieve pressure in the system.

With the addition of thermostatic controls to the heater of the vented system (either by an immersion heater or indirectly heated by a boiler), the system is very effectively protected against a build-up of temperature and pressure.

However, the flow rate from the hot-water taps is restricted by the head of water generated over the tap by the cold-water storage cistern, and can often be quite poor, especially as many of today's users want better performance from systems (e.g. more powerful showers).

The unvented system

In the unvented system, water is stored in the system itself and is available at mains supply pipe pressure. This gives a much higher-performance hot-water system with greater flow rates. By design, therefore, the system is not open to the atmosphere: it works at much higher pressure,

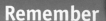
Unvented cylinder

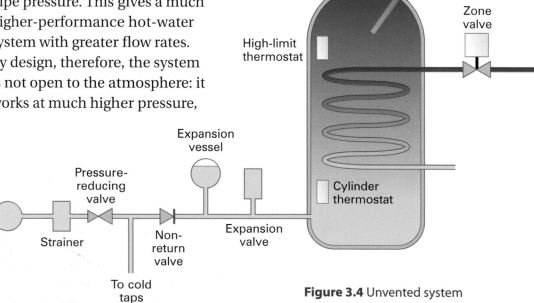
Figure 3.4 Unvented system

and must have a different range of safety control features from those of vented systems. Figure 3.4 shows an example of an unvented system, where you can see that cold water is supplied to the cylinder from the mains supply directly through the cylinder to the hot-water tap. Hence the term 'unvented'; it may be described as a closed circuit.

Cut-away of unvented hot-water cylinder

What would happen if the system was not controlled properly?

This system is operating at a pressure of usually between 2 bar and 3.5 bar, so referring to Figure 3.3 first, at 2 bar the boiling point of water is 133.5°C. If there was an uncontrolled heat source such as a faulty immersion heater feeding an unvented system, then at 2 bar the water would not begin to boil until 133.5°C was reached. As more heat is provided from the uncontrolled source, the temperature of water rises higher. The water is in a confined space, so the volume of water cannot increase by expansion. This increases the pressure in the system, which has the effect on the system that the boiling point of the water is increased. Eventually one of the following happens:

(a) the pressure in the system increases to a point at which a component bursts or ruptures and the pressure in the system goes down, or

(b) someone opens a tap in the property and begins to reduce the pressure.

However, refer to Figure 3.2 again. A reduction of pressure in the system could cause the boiling point in the system to be quickly lowered, causing an immediate change of the cylinder contents from water to steam: as the water changes to steam, its volume can increase up to 1600 times. So a 120-litre cylinder of water expanding up to 1600 times its volume would need to accommodate:

120 x 1600 = 192,000 litres of steam

The explosive effects of a defective unvented system as it flashes to steam are that buildings can be demolished in much the same way as with a gas explosion.

If they are installed and maintained correctly, unvented hot-water systems are perfectly safe – a high degree of control is used to protect them against over-heating and over-pressurisation. There are also key requirements laid down for their installation and maintenance, including the requirement for installers to be competent.

Did you know?

The possibility of this rapid volume increase is commonly known as 'having a bomb on your hands'!

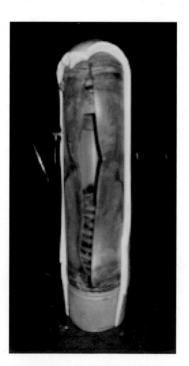

Example of an incorrectly controlled unvented cylinder – this was a mild one!

Advantages	Disadvantages
High-pressure operation At higher pressure, increased flow rate from taps and outlets such as showers is available. If designed correctly, the system can provide a high water-flow rate to multiple outlets at the same time, much higher than a vented system or a standard combination boiler.	**Poor supply/flow rate** The system does not perform correctly where incoming mains supply has a poor pressure and flow rate. Additional system-design options can be applied to overcome such problems, finances permitting!
Balanced pressure at outlets If designed and installed correctly, it is possible to get hot and cold water at the same high pressure. This allows for better performance taps and fittings, such as quarter-turn with ceramic discs. The water supplies to showers are also safer if provided at balanced pressure.	**No stored water** No back-up in the event of mains failure.
Reduction in pipe diameter Pipe diameters supplying plumbing equipment may be reduced, but you need to be careful with this.	**Discharge pipe required** The required discharge pipe can sometimes be difficult to locate.
Quicker to install Less pipework in difficult locations, which usually leads to a reduction in installation time.	**Space requirements** The cylinder and associated components will require more space than a conventional storage cylinder.
Easier siting An unvented cylinder or water heater is usually easier to place. There are some restrictions on siting, mainly in relation to providing an adequate route to outside for a discharge pipe. Freezing of the cistern in the loft space is not experienced with this type of system.	**Purchase price** The price of the unvented system is quite high in relation to the overall costs of installing a traditional vented system.
Reduced water-contamination risk No cold-water storage cistern and less stored water reduces the potential for water contamination.	

Table 3.1 Advantages and disadvantages of unvented hot-water systems

Remember

The improved water-flow rate provides some real benefits to the end user when provided by a well-designed system

Types of unvented hot-water system

The most important factor when unvented systems are designed by the manufacturer and installed by you the plumber is the size of hot-water storage in any vessel in the system.

System	Capacity	Notes	Figure
Unvented instantaneous heater	Usually less than 15 litres	Gas or electric multipoint heater	3.6
Unvented instantaneous-/continuous-flow heaters		Heated by an electrical element and inlet or outlet controlled – more popular on the Continent	
Unvented displacement heaters	Usually up to 15 litres	Over or undersink heaters – very popular in commercial premises	3.7
Unvented water-jacketed tube heaters	Usually under 15 litres capacity (there are exceptions)	Indirectly heated via an internal coil, known as thermal storage and combination boilers	
Unvented direct-fired storage heaters	Over 15 litres capacity	Can be electrically heated or gas fired	3.8 and 3.9
Unvented indirect-fired storage heaters	Over 15 litres capacity	External heat source provided via an internal coil	3.10

Table 3.2 General categories of unvented systems. The operation and installation of these can be applied to every type of unvented installation

Stringent safety controls are essential for any system that has a storage vessel with more than 15 litres storage capacity. Systems with vessels of less than this capacity may have reduced control components. Controls for these systems will be specified by the manufacturer and must be installed as specified by the manufacturer.

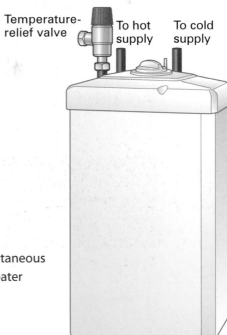

Temperature-relief valve   To hot supply   To cold supply

Figure 3.5 Unvented instantaneous electric multipoint water heater

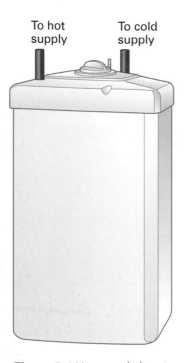

To hot supply   To cold supply

Figure 3.6 Unvented electric displacement heater

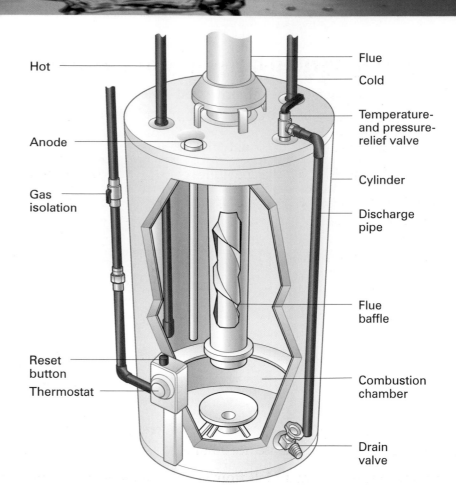

Hot

Anode

Gas isolation

Reset button
Thermostat

Flue

Cold

Temperature- and pressure- relief valve

Cylinder

Discharge pipe

Flue baffle

Combustion chamber

Drain valve

Figure 3.7 Unvented direct-fired gas storage heater

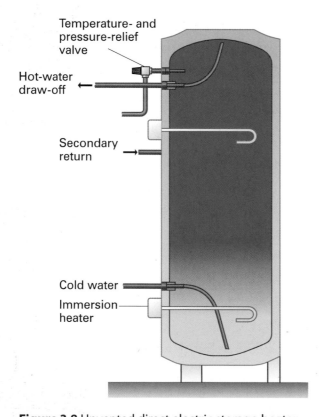

Temperature- and pressure-relief valve

Hot-water draw-off

Secondary return

Cold water

Immersion heater

Figure 3.8 Unvented direct electric storage heater

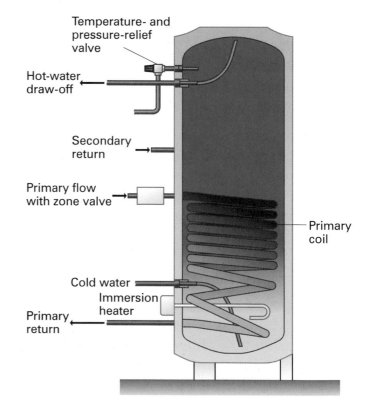

Temperature- and pressure-relief valve

Hot-water draw-off

Secondary return

Primary flow with zone valve

Primary coil

Cold water

Immersion heater

Primary return

Figure 3.9 Unvented indirect storage heater

Unvented system components and their operation

At the end of this section you should be able to:–

- identify the positioning requirements of components within unvented systems
- state the purpose of, and operating principles of, unvented system-control components.

Unvented system components

A key component is the storage cylinder itself.

Storage cylinders

Every new storage cylinder that is installed must be capable of meeting the requirements of Approved Document L1 of the Building Regulations in terms of **thermal insulation** and surface area of heating coil.

Storage cylinders on unvented hot-water systems must be supplied by fully pumped systems (**gravity primaries** are not suitable).

There are a number of materials from which cylinders tend to be manufactured:

- **Copper** – the first unvented storage systems that were installed in the UK were manufactured from copper. They were normally grade 2 cylinders (1.6 mm thick) and operated at a pressure of 2 bar – less than the two following types. The cylinders usually counteract the effects of aggressive water supply by having a **sacrificial anode** installed.

- **Mild steel lined** – Continental manufacturers on entry to the UK market tended to supply cylinders manufactured from mild steel with a glass lining or **polyethylene** coating to the inner surface. Because of the potentially incomplete surface of cylinders with such linings it is essential that a sacrificial anode is fitted to retain the integrity of the cylinder. Most anodes are in the form of a magnesium rod fitted into the top of the cylinder. The anode must be inspected annually as part of the service procedure to check for decay: it may need replacing.

- **Stainless steel** – the material from which the majority of new unvented cylinders are manufactured. Depending on the quality of the stainless steel used, the cylinder may or may not require a sacrificial anode.

Both mild-steel-lined and stainless-steel cylinders tend to be designed to operate at higher operating pressures than their copper counterparts – usually between 3 and 3.5 bar pressure.

Safety tip

Always follow the manufacturer's installation instructions when dealing with unvented hot-water storage systems. Install and maintain as per the instructions, and when replacing components, always replace like for like, with regard to valve pressure settings and flow rates

Remember

Mistakes with unvented systems could cause a danger to life

Did you know?

The advantage of having a higher operating pressure is the potentially higher flow rate through smaller diameter pipes

System controls and devices

Safety devices are included to protect the user and the property. Functional devices are used to protect the supply of water.

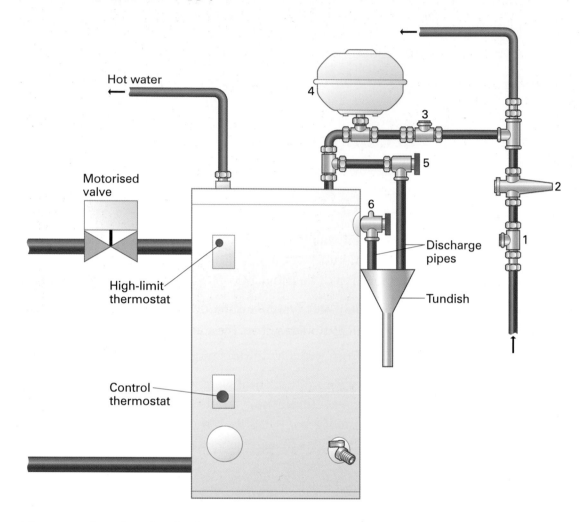

Figure 3.10 Indirect unvented storage system

There are three levels of safety protection to guard against overheating: the control thermostat, the high-limit thermostat (if the control thermostat fails) and the temperature-relief valve (if the high-limit thermostat fails).

Safety item 1	Control thermostat	Controls the water temperature in the cylinder to between 60 and 65°C
Safety item 2	High-limit thermostat (energy cut-out device)	A non-self-resetting device that isolates the heat source at a temperature of around 80–85°C
Safety item 3	Temperature-relief valve (component 6)	Discharges water from the cylinder at a temperature of 90–95°C (water is dumped from the system and replaced by cooler water to prevent boiling)

Table 3.3 Safety devices in an unvented storage system

Line strainer (Component 1)	Prevents grit and debris entering the system from the water supply (causing the controls to malfunction)
Pressure-reducing valve (on older systems may be pressure-limiting valve) (Component 2)	Gives a fixed maximum water pressure
Single check valve (Component 3)	Stops stored hot water entering the cold water supply pipe (a contamination risk)
Expansion vessel or cylinder air gap (Component 4)	Takes up the increased volume of water in the system from the heating process
Expansion valve (Component 5)	Operates if the pressure in the system rises above design limits of the expansion device (i.e. the cylinder air gap or expansion vessel fail)
Isolating (stop) valve (not shown in figure 3.11)	Isolates the water supply from the system (for maintenance)

Table 3.4 Functional devices in an unvented storage system. These are often provided as **composite valves**

> **Definition**
>
> Composite valve – valve that can provide three functions in one unit, e.g. line strainer, check valve and expansion valve

The control components in detail

(a) Stop (isolating) valve

A normal stop valve is required to isolate an unvented cylinder and its control components from the cold-water supply pipe for maintenance purposes etc.

(b) Line strainer

A filter must be provided to prevent particles and grit from the water supply from passing into the system and affecting the correct operation of other expensive controls. The line strainer is usually an integral component of another valve in modern systems. It has been shown as an individual item on the diagram so that you understand where it is positioned in the system.

(c) Pressure-control valve

The pressure control valve may be in the form of either a **pressure-reducing valve** (PRV) or a **pressure-limiting valve**. Pressure-limiting valves tend not to be widely used on modern systems as they are not as effective a control component as pressure-reducing valves. With pressure-limiting valves in high-pressure supply areas, it was found that if the incoming pressure in the system dropped slightly, the valve could provide a higher than required working pressure to the system and cause the expansion valve to discharge water.

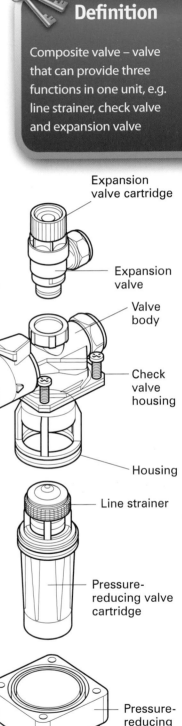

Figure 3.11 Composite valve

Pressure-reducing valve

This valve is designed to control the pressure in the system to a set working pressure. On older copper cylinders, this could have been as low as 2 bar, but on modern steel cylinders this is likely to be set around 3–3.5 bar, depending on the manufacturer's specification.

Under **no-flow conditions** the downstream (outlet) pressure acts on the diaphragm and overcomes the spring pressure. The diaphragm moves up and the linkage that joins the diaphragm to the seat holds the seat closed so that downstream pressure cannot increase.

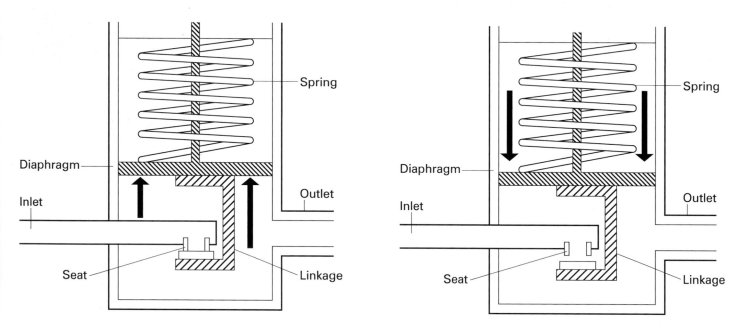

Figure 3.12 Pressure-reducing valve closed

Figure 3.13 Pressure-reducing valve open

Under flow conditions the downstream (outlet) pressure decreases until the spring can overcome the pressure. The diaphragm moves the linkage down and so opens the seat and water flow through the valve. When the outlet is closed, pressure builds up until the spring pressure is overcome and the seat is closed again.

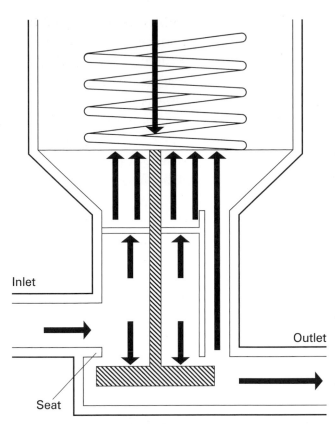

Figure 3.14 Balanced pressure valve

Inlet

Outlet

Seat

Balanced pressure valves operate in basically the same way except that there is an additional 'piston' of the same area as the main seat. This gives better control under low- and high-flow conditions.

Single check valve

(d) Single check valve

A non-return/single check valve is designed to prevent backflow of water from the system into the supply pipe or mains as the system water is heated up. The valve is located on the cold-water inlet upstream of the connection to the expansion vessel, to prevent expanded water being discharged down the cold supply pipe.

(e) Expansion device

There are two methods of accommodating or taking up the expanded water in an unvented hot-water storage system – a cylinder air gap, or an expansion vessel.

Figure 3.15 Cylinder air gap (bubble top)

Expansion vessel

Cylinder air gaps accommodate the expansion of the heated water using an internal air bubble which is generated and trapped at the top of the unit during commissioning. The size of the air bubble is determined by the cylinder manufacturer. (In exceptional circumstances and with systems with extremely long pipe runs the manufacturer may have to be consulted to establish whether the expansion volume is sufficient for the system contents or whether an additional expansion vessel may have to be provided.) The photo above shows the air gap or expansion chamber built into the design of the cylinder. In modern cylinders of this type, the air gap varies in size, based upon the pressure inside the cylinder as a result of a moving or floating **baffle**.

An expansion vessel can be used to take up the volume of expanded water in the system due to the heating process. The vessel contains a flexible diaphragm which separates the stored water in the vessel from a cushion of air or nitrogen.

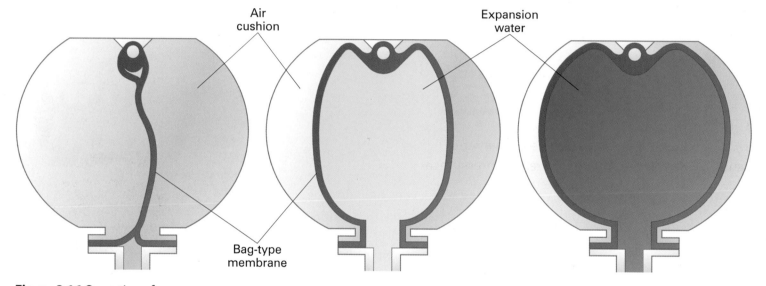

Figure 3.16 Operation of expansion vessel

The vessel is charged or pressurised on the dry side of the flexible diaphragm. The charge pressure is usually determined by the manufacturer in their installation instructions (you will need to check and charge expansion vessels on unvented systems as part of the installation and maintenance procedure).

As the temperature in the cylinder rises, the volume of water increases, and that increase in volume is taken up by the air cushion in the vessel. So when the system is empty the vessel diaphragm is in a collapsed state, as shown in Figure 3.16, left. When the system is at its normal cold-water operating pressure the membrane has taken up the initial pressure built up in the system. As the water heats up, the diaphragm flexes further to take up the increased system volume, as illustrated in Figure 3.16, right, which shows a diaphragm that is fully flexed.

The size of the expansion vessel is usually determined by the cylinder manufacturer, but again, as with the bubble-top version, if there are excessive pipe runs resulting in an excessive increase in system volume, then a larger than normal expansion vessel may be required. The best advice is to consult the manufacturer. The vessel should be fitted on the cold side or inlet side of a storage cylinder to prevent a build-up of scale in hard-water areas and to prevent deterioration of the flexible membrane, as contact with heated water reduces its life expectancy.

The expansion vessel acts as a 'dead leg' in the circuit and the water contained in it is not often changed. In certain conditions the water can begin to stagnate and bacteria grows on the flexible membrane. The WRAS Guide to the Water Regulations identifies two ways to overcome this problem:

- Use a **throughflow expansion vessel** – the cold water is constantly flowing through the vessel as it has an inlet and outlet connection to prevent possible stagnation.

- Provide an **anti-legionella valve** – this is connected to the inlet of an expansion vessel to maintain circulation within it under flow conditions using a **venturi** effect. The valve also incorporates an isolation and drain-down valve facility for easy removal of the vessel for maintenance or replacement.

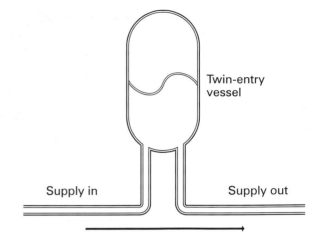

Figure 3.17 Throughflow expansion vessel

Currently there are virtually no manufacturers installing anti-stagnation devices on systems. The installer is left to provide what is specified by the manufacturer.

(f) Expansion valve

The expansion valve is fitted in the system to relieve excess pressure build-up in the system, which is usually due to a component failure. The pressure at which the valve begins to operate is predetermined by the cylinder manufacturer. On modern systems, expansion valves typically tend to operate at pressures ranging from 6 to 8 bar. With earlier systems, the operating pressure for the expansion valve was usually 0.5 bar higher than the system operating pressure.

The two main features of the valve are to protect the system in the event of:

(i) failure of the pressure-reducing valve to maintain the inlet cold-water pressure at the design operating pressure of the system

(ii) failure of the expansion vessel or, in the case of a 'bubble-top cylinder', a loss of expansion volume, i.e. if a loss of air pressure causes an imbalance of pressure between water and air in the device.

No other valve should be fitted between the expansion valve and the storage cylinder. The valve is preset by the appliance manufacturer and should not be adjusted.

Expansion valve and tundish

The valve can be either 'lever-top' pattern (operates by lifting a lever) or 'twist-top' pattern as shown in the photo above.

The safety controls

(a) Temperature-relief valve

Temperature- and pressure-relief valve

This valve is usually supplied and fitted to the storage cylinder. It is preset by the manufacturer to fully discharge at a temperature of 90–95°C, so no adjustment is required.

The temperature activation is by a probe that is immersed in the hottest and highest part of the storage cylinder. The probe is filled with a temperature-sensitive liquid or wax which reacts to temperature change. The valve's key purpose is to protect the system in the event of failure of both the control thermostat and the energy cut-out device.

The valve is often supplied by the manufacturer with an in-built pressure-relief function – called a temperature- and pressure-relief valve. This pressure-relief function in-built in the valve is not a requirement for UK systems as we use an expansion valve, but a temperature- and pressure-relief valve may well be supplied by the manufacturer to eliminate the need for them to produce different valves for the UK market – temperature- and pressure-relief valves are commonly installed on Continental systems that have different installation requirements.

The pressure setting in a combined temperature- and pressure-relief valve is set higher than the expansion valve to ensure that the expansion valve opens first – hence not really providing a useful function.

A means of preventing the formation of a vacuum (**anti-vacuum valve**) was built into some early temperature-relief valves fitted mostly on copper cylinders in the UK market. Anti-vacuum valves tend not to be fitted on modern systems.

Find out

What do you think is the purpose of an anti-vacuum valve? And why would it need to be installed in an unvented hot-water system?

(b) High-limit thermostat (non-self-resetting thermal cut-out)

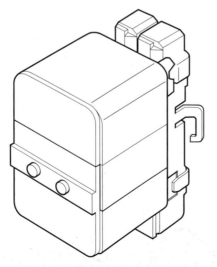

Figure 3.18 High-limit thermostat

An unvented hot-water system must include high-limit thermostats or non-self-resetting thermal cut-out devices. This type of protection must be provided on all heat sources, i.e. hot-water zone feeding an indirect cylinder, or be built into the design of any immersion heaters on direct cylinders (yes – a special immersion heater is required); with direct-fired heaters there must be an overheat thermostat alongside the normal control thermostat. This device must not be capable of automatically resetting itself. It will typically operate at around 80–85°C.

(c) Control thermostat

This is the control thermostat/cylinder thermostat provided on the cylinder which can be adjusted by the end-user. The temperature setting on the thermostat will usually be around 60–65°C.

(d) Tundish

Tundish

Both the expansion valve and the temperature-relief valve are designed to discharge water if a fault occurs in the system. They therefore need to be connected to a discharge pipe(s) to discharge water safely to waste. A blocked discharge pipe could prevent the effective operation of either temperature-relief or expansion valves. To avoid this, an air break is formed in the discharge pipe by means of a tundish.

(e) Composite valves

Most unvented hot-water systems now tend to be supplied with composite control valves. These valves contain two or more of the control functions mentioned earlier. Here is an example.

The composite valve contains a pressure-reducing valve, line strainer, expansion valve and check valve (see Figure 3.11 on page 127).

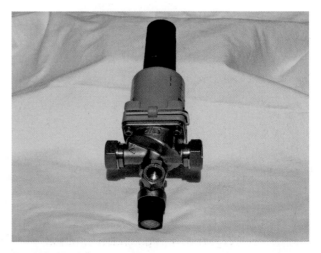

Composite valve

Balanced composite valve

A balanced composite valve contains a pressure-reducing valve, line strainer, expansion valve, check valve and balance cold-water supply connection.

Up until now you have looked at how the components apply to an indirect unvented hot-water storage system; these components and their position in the system (if the storage volume is over 15 litres) will be the same for any system type:

- direct electric storage heater
- direct gas-fired storage heater
- electric multipoint heater.

There are, however, possible differences with displacement heaters below 15 litres capacity, which may have a reduced level of control components. Here, the controls that must be fitted will be as specified and supplied by the manufacturer, but there can be some choices for the installer or specifier.

On the job: Installing a heater

Lucas is going to install a 10-litre capacity unvented displacement heater. He knows there are a number of different installation scenarios and component kits that can be supplied by the manufacturer to suit the installation.

Firstly, he checks out the temperature/pressure-relief valve. As the storage capacity is less than 15 litres he has no obligation to fit this, although the manufacturer recommends it as an additional safety feature. Lucas thinks the best policy, though, is to go for safety at all times!

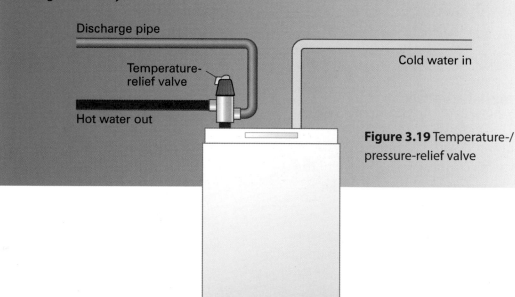

Discharge pipe

Temperature-relief valve

Cold water in

Hot water out

Figure 3.19 Temperature-/pressure-relief valve

Lucas is aware that the functional controls can be different. He looks at the manufacturer's diagram, which shows the minimum level of functional control that may be provided.

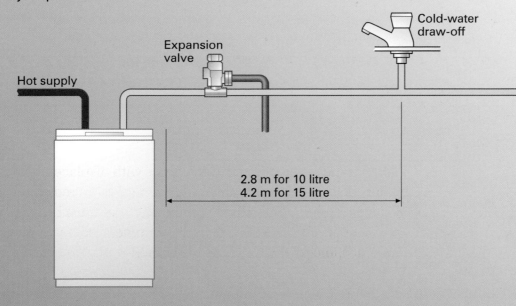

Figure 3.20 Minimum level of functional controls

The only real functional control Lucas has to consider is the expansion valve. This minimum level of control can only be used where the supply inlet water pressure is below 4.1 bar (this may be different for other manufacturers) and where it is possible to accommodate expanded water in the supply pipe as an alternative to providing an expansion space in the storage vessel or an expansion vessel.

For these small heaters it is possible to eliminate the need for an expansion device. The expanded water is taken up in the supply pipe. As the water in the storage vessel is heated, the volume of water increases and is pushed back down the cold supply pipe and ultimately into the mains. It's not as simple as that, though: Lucas has got to make sure that expanded hot-water cannot be drawn-off through the supply pipework, so a minimum distance is specified between the last cold-water draw-off and the point of cold-water connection on the heater. Here it is indicated on the diagram as:

- 2.8 metres for 10-litre capacity
- 4.2 metres for 15-litre capacity.

In addition, the manufacturer requires that the supply pipe right back to the mains must not contain check valves, stop valves with loose jumpers or fittings that prevent reverse flow.

Did you know?

You need to be able to understand the manufacturer's instructions, but to make it easy they provide a simple flowchart to identify which control kits must be fitted to suit the particular installation

Find out

Why do manufacturers require that the supply pipe must not contain check valves, stop valves with loose jumpers or fittings that prevent reverse flow?

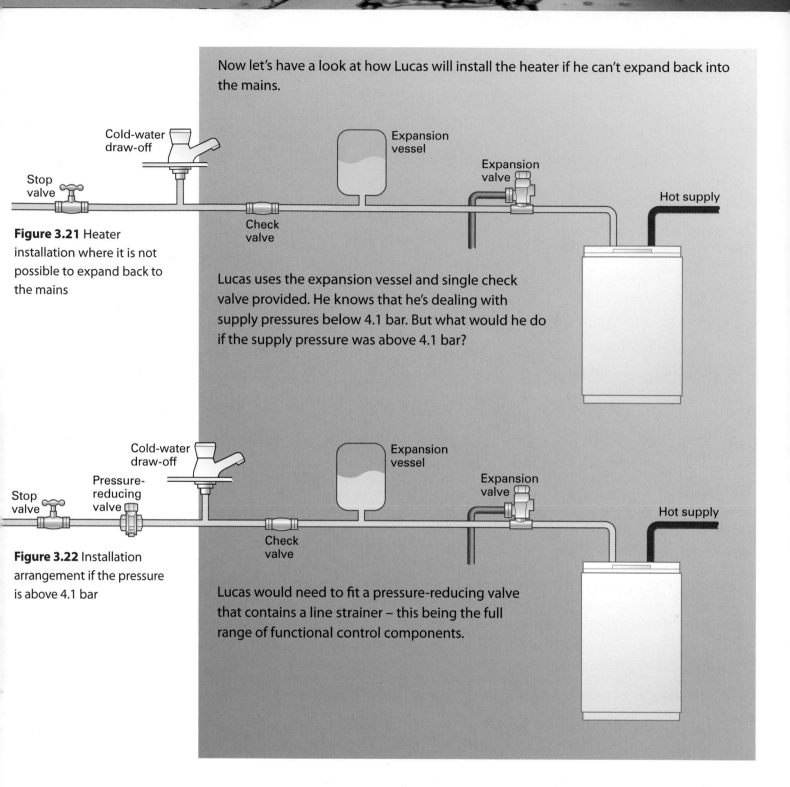

Now let's have a look at how Lucas will install the heater if he can't expand back into the mains.

Cold-water draw-off

Expansion vessel

Expansion valve

Stop valve

Hot supply

Check valve

Figure 3.21 Heater installation where it is not possible to expand back to the mains

Lucas uses the expansion vessel and single check valve provided. He knows that he's dealing with supply pressures below 4.1 bar. But what would he do if the supply pressure was above 4.1 bar?

Cold-water draw-off

Expansion vessel

Pressure-reducing valve

Expansion valve

Stop valve

Hot supply

Check valve

Figure 3.22 Installation arrangement if the pressure is above 4.1 bar

Lucas would need to fit a pressure-reducing valve that contains a line strainer – this being the full range of functional control components.

Unvented hot-water systems legislation

At the end of this section you should be able to:

* state the requirements of legislation relating to the installation of unvented hot-water systems.

You will need a pretty good understanding of the standard to which systems must be manufactured and installed to make them safe.

There are two items of legislation in accordance with which unvented systems must be designed and installed. They are:

* Approved Document G, Section 3 of the Building Regulations (primarily dealing with safety issues), and

* the Water Supply (Water Fittings) Regulations 1999 (primarily dealing with water-supply issues such as contamination and wastage).

This section will concentrate on the detail of the Building Regulations.

Approved Document G3 of the Building Regulations

In the Regulations, an unvented system is defined as 'a hot-water storage system that has a hot-water storage vessel which does not incorporate a vent pipe to the atmosphere'.

G3 does not apply to unvented hot-water storage systems that have storage vessels of 15 litres capacity or less, hence the varying range of controls on the displacement-type heaters. In addition, systems providing space heating do not fall under G3 (sealed heating systems), and systems providing water to industrial process plant are not covered. The Regulations categorise systems into:

* those with a storage capacity of under 500 litres and a heat input of under 45 kW

* those with a storage capacity over 500 litres and a heat input over 45 kW.

There are differences in the installation requirements for these systems, primarily in the application of system controls. In domestic situations, you will concentrate on systems under 500 litre capacity and 45 kW heat input; above this size and it's a pretty big cylinder!

The Regulations identify that there must be precautions to prevent the temperature of stored water exceeding 100°C at any time.

Manufacturers of unvented hot-water systems are required to supply partly assembled systems as described under the Regulations. This is intended to improve safety, as essential control components will not be missed.

Packages and units

Under the Building Regulations G3, systems can be supplied in a package or unit form. Early systems coming on to the market tended to be supplied as units with most of the controls pre-assembled. Modern systems now tend to be supplied as packages with fewer components pre-assembled.

Any unit or package that is installed must be either:

- approved by a member body of the European Organisation for Technical Approval (EOTA) or operating a technical approvals scheme, e.g. the British Board of Agreement (BBA) (you'll find that most manufacturers' systems are BBA approved)

- approved by a certification body belonging to the National Accreditation Council for Certification Bodies (NACCB) and testing to the requirements of an accepted standard such as BS 7206: Specification for Unvented Hot-water Storage Units and Packages

- a proven independent assessment that will clearly demonstrate equivalent performance to the above.

The package

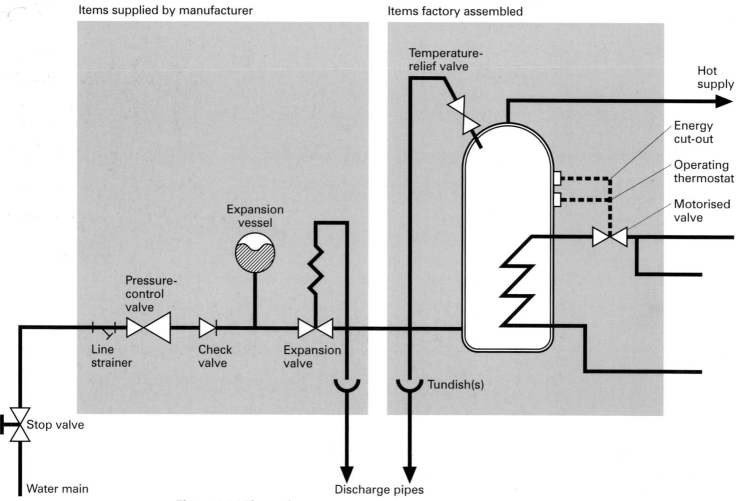

Figure 3.23 The package

All the safety devices are factory-fitted together, with a kit containing other applicable devices supplied by the package manufacturer.

Factory fitted:

- temperature-relief valve
- high-limit thermostat

Supplied by the manufacturer, to be fitted by the installer:

- line strainer
- cylinder primary motorised valve (indirect unit)
- pressure-reducing or -limiting valve
- single check valve
- tundish
- expansion vessel
- expansion valve.

The unit

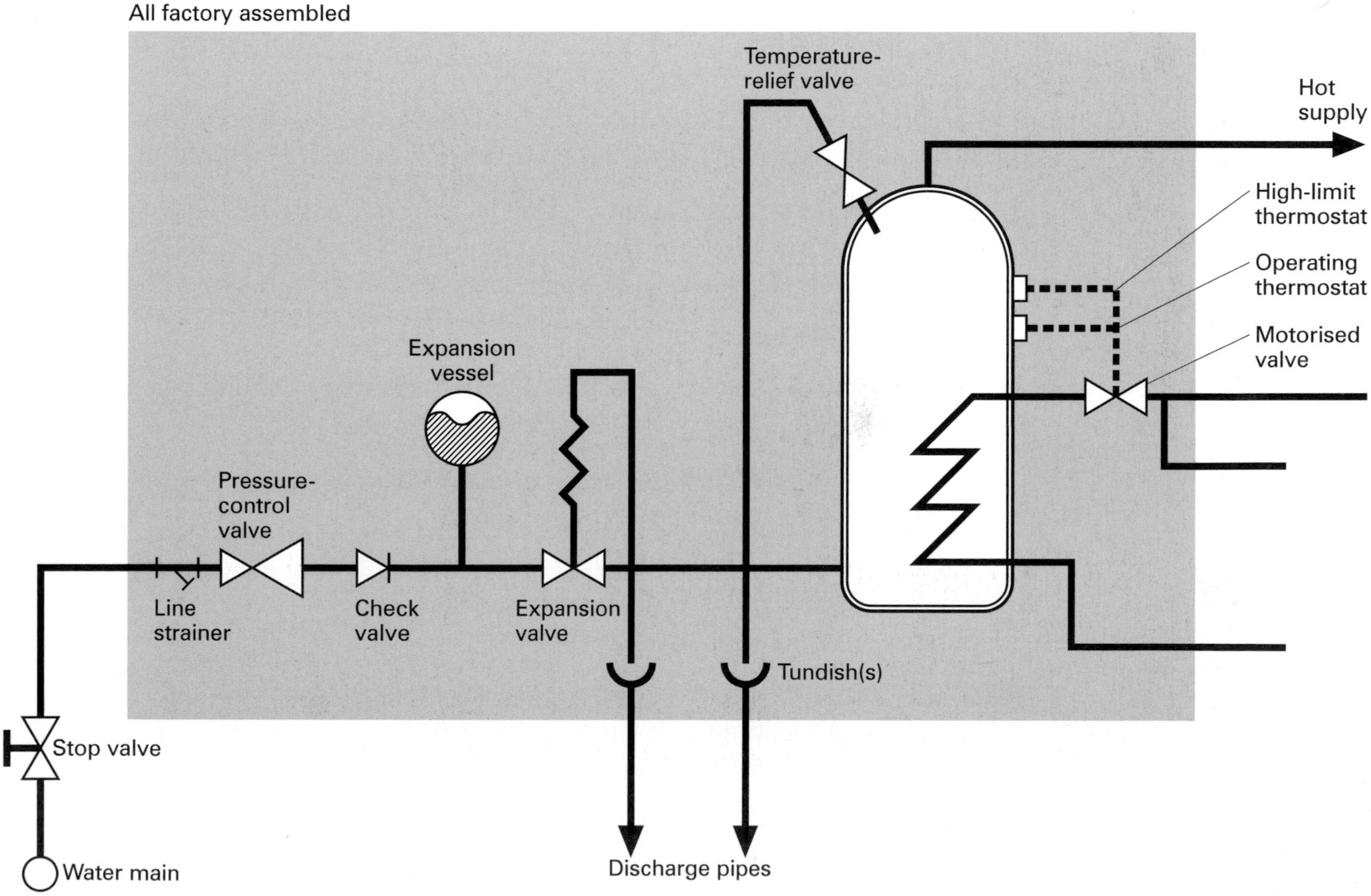

Figure 3.24 The unit

All the safety devices and functional controls are factory-fitted by the manufacturer:

- line strainer
- pressure-reducing or -limiting valve
- single check valve
- tundish
- expansion vessel
- expansion valve
- temperature-relief valve
- high-limit thermostat and cylinder primary motorised valve (indirect unit).

Directly heated systems

Under G3 a directly heated unit or package should have a minimum of two temperature-activated safety devices operating in sequence:

(a) a non-self-resetting thermal cut-out either to BS 3955 for electrical controls or to BS 4201 for thermostats for gas-burning appliances, and

(b) one or more temperature-relief valves manufactured to BS 6283 Part 2 for temperature-relief valves or to BS 6283 Part 3 for combined temperature- and pressure-relief valves.

Both these devices are in addition to the standard control thermostat fitted to maintain the temperature of the stored hot water – hence the three-tier level of protection.

The Regulations do state that it is permissible to use other forms of safety device as alternatives, but they have to be approved by a testing body such as the BBA and provide an equivalent degree of safety, so this isn't really one for the installer.

In both units and packages:

- The temperature-relief valve(s) should be located directly on the storage vessel so that the temperature of the stored water does not exceed 100°C.

- The valve(s) must be properly sized (this one is for the manufacturer) in accordance with the requirements of BS 6283.

- The valves should not be disconnected other than for maintenance or repair (so temporarily capping off a faulty valve on a working system while a new one is obtained is absolutely not allowed).

- Each valve should discharge via a short length of metal pipe (this is referred to as D1 and you'll encounter this later) of a size not less than the outlet size of the temperaturely-relief valve, through an air break over a tundish located vertically as near as possible to the valve(s).

Do you remember that there were special requirements for immersion heaters used with unvented hot-water storage systems? This diagram shows that the immersion heater is a special design and includes an energy cut-out device in addition to the control thermostat. Note that when maintaining these, you'll probably not be able to pick replacement parts straight off the shelf and may well have to wait for them to be ordered.

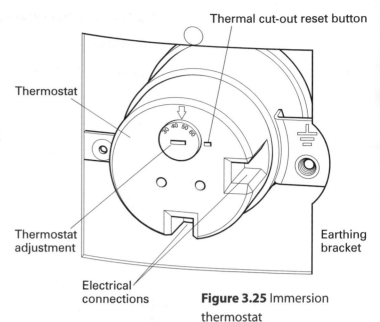

Figure 3.25 Immersion thermostat

Indirectly heated systems

The safety devices listed for direct heating are also required for indirectly heated units and packages, but the following must be considered in addition:

- The non self-resetting thermal cut-out should be wired up to a motorised valve or some other suitable device to shut off the flow to the primary heater.

- If the unit incorporates a boiler, the thermal cut-out may be on the boiler. There are a limited number of manufacturers that provide unvented cylinder and boiler packages that must be fitted together – these take advantage of minimising the number of thermostats by only having one energy cut-out on the boiler.

- The temperature-relief valve should be sized and located, and the discharge pipe provided, as described for direct heating.

- Where an indirect unit or package has any alternative direct method of water heating fitted, e.g. an immersion heater, a non-self-resetting thermal cut-out device will also be needed on the direct source(s).

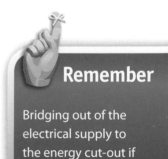

Remember

Bridging out of the electrical supply to the energy cut-out if it has failed is totally irresponsible and highly unsafe

System installation

The unit or package should be installed by a competent person, i.e. one holding a current Registered Operative identity card for the installation of unvented domestic hot-water storage systems issued by one of the following:

(a) the Construction Industry Training Board

(b) the Institute of Plumbing and Heating Engineers

(c) the Association of Installers of Unvented Hot Water Systems (Scotland and Northern Ireland)

(d) individuals who are designated Registered Operatives and employed by companies included on the list of Approved Installers published by the BBA up to 31 December 1991

(e) an equivalent body: as G3 was published in 1992 several newer certification bodies provide certification for unvented systems, namely the British Plumbing Employers Council (BPEC) and Zurich.

Building Regulations approval must be granted to install an unvented hot-water system under G3, and therefore a special notice form must be submitted with a fee, usually to the local building-control office, prior to the installation of each system. Only when approval is granted should the system be installed. This is not usually a big issue as long as the installer can prove that they are competent and the system that is being installed is approved and tested in line with the requirements. If this is the case, it is unlikely that a building-control officer will visit the site to conduct an inspection. However, if the officer is unhappy with any aspect of the application, then a site visit will be the norm.

Discharge points

Unvented systems give great flexibility in terms of where they can be sited, but the main factor to consider is the provision of a discharge pipe from both the temperature-relief valve and the expansion valve to a safe position, and this can sometimes be problematic.

Remember

You aren't approved to install unvented hot-water systems until you have obtained your competency card

Remember

Building control have the right to monitor, inspect and control the installation of unvented systems under G3

Figure 3.26 Discharge pipe arrangements

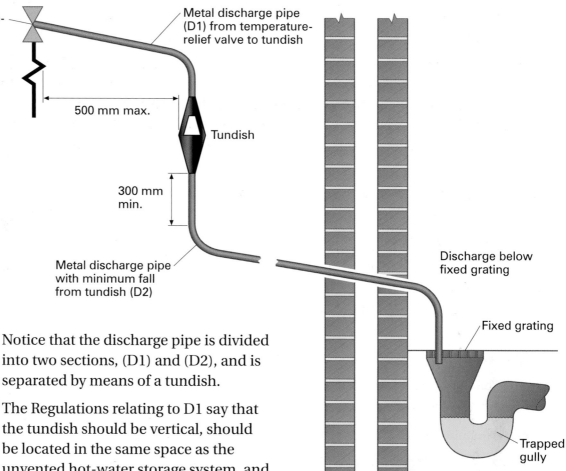

Notice that the discharge pipe is divided into two sections, (D1) and (D2), and is separated by means of a tundish.

The Regulations relating to D1 say that the tundish should be vertical, should be located in the same space as the unvented hot-water storage system, and should be fitted as close as possible to and within 500 mm of the safety device, i.e. temperature-relief valve.

With early systems, the D1 section of pipe used to be manufacturer-supplied; it is now common for you to have to produce this section of pipe yourself: it commonly joins together both the outlets of the temperature-relief and expansion valves, as shown in Figure 3.27.

The requirement here is to make sure that you install to the D1 section, with the pipework falling to the tundish and no more than 500 mm between valve outlet and tundish. Additionally, when joining both valve outlets together, siting the cold-water supply pipe and the expansion valve will also need to be considered, to ensure that the outlet from that valve can fall continuously to the tundish – it is not allowed to rise upwards!

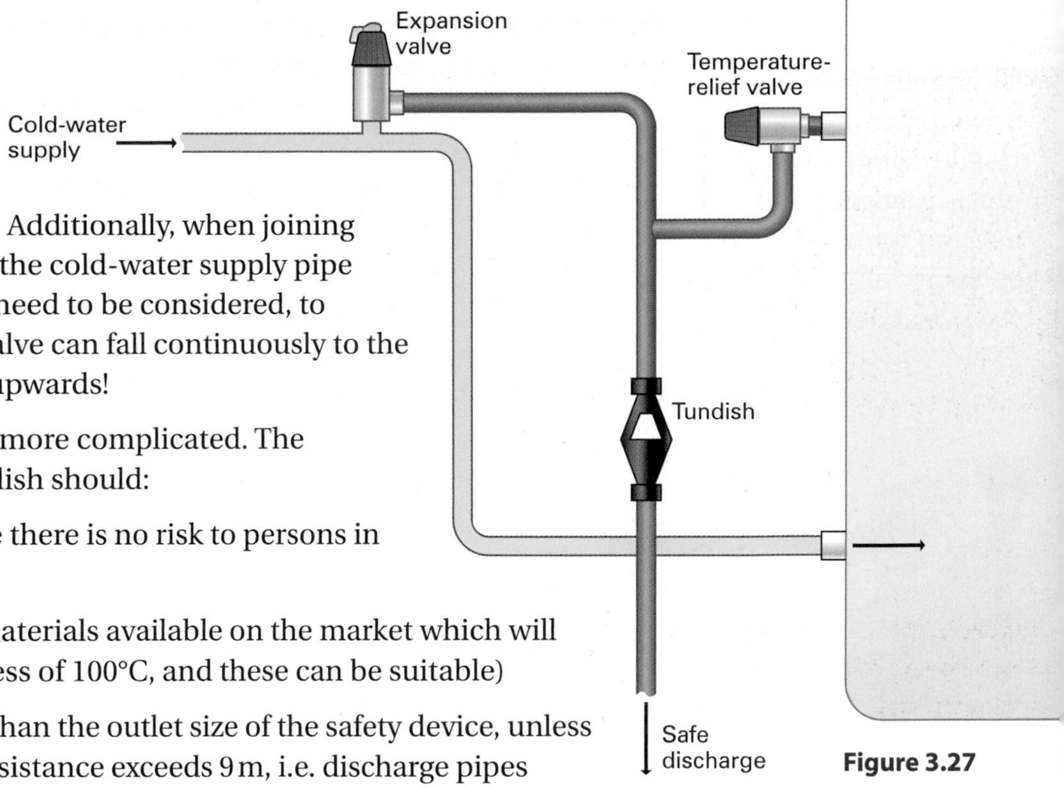

The requirements for D2 are a bit more complicated. The discharge pipe (D2) from the tundish should:

- terminate in a safe place where there is no risk to persons in the vicinity

- be of metal (there are plastic materials available on the market which will cope with temperatures in excess of 100°C, and these can be suitable)

- be at least one pipe size larger than the outlet size of the safety device, unless its total equivalent hydraulic resistance exceeds 9 m, i.e. discharge pipes between 9 m and 18 m equivalent resistance would be at least two sizes larger than the outlet size of the safety device, between 18 m and 27 m at least three sizes larger, and so on. Bends must be taken into account when working out the resistance

- have a vertical section of pipe at least 300 mm long below the tundish, before any elbows or bends in the pipework

- be installed with a continuous fall

- have discharges visible at both the tundish and the final point of discharge (but where this is not possible or is practically difficult there should be clear visibility at one or other of these locations).

Examples of acceptable discharge arrangements are:

(a) low level below a fixed grating is the ideal location as it is highly safe and visible

(b) downward discharge at low level.

Discharges at low level, i.e. up to 100 mm above external surfaces such as car parks, hard standings, grass areas etc. are acceptable, provided that where children may play or otherwise come into contact with discharges a wire cage or similar guard is positioned to prevent contact while maintaining visibility.

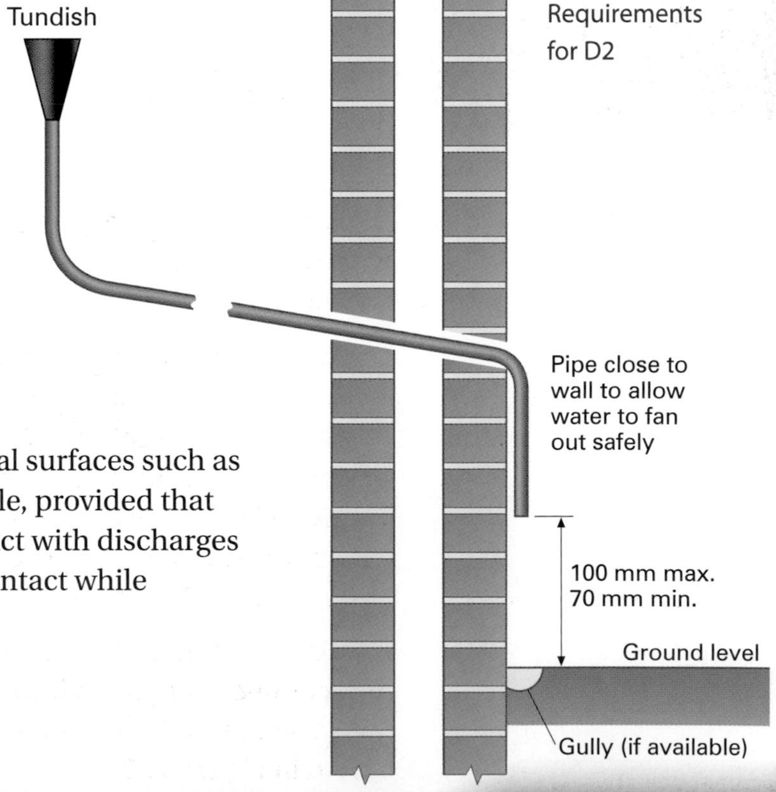

Expansion valve

Cold-water supply

Temperature-relief valve

Tundish

Safe discharge

Figure 3.27
Requirements for D1

Figure 3.28
Requirements for D2

Tundish

Pipe close to wall to allow water to fan out safely

100 mm max. 70 mm min.

Ground level

Gully (if available)

(c) Discharge at high level (into hopper).

Discharge may be made to a metal hopper and metal down pipe with the end of the discharge pipe clearly visible (tundish visible or not) or onto a roof capable of withstanding high-temperature discharges of water and 3 m from any plastic guttering that would collect such discharges (tundish visible).

(d) Discharge at high level (wall termination).

Building Control officers have indicated that in certain cases (and as a last resort) it may be acceptable to discharge at high level if the discharge outlet is terminated in such a way as to direct the flow of water against the external surface of a wall. However, minimum evidence of the height above an accessible surface and the distance required to reduce the discharge to a non-scalding level would have to be established.

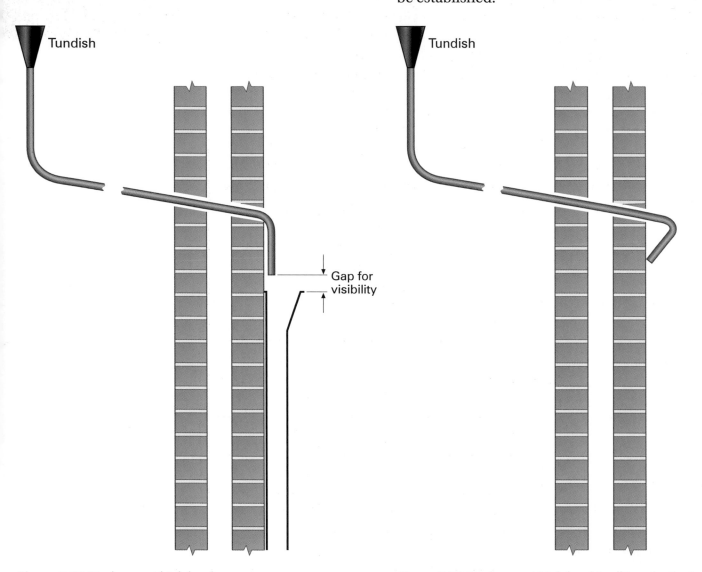

Figure 3.29 Discharge at high level

Figure 3.30 Discharge at high level (wall termination)

Where a single D2 pipe serves a number of discharges, such as in blocks of flats, the number of systems served should be limited to no more than six so that any installation discharging can be traced reasonably easily. The single common discharge pipe should be at least one pipe size larger than the largest individual discharge pipe (D2) to be connected.

If systems are to be installed where discharge is not apparent, i.e. in dwellings occupied by the blind etc., consideration should be given to the installation of an electronically operated device to warn when discharge takes place.

Safety tip

On no account must the pipe be able to discharge onto persons – you're best advised to seek feedback from building control before attempting this one, just to make sure they'll approve it

On the job: Discharge pipe requirements

Remember: the pipe size as a minimum must be one size larger than the valve outlet, up to 9 m in length, and between 9 m and 18 m it must be two sizes larger etc. But on the installation Martin's doing, he can't use straight pipes, and elbows and bends create resistance in the pipe. This pipework resistance has to be catered for when pipe sizing, or the discharge pipe will not meet requirements. Essentially, the length of straight pipe run is reduced for every bend or elbow used, so to simplify the process of sizing the pipe he uses a table like this.

Valve outlet size	Minimum size of discharge pipe D1	Minimum size of discharge pipe D2 from tundish	Maximum resistance allowed, expressed as a length of straight pipe (i.e. no elbows or bends)	Resistance created by each elbow or bend
G 1/2	15 mm	22 mm	Up to 9 m	0.8 m
		28 mm	Up to 18 m	1.0 m
		35 mm	Up to 27 m	1.4 m
G 3/4	22 mm	28 mm	Up to 9 m	1.0 m
		35 mm	Up to 18 m	1.4 m
		42 mm	Up to 27 m	1.7 m
G 1	28 mm	35 mm	Up to 9 m	1.4 m
		42 mm	Up to 18 m	1.7 m
		54 mm	Up to 27 m	2.3 m

Table 3.5 Sizing discharge pipe

Example: A G½ temperature-relief valve has a proposed discharge pipe length (D2) of 5 m from the tundish to its termination. The pipe will include seven elbows.

Martin carries out the calculation in the following way (assuming the minimum pipe size that could be used is 22 mm):

- From Table 3.5, the maximum run allowed for a straight length of 22 mm copper discharge pipe from a G½ temperature-relief valve is 9.0 m.

- Subtract the resistance for seven no. 22 mm elbows at 0.8 m each = 5.6 m.

- Therefore the maximum permitted length equates to 3.4 m.

- 3.4 m is less than the actual length of 5 m, therefore this is unacceptable.

1   Calculate the next largest size

2   Is this acceptable?

Water Supply (Water Fittings) Regulations 1999

The Water Regulations deal with functional (operational) requirements primarily addressing water usage issues. Every unvented storage system (with the exception of those systems with under 15 litres storage, as per Building Regulations), should be fitted with:

1   a temperature-control device

2   either a temperature-relief valve or a combined temperature- and pressure-relief valve

3   an expansion valve

Unless the expanded water is returned to the supply pipe in accordance with Water Regulation requirements, i.e. it should not be capable of being drawn off as drinking water, one of the following should also be fitted:

- an expansion vessel

- an integral expansion system.

The latter only relates to small heaters, as installers can't usually expand back into the mains without drawing off unwholesome water, so some form of expansion device is usually the norm.

Installation and commissioning of unvented systems

At the end of this section you should be able to:

* identify the key installation requirements of unvented hot-water systems
* state the commissioning requirements for unvented hot-water systems.

You now need to consider the practical information required to complete the successful installation of an unvented hot-water system.

Location of the cylinder

When considering the location of the cylinder we need to take account of the following:

* Sufficient clearance should be provided around the cylinder so that we can get at all the parts for maintenance and repairs (clearance details will usually be provided by the manufacturer).

* The floor where the cylinder is to stand should be sufficient to support its weight and the water content – remember, a typical unvented cylinder and controls will be heavier than a standard hot-water storage cylinder. A support for a typical 100-litre cylinder should be capable of withstanding 150 kg. In addition, the floor surface material should be suitable for purpose, i.e. chipboard flooring materials should comply with Building Regulations and capable of withstanding water penetration.

Unvented hot-water storage systems and uncontrollable heat sources

Unvented hot-water storage systems should never be supplied from uncontrollable heat sources such as solid-fuel boilers and solar water-heating circuits. It can be difficult with these types of fuel supply to isolate the supply rapidly in the event of an overheat situation, so they are not regarded as suitable under the Building Regulations.

Incoming water-supply pressure and flow rate

Even though these systems work at higher pressure you cannot install them just anywhere. One of the key failures with this type of system is its installation in properties in which there is insufficient supply flow rate and pressure – remember, one of the advantages of this type of system was high-pressure flow rates. Customers are going to expect this from this type of system, so carry out a few checks first.

The first check is to find out if the supply pressure and flow rate are subject to any sizeable fluctuation throughout a 24-hour cycle, i.e. during peak daytime usage are there any times when the mains supply may be impaired? Your local Water Undertaker (Water Authority) is the best point of contact for this information, and the customer can advise too. The Water Undertaker in situations where there may be fluctuating pressures may advise that a storage system is more suitable.

The second check is to establish the water pressure and flow rate at the supply-pipe entry to the property, and to establish if this meets the minimum requirement as laid down by the unvented system appliance manufacturer – this will usually be quoted in the manufacturer's installation instructions.

As a rule of thumb, a typical domestic storage system will need a minimum pressure of 1.5 bar and 20 litres per minute water-flow rate. Less than this and you should consult the manufacturer for guidance.

If you do not know what the supply pressure and flow rate is, then you need to disconnect the incoming main supply pipe from the rest of the system and:

- pressure test – by attaching a **pressure gauge** to the pipe end
- test for flow rate – using either a flow-measuring device, or a simple procedure is to discharge the supply pipe at full bore into a vessel with a measuring scale for a timed period and calculate the flow rate.

A 15 mm supply pipe can often be sufficient to provide the necessary water supply requirements; however, with new installations it is now common to see a 22 mm supply pipe into properties with unvented hot-water storage systems.

Size of unvented cylinder

As with any system the hot-water storage cylinder should be properly sized. Several manufacturers do provide rule-of-thumb guidance with regard to the size of cylinder, and this can usually be followed for smaller installations; a typical chart is shown in Table 3.6.

Property type and plumbing supply requirements	Indirect (l)	Direct (l)
1-bedroom apartment, 1 shower	125	145
2-bedroom house, 1 bath, 1 shower	145	185
3-bedroom house, 1 bath, 1 shower	185	210
2/3-bedroom house, 2 baths, 1 shower/en suite	210	250
4/5-bedroom house, 2 baths, 1 shower/en suite	250	300
Larger houses	300	300

Table 3.6 Guidance on size of cylinder

Supply pipework size

As with cylinders, pipework should be properly sized. The following points need to be taken into account when looking at the size of pipework in unvented hot-water systems:

- The flow-rate demands on a pipe supplying the property with an unvented system will be higher than with a traditional vented system because all outlets (hot and cold) will be drawing off the supply pipe rather than just the cold and cistern feeds.

- The flow rate after the pressure-reducing valve will typically be reduced due to the need to limit the incoming pressure, and this will have an impact on the required pipe size.

- Outlets situated higher than the pressure-reducing valve will give lower outlet pressures at terminal fittings (taps and valves): a 10 m height difference will result in a 1 bar pressure reduction at the outlet. This can be an issue with tall buildings.

Here are some typical pipe sizes for a smaller domestic property:

- The supply pipe from the pressure-reducing valve (both hot and cold supplies) will need to be 22 mm.

- The pipe feeding the cylinder will need to be 22 mm (cold connection).

- The hot-water draw-off to the point at which it branches to various fittings will need to be 22 mm.

- The cold-water supply from the cylinder to the point at which it branches to various fittings will need to be 22 mm.

- Supplies to individual appliances – baths included – can be 15 mm or less.

Discharge pipes

Remember: the location of the discharge pipe can have a real bearing on the position of the cylinder – it also needs to be properly sized to function correctly and must be positioned safely so as not to cause a potentially serious accident.

Balanced or unbalanced supply

The position of the pressure-reducing valve in the system will have an impact on the pressures within the various parts of the system. If the valve is sited as a composite valve in the airing cupboard in a property (this is the most likely location), then any draw-off points taken from the supply pipe before (not regulated by the valve) may be at a much higher pressure than corresponding cold-supply draw-off points – this is known as an **unbalanced supply**, as shown in Figure 3.31. With many modern taps this will not provide suitable mixing and will be unacceptable for the customer. Under normal circumstances it is usual practice to install a balanced system with all outlets operating at the same supply pressure.

Figure 3.31 Unbalanced supply

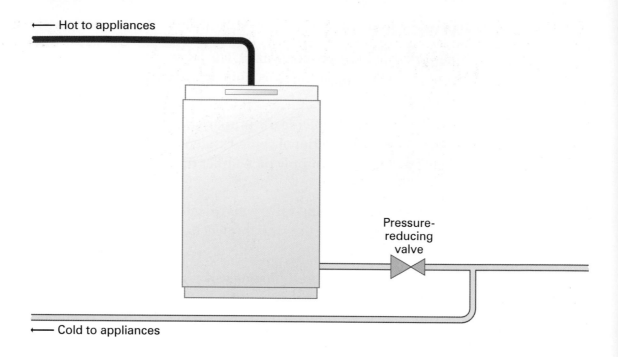

Figure 3.32 Balanced supply

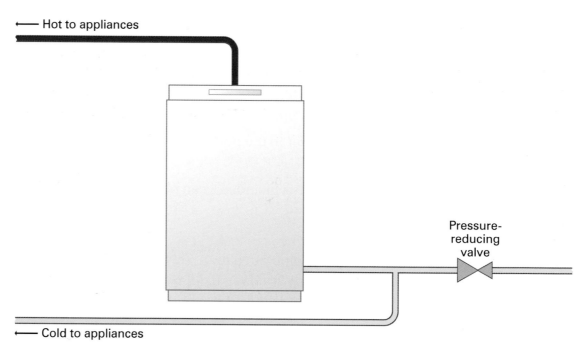

Figure 3.32 shows a balanced supply where all draws-offs are taken after the pressure-reducing valve; in certain cases, and particularly with composite valves, this may mean taking the cold-supply pipe straight from its entry to the building direct to the valve in the airing cupboard and then feeding back to all cold points in the property from this valve.

Composite valves may include a balanced pressure port, as in the following photograph.

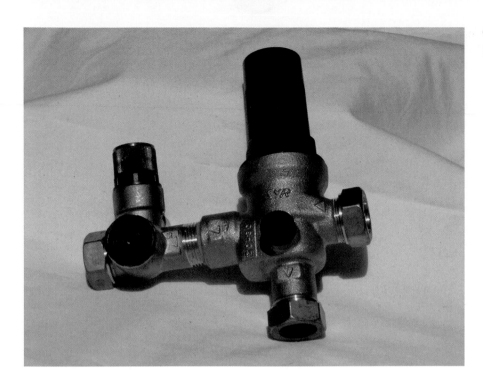

Composite valve with balanced pressure port

You also need to be careful when installing unbalanced systems in which there are mixer taps that allow mixing of hot and cold water in the valve body. The Water Regulations point out the need to provide backflow protection to both hot and cold pipework feeding appliances that incorporate this type of mixer. If you follow these requirements you'll be OK with unvented systems. If you don't, then there's an added problem, as shown in Figure 3.33.

On an unbalanced supply and with mixing in the taps as shown in Figure 3.33, with both taps open, the unbalanced cold supply will back-pressurise the system through the hot supply, causing the expansion valve to **nuisance discharge** and waste water as a likely result.

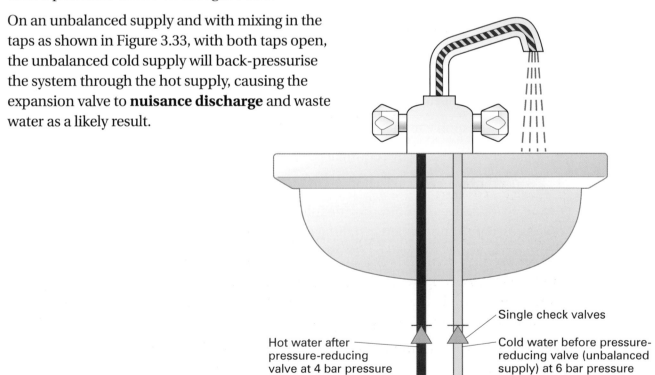

Single check valves

Hot water after pressure-reducing valve at 4 bar pressure

Cold water before pressure-reducing valve (unbalanced supply) at 6 bar pressure

Figure 3.33 Back pressure on an unbalanced supply

Secondary circulation systems

Where secondary circulation systems are to be installed, the secondary return pipe will usually be in 15 mm pipe and incorporate a check valve to prevent backflow. A suitable Water Regulations Advisory Service (WRAS)-approved bronze circulation pump will be required. On large systems incorporating either an expansion vessel or cylinder air gap, it may be necessary to fit an additional expansion vessel to the secondary circuit, due to the increase in system water content. This should be done if the capacity of the secondary circuit exceeds 10 litres.

Use the details below to determine whether an additional vessel is necessary.

Pipe capacity (copper)

- 15 mm = 0.13 l/m (10 litres = 77m)

- 22 mm = 0.38 l/m (10 litres = 26m)

- 28 mm = 0.55 l/m (10 litres = 18m).

Hot-water temperature

Unvented systems provide high flow rates from taps and fittings, and with a water temperature at 60°C there can be a hazard from scalding, especially to children and older people, because of the speed of the hot water emerging from the taps and fittings. **Thermostatic mixing valves** can be added, for example to bath taps.

Control system and electric wiring

Indirectly heated systems must only be connected to fully pumped heating systems. Indirectly heated unvented systems for use in domestic properties require the installation of a motorised valve (usually supplied with the cylinder) to control the temperature of primary water entering the system under normal operating conditions and to isolate the primary heat supply to the cylinder in overheat situations.

The control system can therefore be based on either:

- a 2-port motorised valve configuration – 'S Plan', or

- a 3-port motorised valve configuration – 'Y Plan'.

As the 3-port valve does not provide full isolation of the hot-water port in all operating conditions, when installing a Y-Plan configuration (unless advised by the manufacturer) you will be required to install both a 3-port valve and the 2-port valve supplied by the manufacturer, to ensure that there is effective isolation of the cylinder in an overheat situation. You may also have to order an optional **control relay** to wire both valves.

For new installations it is better to specify the 2-port valve configuration, as it is cheaper and easier to install.

The key point about wiring the electrical connections to the system controls is to follow the manufacturer's installation instructions. The wiring should be made in accordance with the latest edition of the IEE regulations – this goes without saying!

The photo on the right shows a typical dual thermostat provided with most modern unvented hot-water storage systems, where for ease of wiring both the control thermostat and the thermal cut-out device (high-limit thermostat) are housed in one unit; the wiring is made relatively easily with a simple 3-wire connection (live in from the **programmer**, live out to the motorised valve and an earth terminal).

Typical wiring connections to dual thermostat

Commissioning of unvented hot-water systems

The manufacturer usually details a commissioning procedure in their instructions. The following example gives a typical commissioning procedure for unvented hot-water systems.

1   If you are about to commission a system that you haven't installed, then the first check is to establish that all items supplied are those indicated by the manufacturer to be suitable for the system – check valve sizes and discharge rates, and that pipework and wiring connections are correct.

2   Check all system component connections for tightness, especially those made at the factory (they may have come loose in transit).

3   Check that the operating thermostat setting is 60°C to 65°C.

4   Check that the expansion vessel pre-charge pressure is according to manufacturer's specification. The pressure point is usually on the top of the vessel, under a plastic cap. Vessels are normally supplied with the correct charge pressure. Note: Pressure pre-charge details will be found stamped on the expansion vessel or included in the manufacturer's instructions. For bubble tops, the air gap will be established on filling the system; with expansion vessels you'll usually need to have a tyre foot pump and pressure gauge with you as part of the kit!

5   Make sure the drain valve is closed, then open the highest hot and cold water taps and other fittings.

6   As part of the initial filling procedure it is important to ensure that the system is thoroughly flushed. The flushing procedure should be undertaken in two stages:

Stage 1 – Line Strainer (usually part of the pressure-reducing valve)

The line strainer should be removed and water flushed through to ensure that any debris is removed from the pipework up to the strainer. The strainer should then be reassembled.

Stage 2

Open cold-water supply stop valve and allow the cylinder to fill with water.

Continue to fill until water issues from the terminal fittings. Shut these off, then check again for leaks.

7   Manually operate the test device on both relief valves to allow water to flow for about 30 seconds to remove any residue that may have collected on the valve seats. Ensure that water is discharged from the valves and that it runs freely down the discharge pipes. Ensure that relief valves completely shut on testing.

8   The cylinder can now be run up to temperature, confirming operation of control thermostats and that the expansion and temperature-relief valves do not discharge water under normal operating conditions.

9   Check the flow rate and pressure at the outlets (taps) and confirm that design specifications are being met.

10   Complete any commissioning records and leave the manufacturer's instructions on site.

On the job: Commissioning procedure

Martin is installing an unvented hot-water system in a house. The manufacturer of the system he is installing has included a detailed commissioning procedure in its instructions.

1   What are the main things Martin has to do before he can hand the installation over to the customer?

FAQ

Why do I have to do so many tests and checks when I have fitted an unvented cylinder?

The checks will ensure that the cylinder and the hot-water system will be safe, efficient and reliable.

Service and maintenance of unvented systems

At the end of this section you should be able to:

- describe the service and maintenance procedure for unvented systems
- diagnose a range of common faults on unvented systems.

Periodic maintenance and inspection should usually be carried out on an annual basis. Experience of local water conditions may indicate that more frequent inspection is desirable, e.g. when water is particularly hard or scale forming, or where the water supply contains a high proportion of solids such as sand.

The user should, however, be encouraged to report faults such as discharges from relief valves and have them rectified as soon as they occur.

The following steps show a typical servicing procedure for unvented hot-water systems:

1 Check that all approved components are still fitted and are unobstructed. Check to see if all valves are still in position and all thermostats are properly wired. Be careful with the electrics: it can be common for immersion heaters to fail and to be replaced with a standard immersion heater rather than one using the proper thermostats.

2 Check for evidence of recent water discharge from the relief valves, visually and by questioning the customer if possible.

3 Manually check the temperature-relief valve – lift the gear or twist the top on the integral test device on the relief valve for about 30 seconds to remove any residue that may have collected on the valve seat. Check that it reseats and reseals.

4 Manually check the expansion valve – lift gear or twist top on the integral test device on the relief valve for about 30 seconds to remove any residue that may have collected on the valve seat. Check that it reseats – and reseals.

5 Check discharge pipes from both expansion and temperature-relief valves for obstruction and that their termination points have not been obstructed or had building work carried out around them.

6 Check the cylinder operating thermostat setting (e.g. 60–65°C).

7 Isolate gas and electrical supplies to the heating appliance, turn off water supply and relieve water pressure by opening taps:

  (a) Drain inlet pipework where necessary to check, clean or replace the line strainer filter (this may be part of the pressure-reducing valve assembly).

  (b) Check pressure in expansion vessel and top up as necessary while the system is empty or uncharged. Use a tyre gauge and a pump to recharge to operating pressure. If the cylinder includes an air gap as the expansion device, reinstate the air gap as part of the procedure in line with the manufacturer's instructions.

Safety tip

Don't forget the golden rule when maintaining these systems – always use manufacturer-approved replacement parts that meet the original system specification and never make temporary repairs – here it's a real safety issue

8 Reinstate water, electricity and gas supply. Run the system up to temperature and ensure that control thermostats are working effectively and relief valves are not discharging water.

9 Complete any maintenance records for the system.

Fault-finding on unvented systems

Most manufacturers provide pretty good guidance on fault-finding in systems, usually in the form of a flowchart to help diagnose the fault. Here's the chart for a fault where **water is discharging from the temperature-relief valve**.

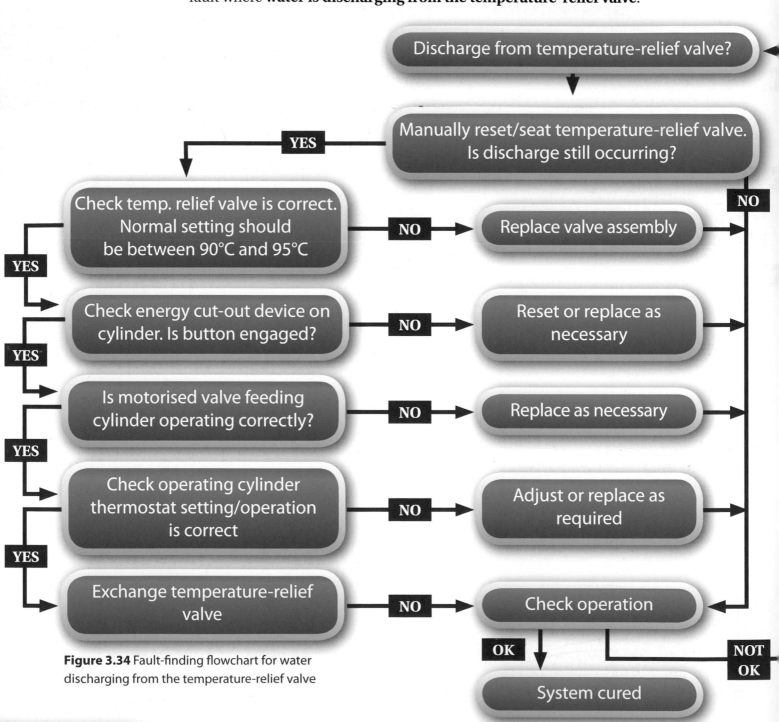

Figure 3.34 Fault-finding flowchart for water discharging from the temperature-relief valve

This is a fault-finding flowchart to be used in situations where there is **poor flow rate at taps or outlets**.

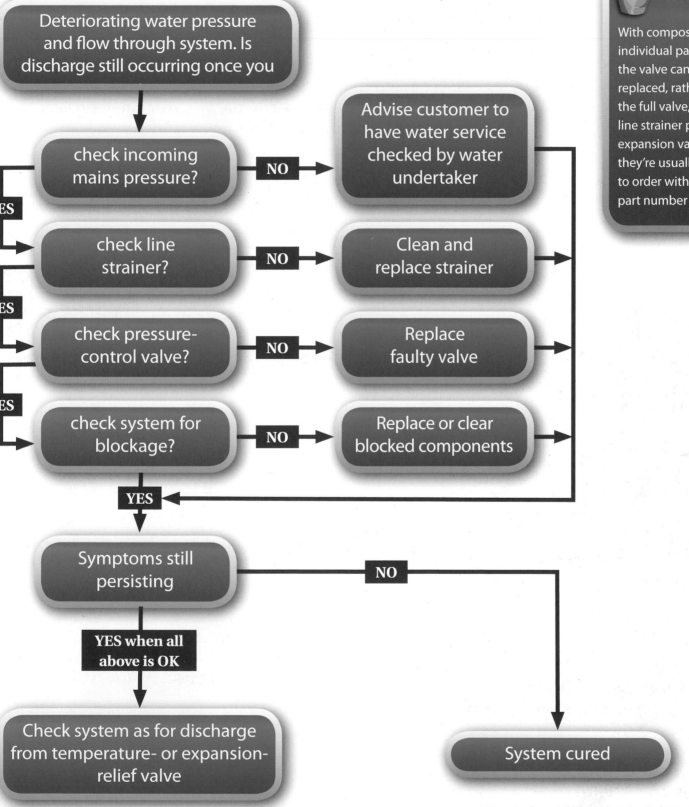

Deteriorating water pressure and flow through system. Is discharge still occurring once you

check incoming mains pressure? — **NO** → Advise customer to have water service checked by water undertaker

YES

check line strainer? — **NO** → Clean and replace strainer

YES

check pressure-control valve? — **NO** → Replace faulty valve

YES

check system for blockage? — **NO** → Replace or clear blocked components

YES

Symptoms still persisting — **NO** → System cured

YES when all above is OK

Check system as for discharge from temperature- or expansion-relief valve

Figure 3.35 Poor flow rate at taps or outlets. Fault-finding flowchart

Another fault is **water discharging from the expansion valve**. Here the chart is for a system that has an expansion vessel.

Figure 3.36 Water discharging from the expansion valve. Flowchart for a system with an expansion vessel

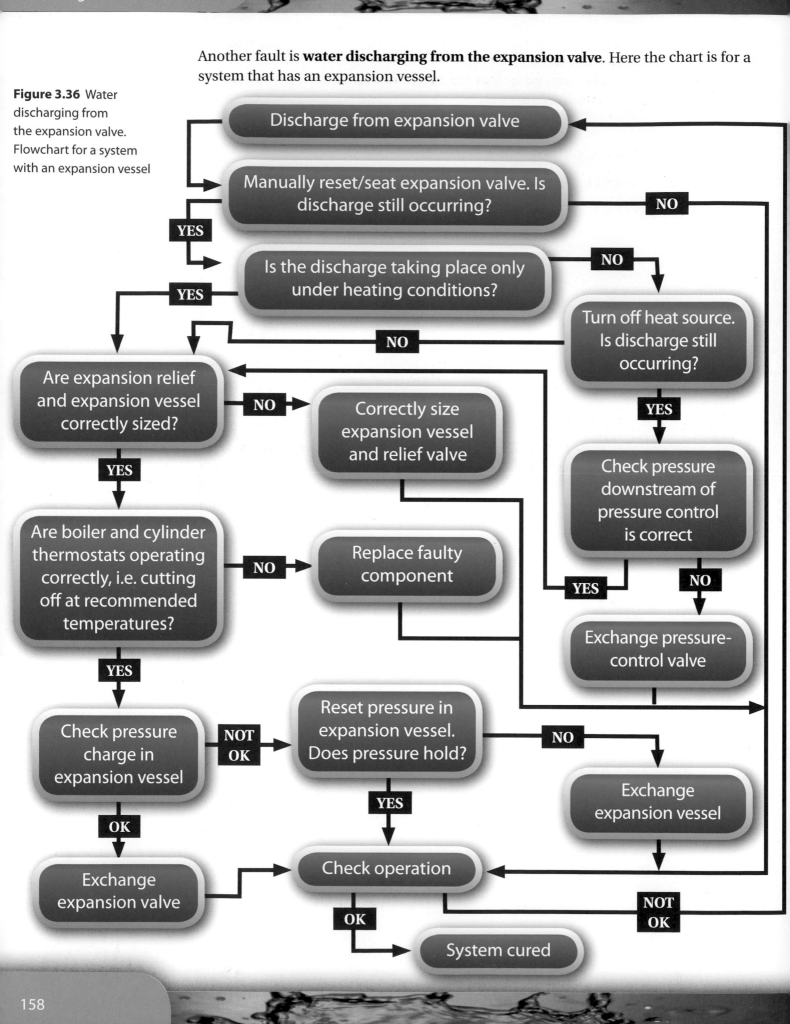

With cylinder air gaps as the expansion device, intermittent discharge from the expansion valve can quite often mean that the air gap or bubble has reduced in volume. The solution to this problem is to:

- turn off the water supply to the cylinder

- open the lowest hot-water tap in the system

- hold open the temperature-relief valve until water ceases to run from the cylinder

- refill the system. The air volume will automatically be recharged as the unit refills.

FAQs

Do unvented cylinders need to be serviced?

Yes, they should be serviced every 12 months.

Why do unvented cylinders need servicing so often?

You need to ensure that all of the safety devices are working correctly and that the cylinder is performing to its full potential.

Knowledge check

1 By how much does water usually expand when heated under normal operating conditions?

2 State two advantages of an unvented hot-water system.

3 Can an unvented appliance with a capacity of less than 15 litres expand water back into the supply pipe if a loose jumper to the stop valve is fitted? Explain your reasons.

4 State two methods of accommodating expanded water in an unvented system.

5 An overheat thermostat on an unvented system will usually operate at what temperature?

6 What components perform a three-tier level of protection on an unvented system?

7 What component prevents potential backflow into the supply pipe in an unvented system?

8 What is the maximum length of the D1 discharge pipe?

9 What is different about the immersion heater fitted to an unvented cylinder?

10 When installing an unvented system with primary flow provided by a mid-position valve, what additional control component is required?

11 How is the operation of the expansion valve checked on system commissioning?

12 The expansion valve from an unvented cylinder with bubble-top arrangement intermittently discharges water. What would be the first action you would try to remedy the problem?

13 A system is suffering from a reduced flow rate at all taps. What are the two most likely causes?

chapter 4

Above ground discharge systems

OVERVIEW

This chapter builds on the learning undertaken at Level 2. It covers the statutory legislation requirements that cover the installation of stack systems in domestic properties. It also looks at macerator WCs, sink waste-disposal units and using valves as alternatives to traps. The chapter concludes with performance testing and fault diagnosis. Key topics covered are:

- **Legislation and system components**
  - Types of stack systems
  - Primary ventilated stack – pipe runs and branch connections
  - Stub stack
  - Air-admittance valves
  - Waste connection to drainage system
  - Urinals
  - Spacing requirements for sanitary equipment
  - Ventilation of sanitary accommodation
  - Disabled access – sanitary accommodation

- **Other system types and components**
  - Macerator WCs
  - Sink waste-disposal units
  - Valves as an alternative to traps (HepvO systems)

- **Performance testing**
  - Soundness testing
  - Performance testing

- **Fault diagnosis**
  - Blockages

Legislation and system components

At the end of this section you should be able to:

- state the relevant legislative requirements relating to sanitary pipework and appliance installation
- identify sanitary accommodation requirements, including disabled access.

The key standards for above ground sanitation systems are: Building Regulations – Part H and British Standard 12056

There are four types of stack system:

- Primary ventilated stack system
- Ventilated discharge branch system
- Secondary modified ventilated stack system
- Stub stacks

Definition

primary ventilated – name for the main waste discharge stack

stub stack – short discharge stack used in limited applications

Primary ventilated stack – pipe runs

The primary ventilated stack system is commonly used in domestic properties. Mistakes are often made in its design, with pipe runs being too long or pipe falls being too shallow or steep.

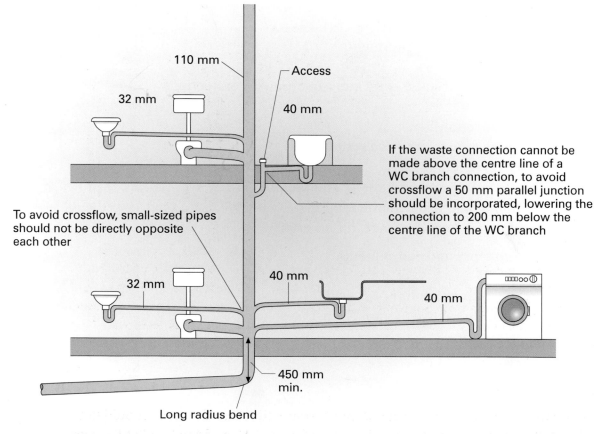

If the waste connection cannot be made above the centre line of a WC branch connection, to avoid crossflow a 50 mm parallel junction should be incorporated, lowering the connection to 200 mm below the centre line of the WC branch

To avoid crossflow, small-sized pipes should not be directly opposite each other

Figure 4.1 Primary ventilated stack system

Pipe size	Maximum length	Slope
32 mm	1.7 m	See design curve (Figure 4.2)
40 mm	3 m	18–90 mm/metre
50 mm	4 m	18–90 mm/metre
WC	6 m	18 mm/metre minimum

Table 4.1 Design specifications for single branch connections

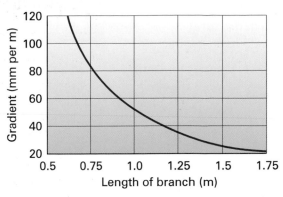

Figure 4.2 Design curve for 32 mm waste pipes

Primary ventilated stack – branch connections

The Regulations lay down specific requirements for the connection of branches into soil stacks.

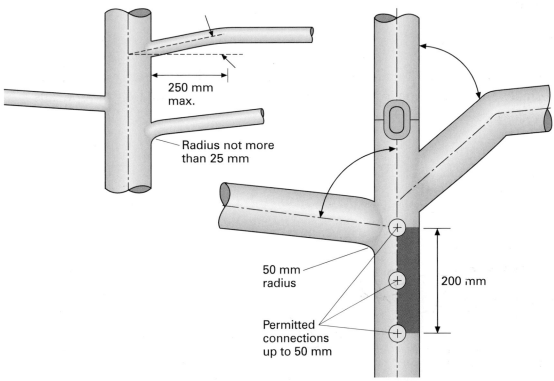

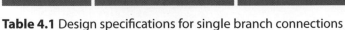

Find out

What is the likely outcome of pipework that is laid with a fall that is too shallow?

Figure 4.3 Branch connections for waste pipes. For the top diagram, junctions, including branch pipe connections of less than 75 mm, should be made at a 45° angle or with a 25 mm bend radius. The bottom diagram shows the prohibited zone distance (opposite the WC connection) in which a branch pipe may not be connected to a distance of 200 mm. Branch connection pipes of over 75 mm diameter must either connect to the stack at a 45° angle or with a minimum bend radius of 50 mm

Foot of the stack

The Regulations state that the lowest point of connection into the stack should not be within 450 mm of the invert of the drain (see Figure 4.1), and the bend at the foot of the stack should have as large a radius as possible but no less than 200 mm.

Find out

If a ventilation pipe from a soil stack discharges within 3 metres of any opening to a building, what height must it rise to above the opening?

What do the Regulations say about maintenance access into above ground discharge pipework systems?

FAQ

Why should a large-radius bend be used at the foot of the stack?

A tight-radius bend will cause turbulence at the foot of the stack, which then causes backpressure. For example, if a WC pan was installed at ground level and connected to the stack, it could cause the trap seal to be lost. This is also why there is a minimum distance of 450 mm for the lowest point of connection to the stack and the invert of the drain.

WCs connected direct to the drain

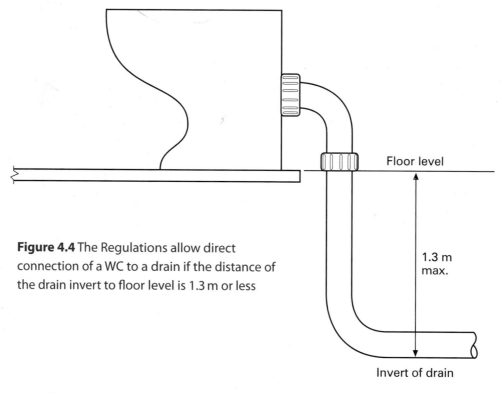

Figure 4.4 The Regulations allow direct connection of a WC to a drain if the distance of the drain invert to floor level is 1.3 m or less

Stub stack

With stub stacks you usually see a short length of pipe rising to above floor level in the room, which is capped. There is no ventilating pipework to the system, but the drain to which it is connected must be ventilated. Stub stacks are usually found in a second bathroom where the other bathroom is at the head of the drain and has a properly ventilated stack. There are specific requirements for the connections into the stack given in Figure 4.5.

Air-admittance valves

Using an air-admittance valve on a primary ventilated stack system removes the need for a stack ventilating pipe. It gives greater flexibility because there are no restrictions on the height of connections above drain invert level.

There are a number of key requirements for installing an air-admittance valve:

- They can only be used inside a building.

- They shouldn't adversely affect the amount of air needed for the below ground drainage system to work.

- They should be placed in a position where air is easily available at their inlet.

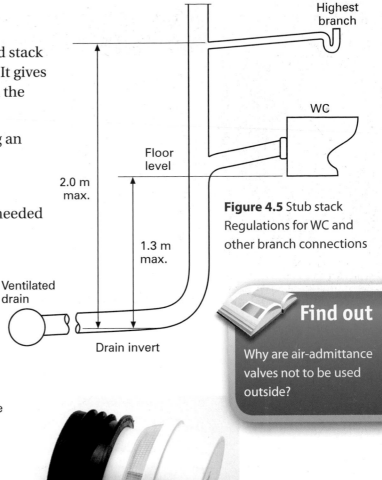

Figure 4.5 Stub stack Regulations for WC and other branch connections

Find out

Why are air-admittance valves not to be used outside?

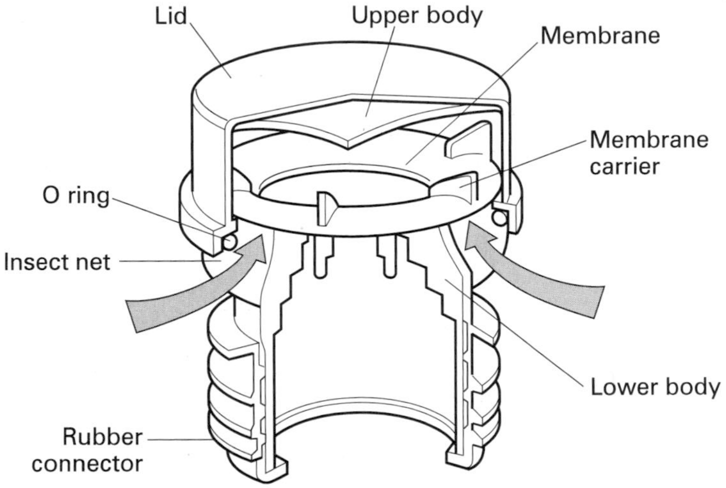

Figure 4.6 Air-admittance valve, showing the internal features

Air-admittance valve

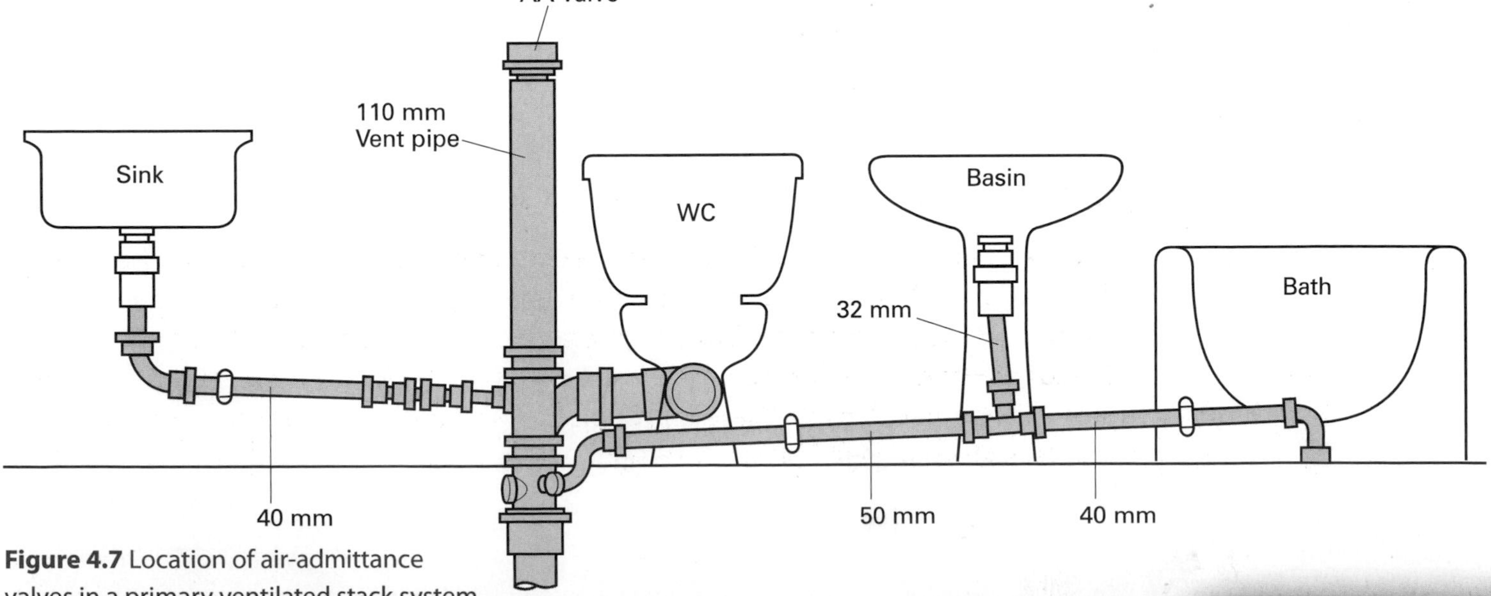

Figure 4.7 Location of air-admittance valves in a primary ventilated stack system

Air-admittance valves are now commonly used because they save costs: you do not have to take the ventilating pipe through the roof. However, you cannot fit them to every property on a site, as the below ground drainage system needs to ventilate itself to the atmosphere to work correctly.

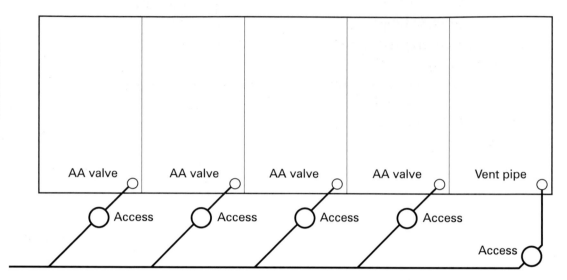

Figure 4.8 A typical layout for ventilating pipes to properties. In this example, every fifth property on a drainage run has a ventilating pipe; other properties are served by air-admittance valves

Waste connection to drainage system

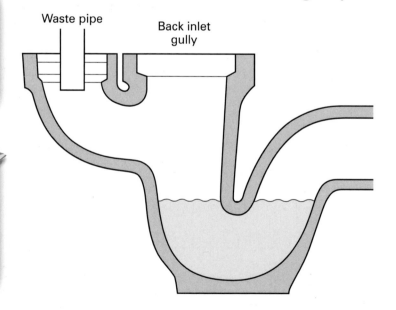

Figure 4.9 Building Regulation requirements for a branch pipe to gully

Urinals

Although urinals are not used in domestic properties, you are likely to install them for small businesses.

Figure 4.10 shows a bowl-type urinal, which is most commonly used. It is directly connected by a branch pipe system to a discharge stack. The outlets are trapped in much the same way as for a basin (32 mm is the minimum size).

For adults the front lip of the bowl is placed about 600 mm above floor level. For children's use, it will be lower (the height will depend on the children's age).

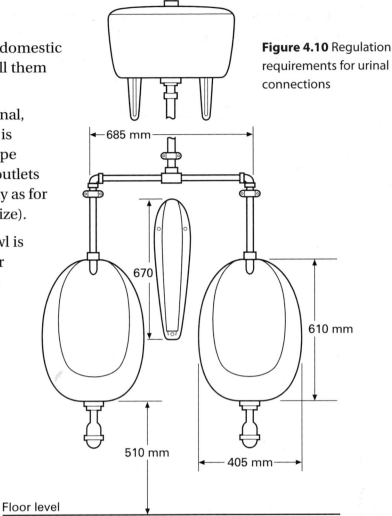

Figure 4.10 Regulation requirements for urinal connections

685 mm

670

610 mm

510 mm

405 mm

Floor level

This is a trough-type urinal. It is used in toilets where there may be a high risk of vandalism. The trough is sized for the maximum number of people that are going to use it and can be made in various lengths. The outlet is put at one end of the trough, which has a slight fall across the base. It then connects to a discharge stack, usually via branch pipework. The size of the branch pipework and trap are determined by the size of the trough and the distance of the branch pipe to the discharge stack.

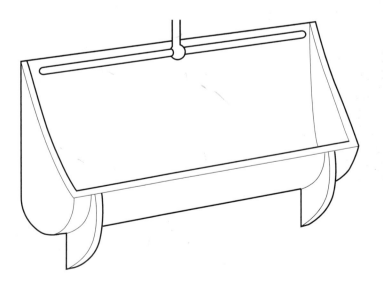

Figure 4.11 Trough-type urinal

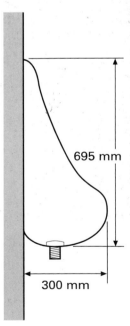

695 mm

300 mm

Below is a slab-type urinal. This urinal is supplied in a number of pieces to assemble on site. If it is on the ground floor of the property the waste connection is made directly to the drainage system via a trapped gully. If it is on the first floor or above the connection is made to a discharge pipe system (also trapped).

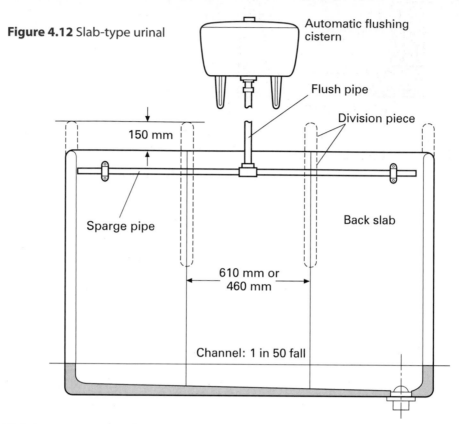

Figure 4.12 Slab-type urinal

Automatic flushing cistern

Flush pipe

Division piece

150 mm

Back slab

Sparge pipe

610 mm or 460 mm

Channel: 1 in 50 fall

This is an example of a slab urinal manufactured in one piece (and usually not able to accommodate more than two people). It is connected in a similar manner to the slab mentioned previously.

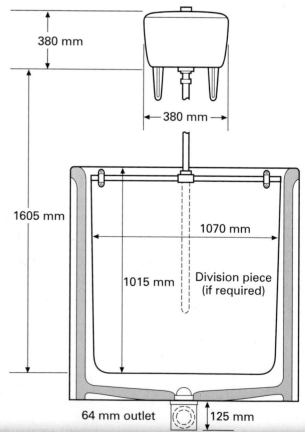

380 mm

380 mm

1605 mm

1070 mm

1015 mm

Division piece (if required)

Figure 4.13

One-piece slab urinal

64 mm outlet

125 mm

Here is an example of a clay trap provided to connect a urinal straight to the drainage system on the ground floor of a property.

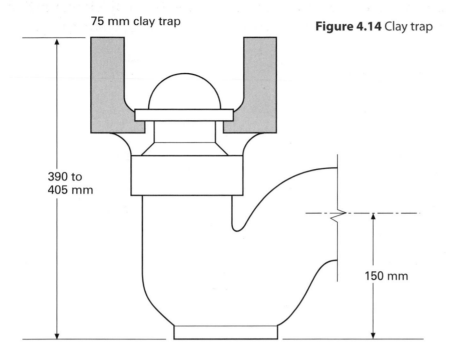

Figure 4.14 Clay trap

Spacing requirements for sanitary equipment

When installing sanitary equipment, it is important to think about the spacing requirement if the householder is to use it properly. Be aware that body size will have an impact on the use of the sanitary equipment.

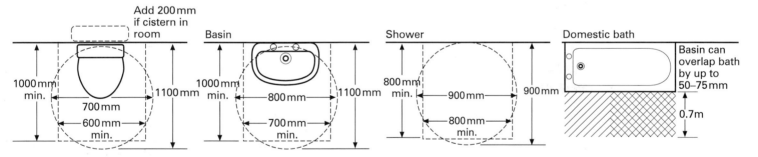

Figure 4.15 Recommended spacing requirements of sanitary equipment for average-sized people

Ventilation of sanitary accommodation

Part F of the Building Regulations details the requirements for ventilation in bathrooms. Installing ventilation will not usually be a part of your job, but you may be asked if ventilation meets Building Regulation requirements.

Disabled access – sanitary accommodation

Approved Document M lays down the requirements for sanitary accommodation for disabled people. It details the size of the accommodation and the layout and positioning of the WC. The sanitary accommodation that is provided will depend on whether it is needed for disabled people who can walk, or for people who use a wheelchair. Wheelchair access requires more space.

Disabled toilets are designed for specific access requirements

On the job: Upgrading toilet facilities

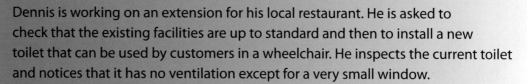

Dennis is working on an extension for his local restaurant. He is asked to check that the existing facilities are up to standard and then to install a new toilet that can be used by customers in a wheelchair. He inspects the current toilet and notices that it has no ventilation except for a very small window.

1   Why is ventilation so important?

2   What might Dennis's layout for the new WC compartment look like?

Other system types and components

At the end of this section you should be able to:

- state the installation and commissioning requirements of a range of extended system components.

For installing **macerator** WCs and **waste-disposal units** you will need to know about more advanced above ground components than were covered at Level 2. In particular you will need to know about using valves as an alternative to traps (a relatively new system design that fixes many of the problems associated with trap-seal loss and the effects of **siphonage**).

Macerator WCs

The most widely used macerator WCs are manufactured by Saniflo.

The key features of a macerator WC are:

- A macerator can be used only in a property where there is another bathroom with a traditional soil connection (second bathrooms only).

- Wiring must be carried out to BS 7671. An unswitched electrical supply point should be provided by a qualified electrician near the point of connection to the unit.

- The unit will usually be connected to 19 mm discharge pipework (depending on the length of pipe run). This gives flexibility as to where the WC can be positioned.

32/40 mm diameter max. 20 m

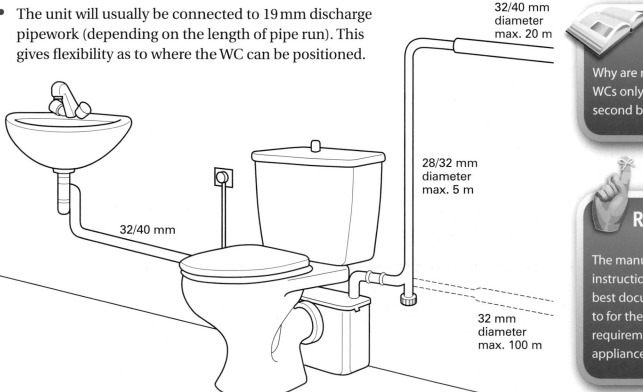

28/32 mm diameter max. 5 m

32/40 mm

32 mm diameter max. 100 m

Figure 4.16 Design components for a macerator unit. The basin is connected to the unit and therefore a single pipe connection is used (this gives greater location options for a cloakroom). The pump can lift the discharge contents vertically as well as move them horizontally

Find out

Why must an air admittance valve be put at the high point of the pipework?

Key requirements for a macerator unit installation are:

- Horizontal pipe runs must have a minimum fall of 5 mm/metre.

- Maximum horizontal pumping distance of 100 m.

- On a horizontal run, the 22 mm pipe should be increased to 32 mm after about 12 metres.

- Any vertical lifting must occur at the beginning of the pipe run, not at the end.

- 90° elbows must not be used – use 2 x 45° bends instead.

- If the horizontal pipe run is significantly below the height of the unit then an air-admittance valve should be fitted to the high point on the pipework.

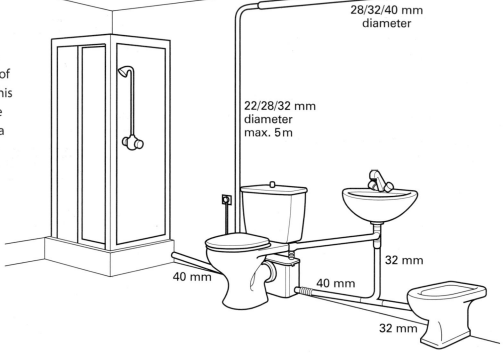

28/32/40 mm diameter

22/28/32 mm diameter max. 5 m

40 mm

32 mm

40 mm

32 mm

Figure 4.17 Design components of another type of macerator unit. This unit is capable of dealing with the outlets from all the appliances in a shower room-type installation.

Commissioning a macerator WC

- Do a visual inspection. Checks should include the soundness of the water supply and the discharge pipe work; also ensure that the electrical supply connection is correct and has been properly tested.

- The electricity supply should be fused at 5 amp and protected by a residual current device (RCD) with a maximum rating of 30 mA.

- Turn on the water and fill the cistern. Check for leaks at the cistern and the supply or distributing pipework. Then set and adjust the float-operated valve to the correct level.

- Turn on the electrical supply and flush the WC once.

- The motor on the unit should run for 10 to 30 seconds (depending on the length and height of the pipe run).

- If the motor runs for longer than this, check whether:
  - there is blockage to the pipe run
  - there are kinks in the flexible hose connections
  - the unit's non-return valve is working properly (this should be sited at its inlet).

- Flush the WC several times to check all the water seals. The discharge pipe should be fully checked for leaks, as should the pipework from other appliances.

- Check the float-operated valve and appliance taps for dripping (annoying short-term activation of the pump can occur).

Sink waste-disposal units

A sink waste-disposal unit is designed to do just what it says – remove waste food and cooking products from the kitchen and discharge them to the drainage system. The waste disposal unit can deal with all food matter, including bone, using a number of cutters to turn the matter into a thin paste. Water is flushed down the unit when it is working and the products are taken out to the drainage system.

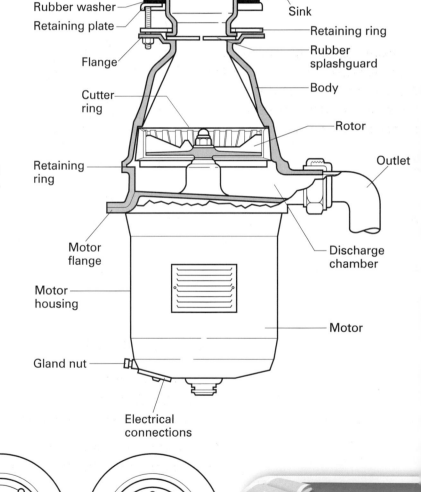

Figure 4.18 A typical waste-disposal unit. Cutter blades are driven by an electric motor. There is usually a rubber splashguard at the inlet to prevent food splashes and debris being thrown back into the room

The drawings show a typical sink mounting arrangement, an example of the rubber splashguard and an insert fitted into the waste, known as a cutlery saver.

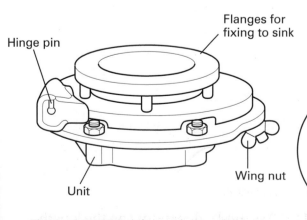

Figure 4.19 Waste-disposal mountings

Rubber splashguard   Cutlery saver

Find out

Find out how many types of waste disposal unit are available

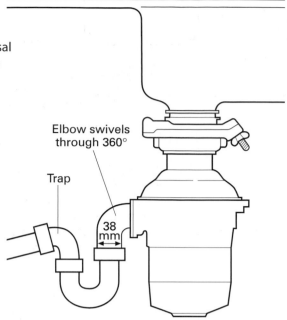

Figure 4.20 Waste-disposal discharge trap and elbow connection

Elbow swivels through 360°

Trap

38 mm

This outline of the unit shows the swivel elbow connection, which can be adjusted to suit the required position. The outlet of the unit is a standard 38 mm or 1½".

A trap should be fitted on the unit to stop smells. The discharge pipe should be laid to a fall of 1 in 12 for horizontal runs (to remove the discharge products effectively). The unit can discharge either to a gully or directly to the soil stack. The unit must not have a **bottle trap** or a grease trap connected to it as they block easily.

On the job: Installing a waste-disposal unit

Pilar is installing a waste-disposal unit. The disposal unit is mounted on the base of a sink with a special-sized waste outlet of approximately 89 mm diameter. Sufficient space must be provided under the sink for the unit. The unit takes its power supply direct from a fused connection unit – the motor in the unit is rated quite high and typically a special 10 amp fuse will be provided. It is also recommended that the circuit be protected with a 30 mA residual current device.

1   What space must Pilar provide under the sink?

2   What things must Pilar check before installing the waste-disposal unit?

3   What must Pilar check when she commissions the unit?

Hepworth valve

Valves as an alternative to traps (the HepvO™ system)

The HepvO valve (made by Hepworth Building Products) can be used as an alternative to the trap. Rather than using a water seal, the HepvO valve uses a tough, collapsible membrane to make the seal. When water is discharged down the valve, the membrane is in the open position. Once the water has discharged, the membrane returns to its normal state, making an airtight seal. Any back pressure on the system forces the membrane into a closed state and no water can be discharged back into the appliance. The valve therefore overcomes the effects of any form of siphonage. There is no trap seal to be lost, and problems from back pressure do not exist.

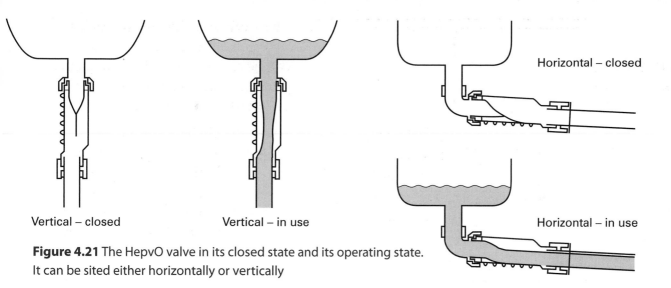

Vertical – closed   **Vertical – in use**

Horizontal – closed

Horizontal – in use

Figure 4.21 The HepvO valve in its closed state and its operating state. It can be sited either horizontally or vertically

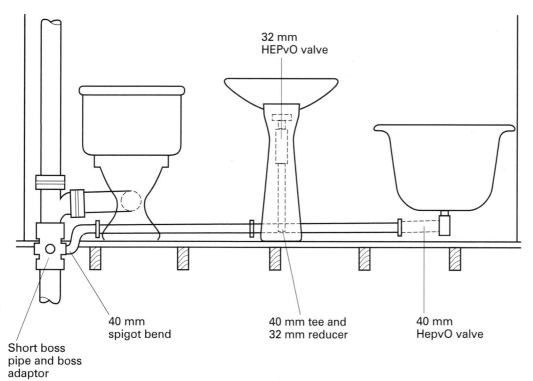

32 mm
HEPvO valve

40 mm
spigot bend

40 mm tee and
32 mm reducer

40 mm
HepvO valve

Short boss
pipe and boss
adaptor

Figure 4.22 A typical example of a HepvO system with combined waste to bath and basin. The basin discharges vertically to branch into the waste pipe, which is unacceptable with a standard trapped system

Did you know?

The HepvO valve can fix problems that occur with standard traps, such as excessively long pipe runs, incorrect pipe falls, incorrect trap seal depth and problems with combining wastes

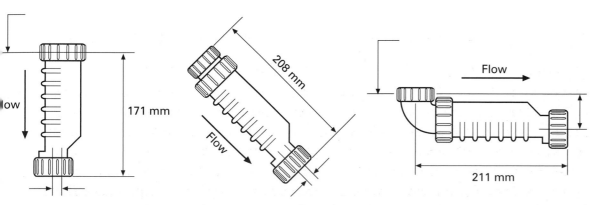

171 mm

208 mm

Flow

Flow

Flow

211 mm

Figure 4.23 HepvO valve dimensions. The valve can be positioned at different angles and comes with adaptors so that it can be put horizontally, under the bath, or fixed in-line, similar to a running trap

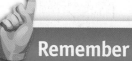

Remember

Don't forget that the pipe must be laid with enough fall to stop deposits forming and blockage occurring

Performance testing

At the end of this section you should be able to:

- state the performance testing requirements of above ground systems.

You need to have a good knowledge of the requirements for soundness testing of above ground discharge system pipework, and for performance testing systems.

Soundness testing

Soundness testing kit in use

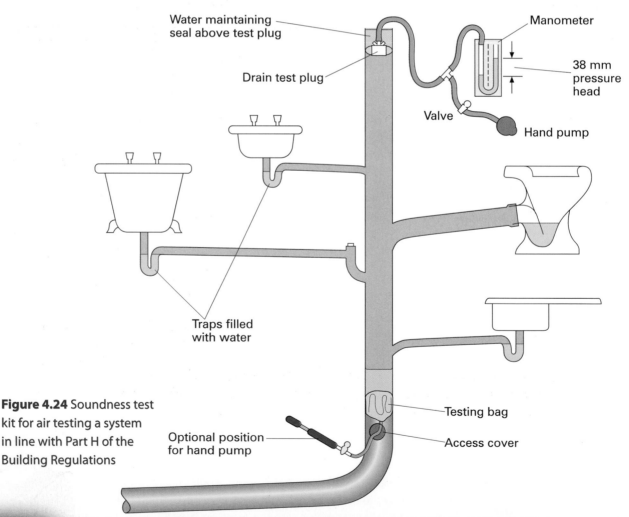

Figure 4.24 Soundness test kit for air testing a system in line with Part H of the Building Regulations

Performance testing

With large systems it is necessary to discharge differing numbers of appliances together. With domestic properties it is slightly easier. If the system has independent waste pipes to each appliance, test each appliance by filling it to the overflow level and discharge the water out of the appliance. Measure, using a dipstick, the amount of water left in the trap.

In installations that have combined waste pipework, e.g. bath and basin connected together, both appliances should be tested at the same time. These systems have a higher risk of trap-seal loss.

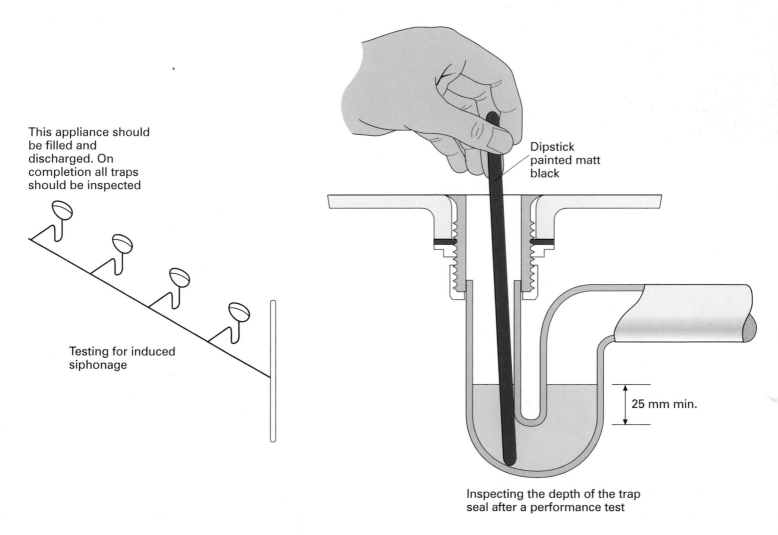

This appliance should be filled and discharged. On completion all traps should be inspected

Testing for induced siphonage

Dipstick painted matt black

25 mm min.

Inspecting the depth of the trap seal after a performance test

Figure 4.25 Best practice is to test each appliance three times (filling the trap to its normal operating level after each test). Use the worst reading as the test reading. It should never be less than 25 mm in depth retained

Fault diagnosis

At the end of this section you should be able to:

• diagnose a range of faults in above ground discharge systems.

There is very little that can go wrong with simple above ground discharge systems, and they are the easiest systems to check and fix. However, there are a number of electro-mechanical devices, such as the macerator WC, that have a higher risk of system breakdown.

Blockages

WCs

Blockages occur every now and then. They are usually due to too much toilet paper being used or a sanitary towel being thrown down the pan. Typically, the material will lodge on the outlet side of the trap, out of view. The blockage can often be removed by discharging a full bucket of water into the pan at the same time that the WC is flushing (creating increased pressure). Alternatively, a disc plunger can be used.

Another cause of poorly performing WCs can be that a detergent-block holder has come loose from the pan. These can make their way to the pan outlet and wedge themselves there, causing a build-up or blockage. The only solution is to remove the WC – then you can get at the offending article!

Traps

Blockages occur in traps because deposits such as hair and soap build up. Traps may need cleaning from time to time. Sink traps have a greater blockage risk due to grease and fat deposits. Because of this, bottle traps should not be used on a sink. Blocked bottle traps will need replacing rather than just cleaning. Smells from the appliance tend to be a common fault.

Remember

If there is an air-admittance valve on the main stack, check it is functioning correctly

FAQ

What is the most common cause of smell complaints and how can these be tackled?

This is usually a result of trap-seal loss. In some cases it can have caused a problem for a long period of time, usually because of system design. It can also be a problem with any air-admittance valve or resealing trap. These have moving parts which can get stuck from time to time. Air does not enter the system, so siphonage occurs. The customer may also report that the overall system is performing badly.

If trap-seal loss results from a problem with the design of the system, then it will need fixing. One option might be to put in a resealing trap or an air-admittance valve, or even a HepvO valve.

Pipework

Blockage in a pipework system should not be a common occurrence. If pipework does block, then it is likely to be in a horizontal run, which is probably not laid to the correct fall. This means you will be faced with a pipework modification job.

Macerator WC

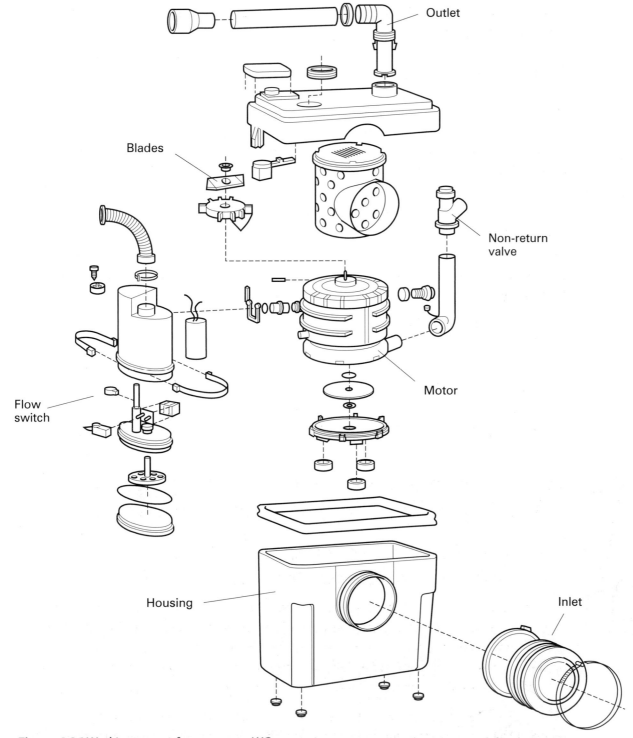

Figure 4.26 Working parts of a macerator WC

A macerator WC has a lot of parts, as you can see from Figure 4.26. Most parts are available as spares, including the motor. These units are pretty reliable but some common faults are listed in Table 4.2.

Fault	Possible cause of fault
The motor operates with an intermittent on/off action	There is a dripping tap or float-operated valve
The water in the pan only discharges slowly	An inner grille is blocked up and needs cleaning
The motor operates but runs for a long time	There are a number of possible problems, including design problems with the pipework system (i.e. the height or length of run is too long), the pump could be blocked or the activating pressure membrane is distorted
The motor does not activate	The electricity may be off, or there may be a defective motor
A rattling or crunching sound is heard	A foreign object such as a toilet block holder has made its way to the grille
The motor hums but does not run	The capacitor or the motor is defective

Table 4.2 Common faults found in macerator WCs

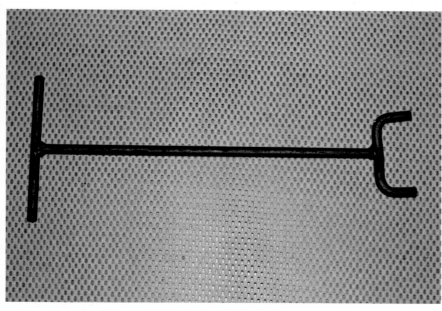

Waste-disposal unit de-jamming tool

Sink waste-disposal unit

There is not much that can go wrong with sink waste-disposal units. The major problem is usually a foreign body in the unit (which tends to be cutlery). This can jam the blades, causing a thermal cut-out device on the electrical supply to stop the motor.

Sink waste-disposal units usually come with a de-jamming tool to help you remove the offending article. Some models have a reverse option to assist with blockages, sending the problem item back into the sink. Again, bottle traps should not be used with the disposal unit.

FAQ

Why shouldn't bottle traps be used with a waste-disposal unit?

A trap to a waste disposal unit tends to block more often than other sanitary appliances. This is because it is moving solid matter, particularly if the unit is being used incorrectly and is not being thoroughly flushed with water. Bottle traps are more likely to block than other traps because of waste.

Knowledge check

1   What is the minimum radius for the bend at the foot of a stack?

2   Sanitary accommodation requirements for disabled access are detailed in which part of the Building Regulations?

3   What might you use to rectify a trap suffering from self-siphonage?

4   A macerator is making a rattling noise. What is the likely fault?

5   Following a performance test to a trap, what should the minimum seal depth be?

6   A ventilating pipe terminating within 3 metres of a building opening should terminate at a minimum of what height above it?

7   In a row of houses with a combination of venting pipes and air-admittance valves, where must a ventilating pipe be placed?

8   What is the current British Standard covering sanitary pipework systems?

9   What is the minimum size of a waste pipe to a sink waste-disposal unit?

10 What is the minimum bend radius for a 100 mm branch connection to a discharge stack?

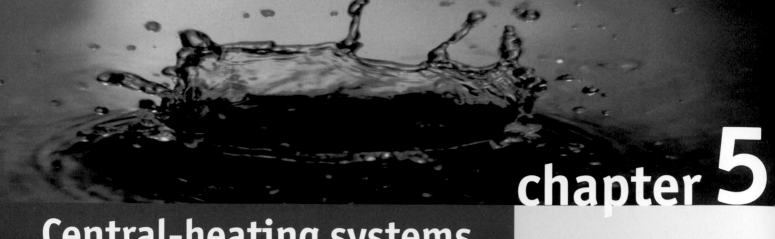

chapter 5

Central-heating systems (including boilers)

OVERVIEW

This section is designed to refresh your memory with the basics of central-heating system layouts, covered at Level 2, so that you get up to speed and progress with the additional work that you will cover in this section at Level 3. By the end of this section you should be able to state the typical layouts for a range of systems. Additional gas-safety and gas-boiler requirements will be detailed when you cover chapters 6 and 7.

What you will learn in this unit:

- Basic system layouts
- Complex system layouts
- Central-heating appliances
- Wiring central-heating system controls
- Commissioning central-heating systems
- Fault diagnosis

Basic system layouts

At the end of this section you should be able to :

- state the typical layouts for a range of systems.

System types

Early heating systems in domestic properties used a one-pipe gravity flow system of pipework layout to the heating circuit.

What do you think was the main disadvantage of this type of circuit layout? Well, the disadvantage is that the flow and return pipes were laid out in a continuous loop from the boiler and round through the radiators, with the return water from the first radiator being mixed with flow water at a lower temperature to supply the flow water to the next radiator, and so on. The end radiator on the circuit was cooler, and if the system was to be designed correctly then radiators needed a larger surface area at the end of the circuit (as compared with those at the beginning of the circuit) to cater for the reduction in water temperature supplied to them.

The one-pipe circuit has been totally replaced in domestic properties by the two-pipe circuit, which, if balanced properly, provides the same flow temperature to each radiator.

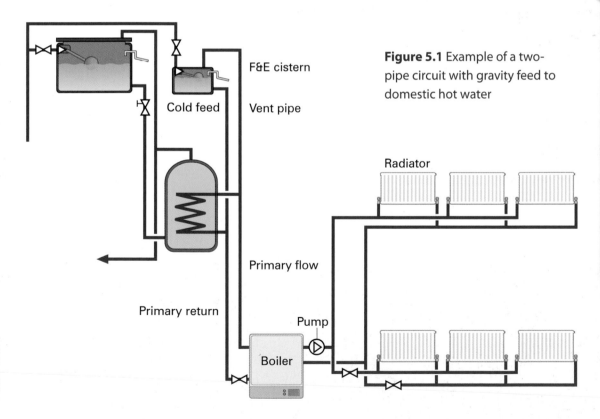

F&E cistern

Cold feed

Vent pipe

Radiator

Primary flow

Primary return

Pump

Boiler

Figure 5.1 Example of a two-pipe circuit with gravity feed to domestic hot water

Find out

At Level 2 you identified that Part L1 of the Building Regulations now requires that traditional central-heating systems (non-combi) be fully pumped for any new installation. What are the two main types of control system that may be used for fully pumped systems in order to comply with the Part L1 Regulation?

The system shown with gravity primaries is now obsolete, following the introduction of Part L1 to the Building Regulations, other than when used by boilers fed by solid fuel.

Remember

The two main types of control system that may be used are:

1 System with mid-position valve
2 System with 2 x 2 port valves

The reason for this is that some solid-fuel systems cannot be controlled to the same extent as their gas or oil counterparts: a certain amount of residual heat is generated in the appliance after the air supply controlling combustion is reduced to its lowest level. Providing boiler interlock controls with positive isolation of circuits would not be safe in this situation.

At Level 2 you identified that Part L1 requires that traditional central-heating systems (non-combi) be fully pumped for any new installation.

The layout of a system with 2 x 2 port valves should look something like Figure 5.2.

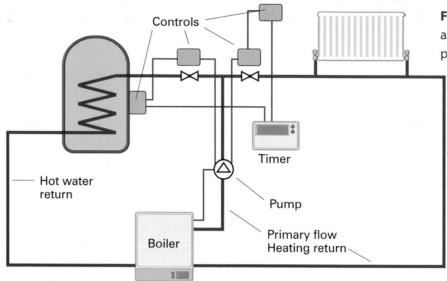

Figure 5.2 Layout of a system with 2 x 2 port valves

With fully pumped systems the return from the cylinder must be connected directly into the boiler or be the last connection into the heating return pipe from the boiler. This is to avoid reverse circulation in the system and heating of the radiators when the heating circuit is isolated.

To operate correctly, a fully pumped system needs to be provided with the following electrical control components:

- room thermostat or programmable room thermostat
- cylinder thermostat
- boiler connection
- pump connection
- zone valves – mid position or 2 x 2 port valves
- programmer.

With zoned heating circuits you can come across problems with return connections to the circuit, so the returns need to be kept separate from one another, with both being taken back to form a manifold arrangement usually near the boiler; again the hot water return should be the last connection before the boiler.

In certain circumstances the returns cannot be kept separate and it may be necessary to install non-return valves to avoid reverse circulation. These should

Remember

In a larger house Part L1 of the Building Regulations identifies the need to zone different spaces of the building. The maximum recommended floor area of a heating zone is 150 m². To control the zones, two or more 2-port valves would be required on the heating circuit, each valve controlled by its own room thermostat with separate timing control for each zone (circuit)

be avoided if possible as they can be prone to failure, but if they prove necessary they must be sited where they are readily accessible, and clear instruction must be left on the job as to their positioning in the system.

An air separator is commonly used on open-vented fully pumped systems, as shown in Figure 5.3.

The purpose of the air separator is to enable the cold feed and vent pipe to be joined closely together, forming the system neutral point, with the majority of the system operating under positive pressure. The grouping of the connections causes turbulence of water flow in the separator, which in turn removes air from the system. This reduces noise in the system and reduces the risk of corrosion.

Part L1 of the Building Regulations now requires fully pumped systems to be installed on new installations. Part L1 says that a system that has a DHW gravity circuit (with no control) and pumped heating circuit that is having a boiler replaced should preferably be converted to a fully pumped system with full controls/boiler interlock. As a minimum, controls should be provided to the gravity DHW circuit including temperature and time control, and full boiler interlock.

Method of operation

This system provides independent temperature control of both the heating and hot-water circuits in a pumped heating and gravity domestic hot-water heating system. The pump and the boiler are switched off when space and hot-water temperature requirements are met.

Time control can be through either a time switch or a programmer. **Thermostatic radiator valves (TRVs)** could also be fitted to provide overriding temperature controls in individual rooms. The pump and circulation to the cylinder can be turned off independently by the thermostats and there is an interlock to ensure that when the final circuit turns off so do the pump and boiler to avoid wasting energy.

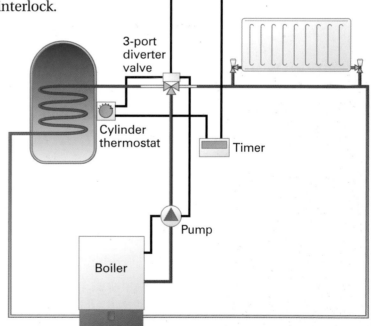

Figure 5.3 Air separator used on open-vented fully pumped system

Figure 5.4 Minimum specification layout for conversion of fully pumped system

Thermostatic radiator valve

Complex system layouts

At the end of this section you should be able to:

- state typical pipe sizes for domestic applications
- state the detailed requirements for pump positioning
- identify the key requirements of sealed systems, their components and positioning requirements in systems
- recognise typical microbore system layouts
- describe other system considerations.

Typical pipe sizes for central-heating systems

In sizing pipework for domestic heating circuits you are attempting to identify the requirements for:

- getting a required amount of water from point A to B (this is the flow rate of water that you require to be delivered around the circuit to get the necessary heat output from your radiators and/or the coil in your cylinder)
- overcoming the resistance to the flow of water (frictional resistance) created by the fittings that make up that system.

So the two key issues here are pressure and flow.

To be absolutely safe each job should be the subject of proper system design and pipe sizing, and we're going to look at some of these principles a little later. As the systems that we install in domestic properties are relatively small, rules of thumb exist about the size of pipework.

Gravity primaries

The usual size for gravity primary circulators to an indirect cylinder is 28 mm and not 22 mm, as is often found on many older systems.

There are a number of key factors that need to be taken into account when dealing with gravity primaries (you probably won't come across them much, other than perhaps with solid-fuel installations, but you might be called on to look at problems with them):

- The pipework must rise continuously from the boiler to the cylinder. This is to ensure that all the air is vented from the pipework. The system works on the basis of gravity circulation created by a difference of temperature, and hence density of water, between the flow and return pipes. That difference in density creates the circulation effect in the pipework. The pressures that are created to make this happen are very low, hence the need to make sure that the pipework is laid to proper falls, as in the event of an air lock there is insufficient pressure to remove the air bubble, and circulation will not occur. This can be a common problem.
- The **circulating head** – this is the distance between the centre line of the cylinder coil and the centre line of the boiler and must usually be a minimum of one metre or the minimum specified in the manufacturer's installation instructions.

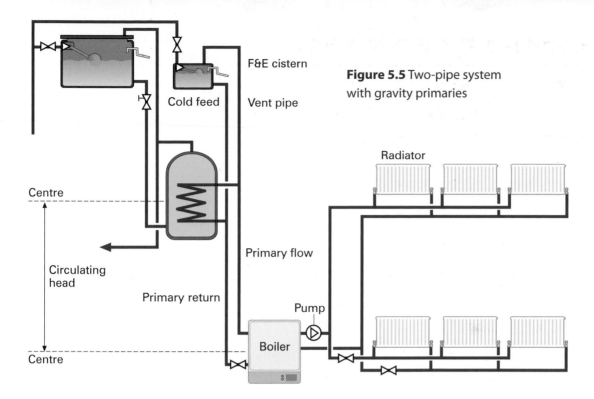

Figure 5.5 Two-pipe system with gravity primaries

- Long horizontal runs should be avoided: as a guide, for every 3 metres of horizontal run there should be at least 1 metre of vertical rise in the pipework.

- Changes of direction should be avoided wherever possible, and bends and swept tees are preferred to elbows and tight bends as they do not create as much frictional resistance, which can limit the circulation in the pipework.

- Larger solid-fuel appliances such as wood-burning stoves may require a minimum 35 mm diameter pipework; this may also be the case in situations where excessive changes of direction are required or the length of horizontal run is excessive.

The pipework circuit should be based on either:

- both flow and return pipework rising across roof space laid to appropriate falls, or

- flow pipe rising to run across roof space and return from cylinder down through ground floor, falling across floor void to connection point at boiler.

Pumped circuits

In the heating circuit, the following pipe sizes will carry the heat loads shown in Table 5.1.

With experience it is usually possible to apply a rule-of-thumb sizing to heating-system pipework. However, there is no getting away from the fact that the radiators and heat requirement to the cylinder need to be properly established by correct sizing in the first place.

Copper pipe (OD)	Approx. loading
8	1.5 kW
10	2.5 kW
15	6 kW
22	13 kW
28	22 kW

Table 5.1 Pipe sizes for heating circuit

Remember

Key discussions with the customer should centre on accommodating the pipework, how it may be hidden or boxed in etc.

Remember

You can't guess at radiator sizes!

Experience tends to play a great part when using rules of thumb such as this, and when deciding on a pipe size an experienced plumber will take account of:

- the length of pipe run required
- the number of fittings or changes of direction that are required in the pipe run.

If the pipe run has excessive length or a large number of changes of direction and it is required to deliver close to the maximum loading as indicated in Table 5.1, then the experienced plumber is likely to go to the next pipe size up to ensure the correct flow rate through that part of the system.

With the domestic hot-water circuit this is likely to need to be fed by 22 mm for most domestic installations. A number of conversion kits were widely used in the 1970s and 1980s to convert a direct cylinder to an indirect cylinder using a heat-exchanger device inserted in the immersion heater boss – these tended to be fed by 15 mm tube.

Boiler mains

The usual pipework size for the boiler mains circuit is 22 mm (to the point at which the split between hot water and heating occurs) for boiler sizes up to 15 kW. For boilers between 15 kW and 18 kW the boiler mains flow and return (to the point at which the split between HW and heating occurs) is usually 28 mm pipework with both heating and hot-water circuits then 22 mm. Between 18 kW and 30 kW, 28 mm boiler mains tend to be the norm, with 22 mm feeding the cylinder and aspects of the heating circuit needing to be fed by 28 mm. Over 30 kW and the circuits really need to be properly sized.

The size of cold feed and open-vent pipe for systems up to 25 kW is 15 mm and 22 mm respectively, with an F&E cistern required with a nominal capacity of 45 litres.

Between 25 kW and 45 kW the vent pipe should be increased to 28 mm, the cold feed to 22 mm and the nominal capacity of the F&E cistern to 70 litres.

Pump positioning – open-vented systems

The first thing that we need to be aware of is that the system operates best if it is under positive (above atmospheric) pressure. There are three reasons for this:

1 Pumps operate much more successfully when they are pushing rather than sucking, and this really links in with the following point.

2 This follows on from the temperature and pressure effects detail that we looked at when we covered unvented hot-water systems. When you reduce the pressure in a system to below atmospheric pressure, water will begin to boil at a lower temperature than the normal boiling point of 100°C. If there is sufficient negative pressure it can cause an effect in the pipes known as **cavitation**. When a system begins to cavitate, excessive noise is produced and the mini pockets of air and steam that are produced in the pipework have the ability to damage internal component parts of the system such as valve seatings – this is a problem that needs to be avoided.

3 The parts of the system that operate below atmospheric pressure have the ability to draw air into the pipework through microleak points from the surrounding air that is at the higher atmospheric pressure. So there can be very little sign of leakage, but air can get into the system.

Our main aim therefore should be to keep as much of the system as possible working under positive (above atmospheric) pressure. The key to achieving this is the position of the pump in the system in relation to the connection of the cold-feed pipe.

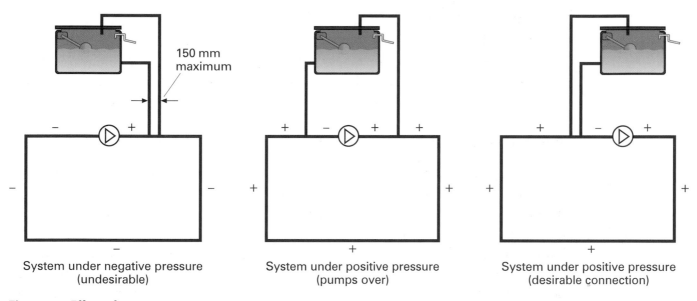

Figure 5.6 Effect of pump position

The drawings in Figure 5.6 represent a system layout that is essentially a loop. You heat water at the boiler; it is pumped to the radiators and returned from the radiators to the boiler to be reheated. The key point to notice is that the connection point of the cold feed in the system is the point at which the pressure in the system changes from positive to negative. This is called the **neutral point**.

The reason why the pressure changes at the neutral point is that the static pressure created by the head of water in the cold-feed pipe, i.e. the base of the F&E cistern to the point of cold-feed connection into the system, will be higher than the pressure created in the circuit by the pump at that point, so an external pressure is exerted at that point which causes a pressure change from positive to negative.

Fitting a central-heating pump

There's one other thing that we need to consider, and this is where we place the open-vent pipe.

Remember

The open-vent pipe has two key functions:

- to remove air from the system on filling
- to relieve pressure build-up in the event of overheating

The vent pipe usually needs to be placed at a point in the system where it can provide these functions. This is usually directly into the boiler or at a high point in the mains pipework, as shown in diagram 3 (right) in Figure 5.6.

Diagram 1 (left):

- The system is virtually all operating under negative pressure, so is not a desirable layout.

Diagram 2 (middle):

- The system is operating under positive pressure, which is desirable. However, the open-vent pipe is connected into the system at its point of highest positive pressure (near the pump). The likely result is that on system start-up water will be discharged into the cistern, aerating it and increasing the potential for corrosion or, in the worst-case scenario (particularly where the head on the system generated by the F&E cistern is low), then the F&E cistern acts as part of the circuit, i.e. hot water is discharged via the vent pipe and returns to the system via the cold feed (in a loop).

Diagram 3 (right) shows the preferred method of connection employed with most open-vented systems that are now installed. Key features are:

- There is positive pressure across virtually every aspect of the system.
- The vent pipe is connected at the point of lowest possible pressure (must be within 150 mm of the cold-feed pipe – but before it).
- It can be used where there is a relatively low head of pressure generated over the system by the F&E cistern – but never less than the head that is developed by the pump, divided by 3.

So if a pump is used that is set to overcome the frictional resistance in a circuit at 20 kPa, which is equal to 2 metres head, then the minimum head is 2000 mm/3 = approx. 700 mm. This method of feed and vent arrangement is known as **close coupling**.

Remember

The best position in a system for a pump is sited vertically, pumping upwards – you can't always achieve this, though, with a close-coupled arrangement

Other key features of close coupling are:

- it is preferable for the connections to be formed using an air separator – remember we saw that earlier
- if an air separator is not used, then:
  - the vent pipe should be taken horizontally from the main flow pipe
  - it is preferable to increase the pipe size of the main flow pipe (either side of the vent pipe) to the next pipe size up; this is to promote air separation in the main flow pipe.

Sealed central-heating systems

Figure 5.7 shows a typical layout for a sealed system.

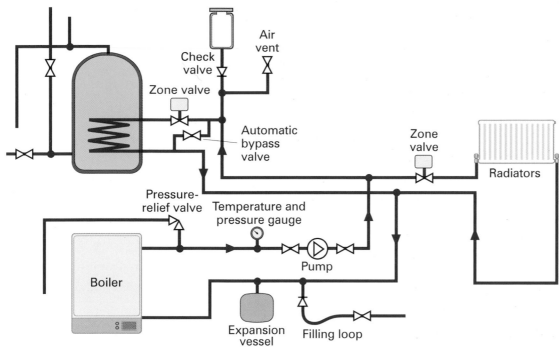

Figure 5.7 Sealed heating system

Sealed system components

Pressure-relief valve

This replaces the open-vent pipe (as seen on open-vented systems). It must therefore be installed on all types of sealed system. For domestic systems it must:

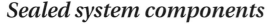

Figure 5.8 Sealed-system valve

- be non-adjustable

- be preset to discharge when the system reaches a maximum pressure of 3 bar

- have a manual test device

- have a connection for a full-bore discharge pipe

- be connected either into the boiler or on the flow pipe close to the boiler

- have a metal discharge pipe, which as a minimum should be the same size as the valve outlet, discharging to a tundish which in turn discharges in a safe, visible, low-level location.

Sealed vessel components, valve and gauge

Figure 5.9 Sealed-system gauge

Find out

What is wrong if a system needs to be constantly topped up with a small amount of water on a weekly basis?

Find out

Why is it that the filling loop should always be disconnected from the system during normal operation?

Pressure gauge

A pressure gauge must be provided to assist with system filling and top-up. The pressure gauge should be capable of giving a reading between 0 and 4 bar. It is also preferable that a temperature gauge is provided to measure the boiler flow temperature (this will indicate any potential faults in the system). The temperature gauge should be capable of giving readings up to 100°C.

The pressure gauge will usually be sited in the vicinity of the expansion vessel.

Method of filling

There are two main methods of filling the system.

1 Filling and pressurising by filling loop

We dealt with this when we looked at the Water Regulations in Section 2. Essentially a temporary connection is made to the system via a flexible filling loop, which includes a flexible hose, stop valve and double check valve. The connection is used:

- to fill the system
- to pressurise the system
- for the purposes of future top-up.

2 Refilling through a top-up unit

You probably won't come across this much but you'll need to know about it if you do. The unit is essentially a small bottle sited at the highest part of the system and fitted with a double check valve assembly and automatic air eliminator (all provided as part of the package). The water in the system is topped up manually by the unit, which contains approximately three litres, when the water level drops. The connection to the unit should be made into the return side of the distribution pipework or the domestic return from the cylinder.

The system still requires a filling loop for initial filling purposes. However, the system is not pressurised on fill-up; the pressure acting on the system is the head of water exerted by the top-up unit. Care needs to be taken when determining the initial charge pressure of the expansion vessel for it to function correctly.

There are two other potential methods of filling that could be utilised:

- automatic filling through a make-up cistern
- permanent supply connection using a filling loop that can be disconnected.

The use of an automatic filling cistern for domestic applications is very unlikely owing to its cost, so we can essentially rule this one out. The use of a Type CA disconnector came in with the Water Regulations, and as we've seen it is possible now to have a permanent connection to the main supply pipe. However, a well-installed sealed system should not require much in the way of top-up as evaporation is minimal, and the need to periodically top up a system is a sign that it contains leaks. As with any other system, constantly introducing fresh water to the system increases the system's exposure to aerated water and promotes corrosion.

Both of these methods of fill connection, pressurisation and top-up should therefore be avoided for domestic systems.

Boiler

The boiler used with a sealed system must have a high-limit thermostat (energy cut-out device) fitted to it so that, in the event of control-thermostat failure, there is the added protection.

Expansion vessel

An expansion vessel is used in the sealed system, and it replaces the F&E cistern in an open-vented system, i.e. it takes up the increase in system volume when the system is heated.

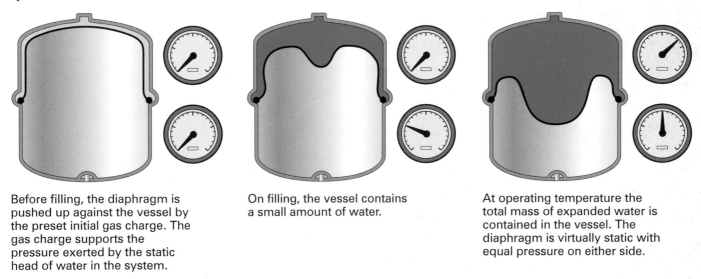

Before filling, the diaphragm is pushed up against the vessel by the preset initial gas charge. The gas charge supports the pressure exerted by the static head of water in the system.

On filling, the vessel contains a small amount of water.

At operating temperature the total mass of expanded water is contained in the vessel. The diaphragm is virtually static with equal pressure on either side.

Figure 5.10 Expansion vessel

The vessel should be located close to the suction side of the pump to ensure that the system operates under positive pressure.

The point of connection of the expansion vessel to the system is the neutral point of the system. It is preferable to install the vessel at the coolest part of the system to maximise the life span of the flexible diaphragm. The pipe connecting the vessel to the system should be the same size as the vessel outlet, and there must be no isolating valve installed between the vessel and its point of connection to the system.

Different arrangements are available for mounting expansion vessels, including a manufacturer-produced mounting bracket, as shown in Figure 5.11.

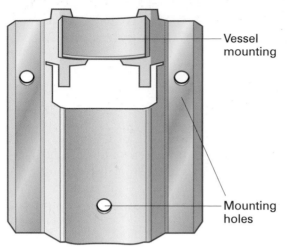

Figure 5.11 Expansion-vessel bracket

Expansion vessels are normally red but are now being manufactured in different colours

Sizing the vessel

When it comes to sizing the vessel there are two points that need to be considered.

1. Vessel charge pressure

Vessels for domestic use tend to be available with an initial charge pressure of 0.5 bar, 1 bar or 1.5 bar. The initial charge pressure should be in accordance with manufacturer instructions and must always exceed the static pressure of the heating system at the level of the vessel.

Essentially, for the vessel to work correctly, the air or nitrogen charge pressure on the dry side of the vessel should be slightly higher than the static pressure on the wet side of the system when cold-filled.

So it follows that if a vessel was required to be fitted in the cellar of a three-storey property with 8 metres static head of water in the system above it:

1 metre = 0.1 bar, therefore 8 metres = 0.8 bars

We would select a 1 bar vessel and would charge the system to around 0.9 bar (less than the vessel charge of 1 bar).

2. Volume of the vessel

The other factor that needs considering is the volume (size) of vessel that we require, and this is based on the amount of water contained in the system, which can vary dramatically.

The amount of water contained in the major system components can usually be obtained from the manufacturer (usually boiler, cylinder and radiators).

Table 5.2 can be used to determine the amount of water in the system per metre run of pipe.

So let's look at a worked example. You have identified from manufacturers' catalogues that a system you are to install contains the following:

- Boiler – 1 litre
- Cylinder – 2.5 litres
- Radiators – 41 litres
- 15 mm pipe – 78 metres
- 22 mm pipe – 60 metres.

Using Table 5.2:

60 metres of 22 mm pipe contains – $60 \times 0.320 = 19.2$ litres

78 metres of 15 mm pipe contains – $78 \times 0.145 = 11.3$ litres

Total content = 75 litres. We then apply the information to Table 5.3 (assuming we already know the vessel charge pressure).

So, if our system utilises a 0.5 bar pressure vessel and the pressure valve setting is 3 bar, for a system containing 75 litres, the vessel volume will be 6.3 litres.

The next vessel size up usually available is 8 litres at 0.5 bar pressure; this is the specification for the vessel.

Table 5.2 Water in system per metre run of pipe

Pipe OD (mm)	Water content (litres)
8	0.036
10	0.055
15	0.145
22	0.320
28	0.539

Safety valve setting (bar)	3.0		
Vessel charge and initial system pressure (bar)	0.5	1.0	1.5
Total water content of system (litres)	Vessel volume (litres)		
25	2.1	2.7	3.9
50	4.2	5.4	7.8
75	6.3	8.2	11.7
100	8.3	10.9	15.6
125	10.4	13.6	19.5
150	12.5	16.3	23.4
175	14.6	19.1	27.3
200	16.7	21.8	31.2
225	18.7	24.5	35.1
250	20.8	27.2	39.0
275	22.9	30.0	42.9
300	25.0	32.7	46.8
Multiplying factors for other system volumes	0.0833	0.109	0.156

Table 5.3 Calculating the volume of the vessel

Composite valves

A number of composite valve arrangements are available, such as a combined pressure-relief valve and pressure gauge, as shown in Figure 5.12.

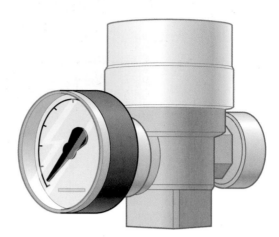

Figure 5.12 Combined-pressure relief valve and pressure gauge

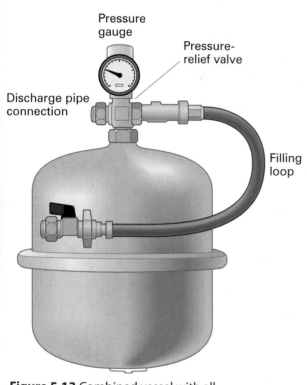

Pressure gauge

Pressure-relief valve

Discharge pipe connection

Filling loop

Figure 5.13 Combined vessel with all components attached

In Figure 5.13 we see an example of a combined expansion vessel filling loop, pressure-relief valve and pressure gauge all in one unit.

The main point that you need to consider when deciding on a composite valve arrangement is whether placing components together in a composite valve meets the key requirements that must be met when installing the system, e.g. proximity of the pressure-relief valve to the boiler and the availability of the discharge pipe connection.

Combination boilers and system boilers

These usually contain all the key components that make up a sealed system, and can include the filling loop as well. One key issue that you may need to take into account is the size of the expansion vessel, which is based on an average system size as determined by the boiler manufacturer. If the system has a higher water content then an additional vessel may be necessary.

Microbore systems

These systems use smaller than normal pipe diameters (8 mm and 10 mm) to feed the radiator circuits.

The microbore system was commonly used in the 1970s and early 1980s, and is still used in some cases today. It was based on the supply to the radiators provided from the heating mains via a number of manifolds, usually one downstairs and one upstairs, which in turn fed the radiators via individual 8 mm or 10 mm flow and return pipes.

Figure 5.14 shows a typical layout for the system, which could be installed with open-vented or sealed systems.

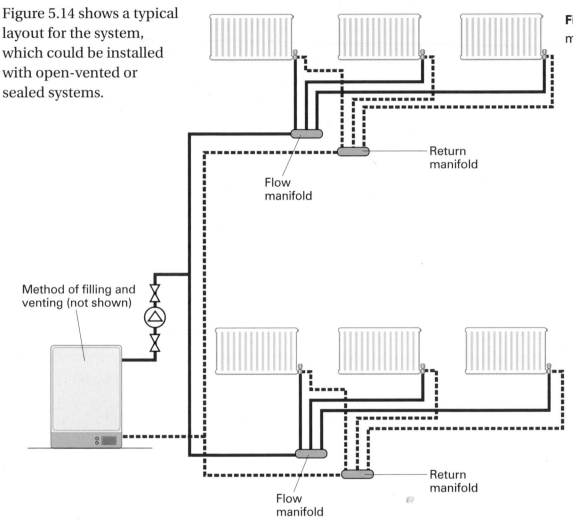

Figure 5.14 Layout of microbore system

The manifold could be made up in sections to suit the number of connections required, as shown in Figure 5.15.

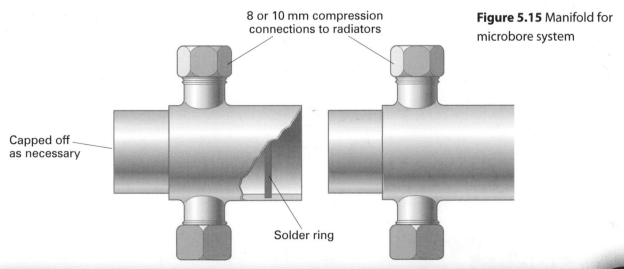

Figure 5.15 Manifold for microbore system

Microbore manifold arrangement

An alternative to using the manifold was to use a '**spider**' – this device does exactly the same job.

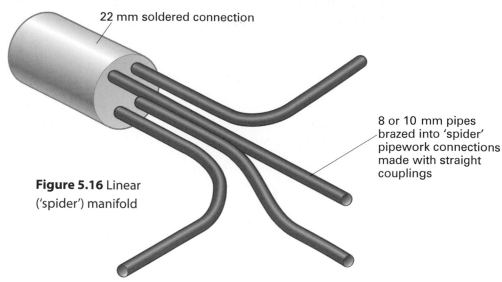

22 mm soldered connection

8 or 10 mm pipes brazed into 'spider' pipework connections made with straight couplings

Figure 5.16 Linear ('spider') manifold

The connections to the radiators could be different with this system, as twin-entry valves could be used.

The twin-entry valve, available as a manual or thermostatic valve, required a tube to be used to extend the return connection some way across the base of the radiator to ensure proper circulation across the radiator. With the introduction of radiators that have (largely) back-entry tappings, the twin-entry valve became redundant.

The main issue with the system layout as shown in Figure 5.14 was that it was only really suitable for smaller domestic properties, owing to possible longer pipe runs (greater frictional resistance) and the limitation on available flow rate and hence heat output with 8 mm and 10 mm pipework. The type of property suitable tended to be two-bedroom starter homes.

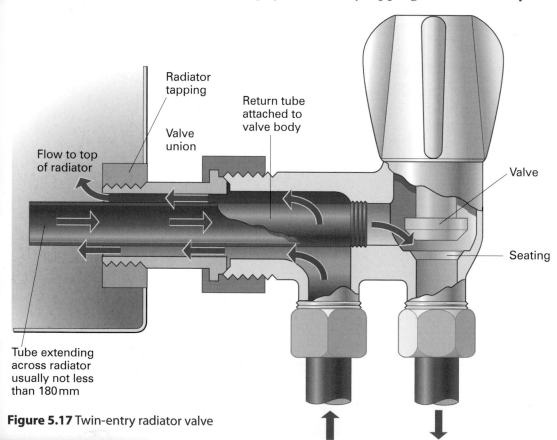

Radiator tapping

Return tube attached to valve body

Valve union

Flow to top of radiator

Valve

Seating

Tube extending across radiator usually not less than 180 mm

Figure 5.17 Twin-entry radiator valve

There were some real advantages to the system if it was installed correctly, particularly as it saved time, with pipe runs installed more quickly on new-build properties, especially if twin-entry valves were used. The system tended, however, to suffer from poor installation standards, including:

- pipe runs to radiators that were too long and overloaded

- unsupported pipe runs not laid to proper falls, causing severe air locks.

8 mm and 10 mm pipework have not disappeared from the scene, and the tendency is to use these pipe sizes where appropriate on a job but to pipe up as a normal small-bore installation without the use of manifolds.

One of the key issues that you need to be aware of when using 8 mm and 10 mm is that it is supplied in soft copper rolls which are malleable and easily flattened, so you need to be careful about how you make the connections up to radiators. Installed as normal 15 mm pipework they can be subject to flattening by vacuum cleaners, so the best approach is to rise up behind the skirting board and bring the pipe out to the radiator above 'knocking height'.

Other system considerations

Bypass valves

This applies largely to boilers installed in close-coupled systems that don't specify that a bypass is required as part of the system installation. Current Regulations require this valve to be fitted onto systems.

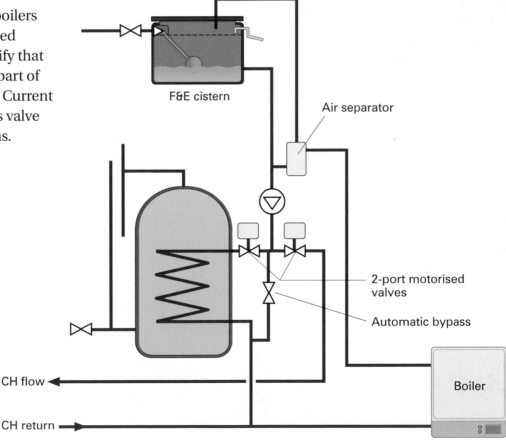

Figure 5.18 Layout of a bypass with 2 x 2 port valve close-coupled system

Referring to Figure 5.18, and assuming that the thermostat has failed and the boiler has begun to boil, the in-built safety mechanisms in the system should 'kick in' as the boiling water is discharged via the flow pipe and up over the vent pipe into the F&E cistern, where cooler water should be drawn down the cold-feed pipe to return to the boiler. The issue with the system shown above is that normally closed zone valves provide a positive shut-off to both heating and hot-water circuits, so if the boiler begins to boil when the electrics are not calling for heat (this is quite possible if the gas valve on the boiler fails in the open position), then there can be no cooling circulation of water, giving rise to a potentially dangerous situation.

Essentially, it's for the same reason that we don't put a gate valve on the cold-feed pipe.

Low water-content boilers

Most boilers now tend to be low water-content boilers. These boilers provide a great number of advantages to us when installing systems, mainly related to space-saving.

One disadvantage they have is that, in order to save space, the heat exchanger size is radically reduced, meaning that quite often we have to circulate more water through the boiler than the heating system requires. If the necessary circulation is not provided then the boiler tends to produce an unpleasant noise known as kettling, often causing a tripping of overheat thermostats and, in the worst cases, damage to the boiler heat exchanger.

Boilers with pump over-run thermostats

Because of the residual heat contained in the heat exchanger when the boiler had shut off, which enabled the temperature in the heat exchanger to rise to a higher level, a pump overrun thermostat was sometimes included – this needed a slightly different wiring arrangement to conventional boilers, whereby the operation of the pump was directly controlled by electrical circuitry in the boiler. This circuitry responded to the opening of the motorised valves and heat requirement to a particular circuit, and also to the activation of the overrun thermostat, causing the pump to keep operating for a few minutes after the last circuit had closed to remove the residual heat from the boiler. The residual heat from the boiler was usually removed via circulation around a bypass circuit.

One other system design feature indicating that a bypass is necessary is a system that is installed with thermostatic radiator valves (TRVs). Current practice is to install TRVs to all radiators in a property to meet energy-efficiency requirements, other than the room in which the room thermostat is sited (boiler interlocking control). In the event that the heating circuit is operating in mild weather, there is a likelihood that all the TRVs will close, leaving only the room in which the room thermostat is sited operational. This will often be the hallway, which is a relatively small radiator and will not give an adequate flow rate through the boiler alone.

The bypass will often be in-built into system boilers and combination boilers. Figure 5.19 shows possible bypass positions in a system (bypass not built into the boiler).

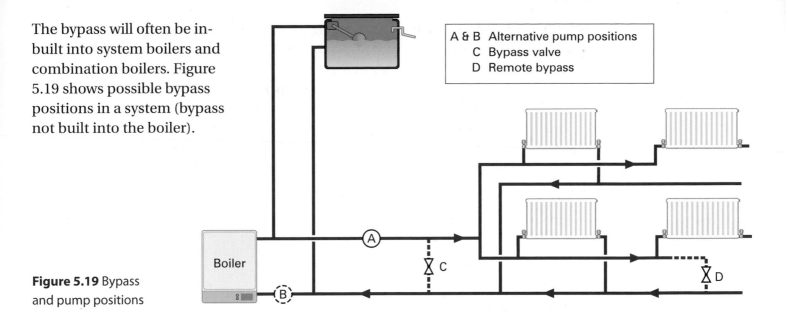

A & B Alternative pump positions
C Bypass valve
D Remote bypass

Boiler

Figure 5.19 Bypass and pump positions

It is preferable for the bypass to be an automatic valve. The automatic valve is able to vary the flow rate through the bypass in relation to the differing flow rates occurring in the system.

Frost protection

At Level 2 we introduced the need to provide frost protection to cover parts of the circuit that may be exposed to freezing, such as external boilerhouses. Protection against frost is obtained by installing a frost thermostat, which overrides the operation of other controls – room thermostat and programmer – to bring the circuit into operation to keep it from freezing.

The issue here is that there may be prolonged periods of frost when the boiler is firing and the system is operational but when heat is not required, so presenting a waste of energy. A frost thermostat should therefore be used together with a pipe thermostat (wired in series) to control the operation of the circuit.

So, essentially, when the frost thermostat calls for heat the circuit is energised and boiler and pump operate until the preset temperature of the pipe thermostat is reached (usually set at 50°C); the boiler then begins to cycle on and off but does not use as much energy.

FAQ

What temperature should you set the frost thermostat to?

Between 4°C and 8°C, depending on the location of the thermostat. Manufacturers' information will give you some guidelines.

Central-heating appliances

At the end of this section you should be able to:

- state the working principles of flue systems
- identify a range of appliance types and flueing options for appliances fuelled by solid fuel, gas and oil
- state key flueing options for a range of gas-fired boilers.

This section is designed to provide you with an overview of the types of central-heating appliance that are available for use in central-heating systems. Gas training is dealt with in Chapters 7 and 8; appreciation of oil-fired boilers is provided, although oil systems are fundamentally outside the scope of this book. If at some point in the future you are required to work on oil appliances, additional training will be required, plus completion of industry-recognised assessments; these are provided by an organisation known as OFTEC, which provides a similar service to the oil industry as CORGI does for gas.

Basic flue-operating principles

To complete this session you'll need a basic understanding of how the two main types of flue system work.

Open flues

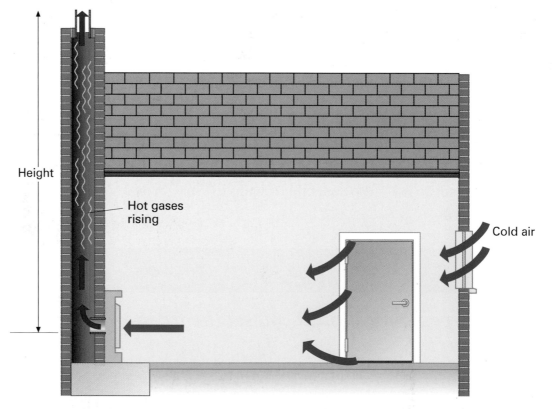

Figure 5.20 Convection currents in a room

Essentially, air for combustion at the boiler is drawn from the room in which it is situated, usually by air vents placed in the outside wall. As part of the combustion process the air is mixed with the fuel and is heated. The heated air and products of combustion from the burning process are drawn through the flue by convection – cool air inside the room, higher temperature air/products of combustion at the flue terminal. It's exactly the same process as gravity circulation in pipework systems and, as with gravity circulation, the greater the flue height (circulating) head, the greater the flue draught or speed at which the products of combustion will be taken through the flue.

Balanced flue boiler/terminal

Room-sealed flues

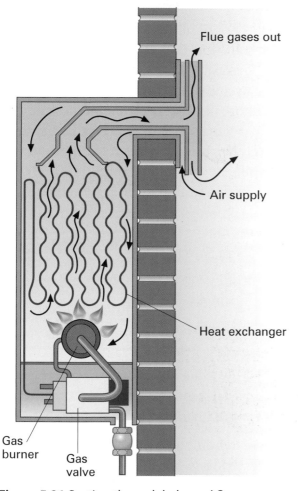

Flue gases out

Air supply

Heat exchanger

Gas burner

Gas valve

Figure 5.21 Section through balanced flue

Here the air for combustion is drawn in from outside, making the boiler much safer by design. The operating principle from this point on is very similar to open flues. Cool air is drawn in through the outside terminal and is supplied up through the burner, where it is mixed with the fuel and combustion takes place. The warm air/products of combustion rise through the heat exchanger and are collected in the flue hood, where they discharge to outside via the terminal. Essentially the difference in temperature between the warm flue outlet and the cool air inlet causes the convection current effect, which moves the flue products. Fanned flues are commonly added to this type of appliance to reduce the size of flue required and to provide a number of varied installation options, which we'll look at later in this section.

Appliance types and flueing options

Solid fuel

Appliances generally fall into four categories:

- 'wrap-around' back boilers
- room heaters
- independent boilers
- cookers – these generally work on the principle of room heaters so we will not deal with these in any great detail.

Solid-fuel appliances all operate on open-flue arrangements. As it is not possible to directly turn the fuel supply on with solid-fuel boilers, as it is with gas and oil, the rate of burning and heat output to the water is largely controlled by adjusting the air supply to the appliance, and therefore installation standards with these appliances (particularly the connection to the flue system and the seals around the appliance) need to be to a high standard for the appliance to operate correctly. If you get an air leak into the appliance or the flue system then you have a problem.

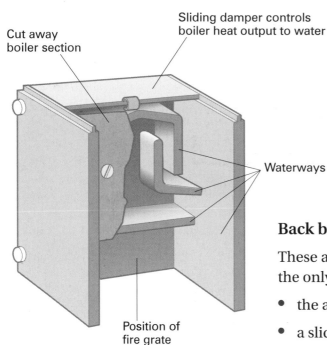

Figure 5.22 Solid-fuel high-output back boiler

(labels: Cut away boiler section; Sliding damper controls boiler heat output to water; Waterways; Position of fire grate)

Back boilers

These are the least effectively controlled of any solid-fuel appliance, as the only control over the rate of burning is:

- the amount of fuel placed on the fire grate
- a sliding damper positioned on the top of the boiler, reducing the heat output from the boiler by restricting the passage of the products of combustion from passing the rear surface of the boiler; the front surface of the boiler is still uncontrolled, however.

It can therefore be possible to get the appliance to boiling point in certain circumstances. The boiler is supplied with four tappings (connections), to which a system of gravity primaries with a pumped heating circuit are connected. Boilers are available up to a heat output of 12 kW (based on a high-output appliance with several flueway/heat-exchanger passes).

Control systems installed with this type of appliance tend to be relatively primitive:

- Thermostatic radiator valves can be installed.
- A room thermostat can be fitted (radiator in the room in which the thermostat is fitted is not to have a TRV).

- There is little use in fitting a programmer, as the fire will have 'died' to a relatively low level during the night, so there will be little heat available in the morning until the fuel is replenished. Therefore the pump tends to be operated by a simple double-pole switch.

- As an added safeguard a pipe thermostat should be supplied and attached to the domestic hot-water primary flow pipe: the thermostat should be wired with a permanent live connection so that in the event of an overheat situation the pump is activated, irrespective of whether it is turned on at the double-pole switch. This will dissipate excess heat quickly through the radiator circuit and prevent boiling.

All solid-fuel appliances also require what is known as a heat-leak (sink) radiator.

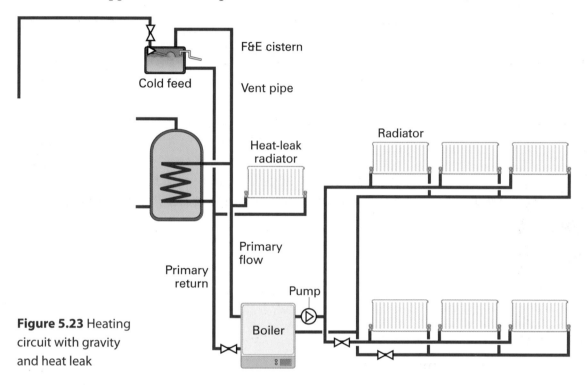

Figure 5.23 Heating circuit with gravity and heat leak

A gravity-fed radiator is installed across the flow- and return-pipework to the cylinder. Gravity circulation works on the basis of the temperature difference between flow and return pipes. However, as the cylinder begins to reach temperature, the return temperature will rise and circulation will begin to slow down to a point where, if little hot water is drawn off – it will virtually cease. The rate of burning in a solid-fuel boiler is controlled by the air intake to the appliance. To keep it burning we cannot fully stop that air supply, so a small amount of air will always be required. This means that there is not a positive shut-down of the heating process, and if heat is not taken away from the boiler when the cylinder reaches its operating temperature then the boiler will overheat.

To overcome this problem with solid-fuel appliances, a heat-leak radiator is connected to the gravity primaries off the cylinder to ensure a minimum amount of heat circulation though the gravity primaries, which in turn prevents the boiler overheating.

Safety tip

A heat leak must be fitted to all solid-fuel appliances feeding central-heating systems. Also linked to this, the cylinder should be no less than 100 litre capacity

Key requirements for the heat leak are:

- It must be connected thus: flow to the top of the radiator, return to the bottom, to work effectively.

- ¾" tappings on the radiator, and 22 mm pipe will normally be required to provide adequate gravity circulation through the radiator.

- If valves are fitted to the radiator they should be lockshield valves to prevent isolation by the customer.

- The radiator is usually positioned in the bathroom or a cupboard adjacent to the cylinder.

Room heaters

These are slightly more sophisticated than back boilers in that there is thermostatic control of the water in the boiler. This is provided by means of a boiler thermostat, which has a phial inserted in the boiler waterways. The phial is connected to a bellows which moves in response to changes in the boiler water temperature. The bellows is connected to a lever arm attached to a metal plate, which increases or decreases the air supply to the boiler – the thermostat is shown in Figure 5.24.

Room heaters are available as built-in models or free-standing 'sitting on a hearth' arrangements and are connected to a flue pipe. They generally provide heat outputs up to 15 kW.

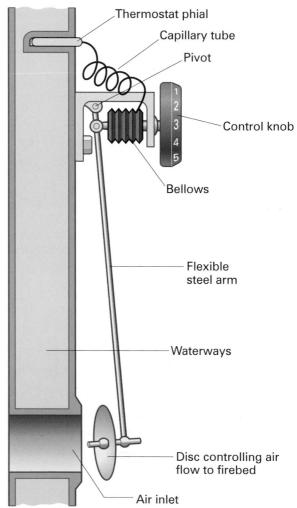

Figure 5.24 Thermostat to solid-fuel boiler

Solid-fuel room-heater

In terms of level of controls provided, these are very similar to back boilers, with the exception that some plumbers do install them with timing devices to control the operation of the pump, as an amount of fuel can be available to provide some form of heating in the morning. A small hole, usually in the ash-pit door, provides a minimum air supply to prevent the fire from going out when boiler temperature has been reached.

A heat-leak radiator is required and an overheat pipe thermostat is considered desirable but not absolutely essential with this type of appliance.

Independent boilers

This tends to be the most sophisticated form of solid-fuel heating appliance. It is a free-standing boiler similar to a gas or oil boiler. It has a fuel store which can contain fuel for several days' operation, making it relatively automatic. The boiler incorporates electrical controls.

Air is supplied to the fire-bed by means of a fan. The operation of the fan is controlled by an electrically operated boiler-control thermostat; an overheat thermostat is also usually provided as added protection in the event of control-thermostat failure. An independent boiler will also now usually include a flue-blockage thermostat. This detects blockages in the flue system and again prevents the fan from operating.

This type of boiler should always be fitted with a timeclock to control heating-circuit operation. It is also possible, using a manufacturer-produced control kit, to fully pump the boiler, giving improved circulation and temperature control of the hot-water cylinder. In all cases, however, the heat-leak radiator must be provided.

Additional points that we need to consider in relation to solid-fuel systems are:

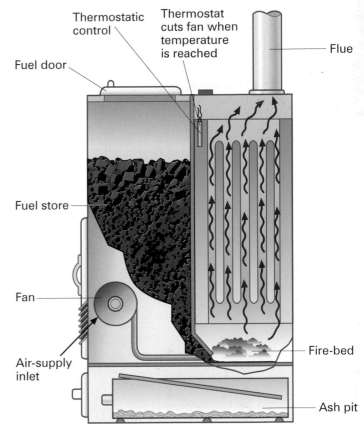

Figure 5.25 High-output independent solid-fuel boiler

(a) They can't be used with unvented systems and are not suitable for sealed systems – open-vented only.

(b) It is considered best practice to fit a temperature-relief valve to the cylinder with tundish and discharge pipework as additional protection to the open-vent pipe.

(c) As the water temperature in the cylinder is largely uncontrollable and usually reaches boiler temperature at the top, the installation of thermostatic mixing valves at point of use is strongly recommended to minimise the risk of scalding, particularly in properties in which young children, the elderly or the infirm live.

Flues for solid-fuel appliances

Part J of the Building Regulations details the requirements for flue systems that serve solid-fuel appliances. You encounter quite onerous requirements with solid fuel, as the heat generated by this type of appliance is much greater than their gas or oil counterparts, and the products of combustion tend to be more corrosive.

Appliances may be discharged into existing brick-built chimneys (usually 225 mm × 225 mm), provided that they are sound. Existing clay liners are also suitable, subject to their correct initial installation. The connection between the appliance and the flue should be properly made, with any flue being properly gathered to connect to the flue pipe. Provision via soot doors for sweeping is essential, as shown in Figure 5.26.

Where defects are found in existing brick-built chimneys, it is possible to line them either:

- using a specialist company which provides a new concrete lining to the flue system, or

- using a flexible stainless-steel flue liner (specially manufactured for solid-fuel use).

Great care should be taken when connecting flue pipes to the flue system, particularly in relation to any combustible material. Part J provides full requirements.

In situations where a new metal flue pipe is considered, Part J lays down very rigid and detailed requirements for how close the flue may be sited to any combustible material. The manufacturer will provide details on this; great care needs to be taken where flues pass through floors and ceilings in relation to combustible material, and there is a need to provide adequate fire-stop measures at these points.

Where an internal flue pipe passes through a cupboard it must be properly guarded from contact. This is usually via a plasterboard shield fixed with metallic fixings rather than a combustible timber frame.

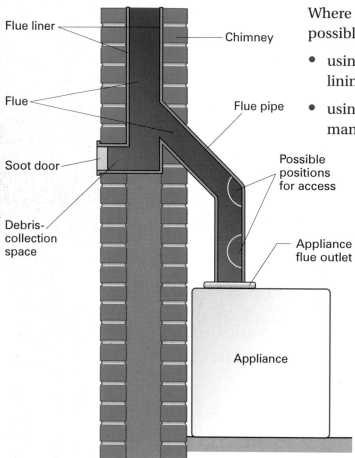

Figure 5.26 Flue connection from boiler

An external flue pipe must be twin-wall-insulated, and suitable for internal use with solid fuel. It is also possible to site metal flue pipes external to the building. Key points are:

- The flue pipe should be sleeved at the point where it passes through the external wall.

- Suitable provision must be provided for sweeping.

- The flue pipe must be suitable for external use with solid fuel. With the increased risk of cooling in the flue pipe and the possibility of an increased amount of acidic water condensate being produced, the flue pipe will usually be twin-wall-insulated with a ceramic lining to the inner surface of the flue pipe – quite an expensive product to purchase.

Safety tip

Flexible liners designed for gas systems are not suitable and are wholly unacceptable

Flues for use with solid fuel (chimneys and metal flue pipes) must, if they discharge at or near the ridge of the building, rise to a minimum height of 600 mm above the ridge. Metal flue pipes discharging at the eaves point must usually rise to a minimum height as well.

Ventilation for solid-fuel appliances

Solid-fuel appliances must be provided with a permanent supply of fresh air for combustion to take place properly. If this fresh air is not provided correctly then, as a result of the burning process, a poisonous gas known as carbon monoxide may be produced – this is a killer.

To overcome this we need to install permanent ventilators in the room in which the appliance is sited. The size of the ventilators is based on the following:

- **back boiler with a throat** – the free area of the air vent should be at least 50 per cent of the area of the throat opening

- **independent boiler/room heater (with no draught stabiliser)** – 550 mm²/kW free area of appliance heat output above 5 kW

- **independent boiler/room heater (with draught stabiliser)** – 300 mm²/kW free area for the first 5 kW of heat output with 850 mm²/kW for the balance of the output.

The rate of burning in solid-fuel appliances is controlled by the air supply, with the burning process being highly dependent on the flue draught (flue pull) created in the chimney. Flue draught is affected by:

- length of the flue (this should usually be a minimum of 4.5 m)

- number of bends in the flue

- angle of bends

- point at which the flue pipe terminates.

Too little draught may well be caused by any of the factors above and would result in two possible outcomes – the fire constantly 'goes out', and more seriously, incomplete combustion and the production of carbon monoxide. A solid-fuel appliance with poor flue draught should therefore be fully investigated and rectified.

It is usually the case, however, that we are in a situation where too much draught is created by the flue system, resulting in overburning/overheating and inefficient performance of the appliance. The solution here is to use a draught stabiliser.

The draught stabiliser responds to fluctuations in flue draught by opening a hinged flap and allowing air to enter directly into the flue system to reduce the flue draught. The stabiliser will ensure a constant draught through the appliance but must be properly set up and the counter-balance weight properly adjusted on commissioning.

Find out

What is the minimum height above the eaves for a flue?

Did you know?

You may have heard a lot about carbon monoxide with gas appliances, but you may be surprised to know that, based on the number of appliances installed, solid fuel accounts for more deaths than gas (statistics provided by the HSE)

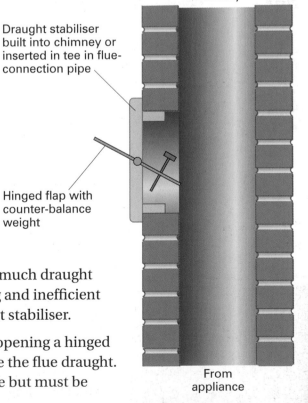

Chimney

Draught stabiliser built into chimney or inserted in tee in flue-connection pipe

Hinged flap with counter-balance weight

From appliance

Figure 5.27 Draught stabiliser to flue

Hearths for solid-fuel appliances

This generally involves building work. A solid-fuel boiler cannot just be sat on a concrete slab on a timber floor – it is required to sit on a full constructional hearth that projects for fire safety with sufficient clearance to all sides of the appliance. This is fundamental building work – outside the scope of the plumbing qualification.

If you're asked to install a solid-fuel appliance you'll need to check the dimensions of the constructional hearth against Part J and determine its suitability – building work may be required.

Commissioning solid-fuel appliances

With room-heaters and independent boilers we are looking at thermostat control and therefore a higher degree of control over the appliance. That control is heavily influenced by the flue draught created in the flue system. With this type of appliance we'll usually need to check the flue draught and flue gas temperature (the latter may not be possible on a room-heater).

Tests and checks will be carried out through a test port located in the appliance or by following manufacturer instructions. The manufacturer will usually quote the required flue gas temperature and flue-draught requirements. A special flue thermometer is used to measure temperature. Flue-draught readings are usually taken with a device known as an **inclined manometer** or draught gauge.

Manufacturer instructions should be followed when using the manometer, which is connected via tubing to take flue-draught readings, usually just over the fire-bed. Readings are usually taken with the fire idling (no heat demand) to determine whether the boiler will go out (if insufficient draught or air supply is available) or whether it will overheat if too much draught or air is available. Similarly, readings are taken with the boiler operating under full load to adjust any air-supply requirements. The draught stabiliser (if fitted) is adjusted as part of this boiler-commissioning process – the stabiliser being adjusted based on the readings of the manometer.

The customer must be fully advised about the flue and appliance cleaning requirements as per manufacturer instructions. This will cover:

- intervals at which the flue must be swept
- intervals for cleaning boiler flueways.

Oil-fired appliances

Here we're going to cover the basics of the key oil-fired appliance – pressure-jet boilers – so that you can get an appreciation of the type of system that may be connected to them. We will not be covering the requirements for fuel-supply systems, which you'll need to know if you are to fit them, or the detailed flue standards that must be met, as detailed in Part J of the Building Regulations. We will, however, look at various pressure-jet boiler flueing options.

Remember

The most important part of the commissioning process with solid-fuel appliances is the hand-over stage

There are essentially two types of oil-fired appliance available:

- Those with **vaporising burners** (quite typically these are appliances such as AGA cookers). This is a relatively unsophisticated type of appliance, as the controls are quite primitive. These types of appliance are usually hot-water only systems or systems with hot-water gravity primaries and pumped heating circuits: their installation tends to be a relatively specialist activity.

- Those appliances fired by **pressure-jet burners** – most modern domestic boilers will be 'fired' by a pressure-jet burner.

Remember

Oil-fired appliances are, strictly speaking, not a part of the range of systems covered by the plumbing technical certificate, so if you wish to progress to cover them fully then you'll have to complete a course outside your apprenticeship

Atomising oil burner

The pressure-jet burner

The operation of this type of burner used to be regarded as rather complex. However, with the introduction of many of the components that are now seen in gas appliances, it is relatively straightforward.

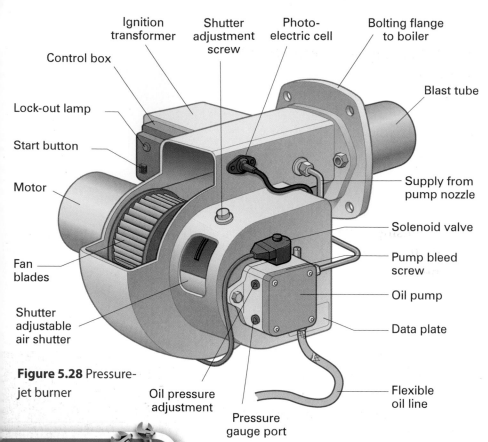

Figure 5.28 Pressure-jet burner

The boiler works on the principle of forcing oil at high pressure through a very fine nozzle ('atomiser'), resulting in the production of a fine spray of oil – as shown in Figure 5.29.

The process of forcing the oil through the nozzle changes its state and makes it easily ignitable, particularly when the correct amount of air is provided in the combustion chamber by a fan attached to the burner. The jet of oil from the nozzle is initially lit by spark electrodes (the heat source) and once lit it continues to burn, as long as the air supply and fuel continues to be provided through the nozzle.

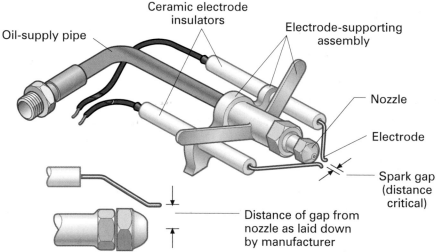

Figure 5.29 Electrodes/nozzle assembly for pressure jet

There are a number of critical points in making the burner operate successfully:

- The amount of air required is set by adjusting a flap on the burner.

- The nozzle must be the correct size for the boiler it is serving, and the nozzle angle must be correct (different boilers have a different angle of nozzle for different spreads of oil coming from the nozzle).

- The distance between the nozzle and the spark electrodes must be correct, as should the gap between them – get this wrong and it won't light properly.

- Oil must be delivered at the right pressure and hence quantity (flow rate) at the nozzle.

The lighting sequence

The lighting sequence is as follows. (It is completed in a matter of seconds and varies slightly from manufacturer to manufacturer.)

Stage 1

With the boiler-control thermostat calling for heat from the boiler, the control box activates the burner motor; this drives the fan and oil pump. Pressure builds in the pump outlet.

Stage 2

The control box activates the ignition transformer, causing sparking at the ignition electrodes – the solenoid valve opens and oil is sent to the nozzle, where it discharges and is mixed with the air supplied by the fan. In the presence of the spark at the electrodes, it ignites.

Stage 3

On lighting, the presence of a flame is detected by the photo-electric cell (this senses the presence of light). If the cell identifies that the flame has been lit then the electrodes are shut down. The fan and motor continue to operate, supplying air and fuel to heat the boiler. In the event that the cell does not detect the presence of a flame, it goes into a restart procedure whereby the solenoid valve closes, preventing oil from entering the boiler. However, the fan and motor remain operational for a predetermined period to clear any vapour that has collected in the boiler.

Following this period the boiler then re-attempts its lighting sequence – if the presence of a flame is still not detected then the boiler proceeds to lock-out: the control box shuts down all boiler operations and it cannot be reactivated until the cause of the fault has been addressed and the lock-out button reset.

Remember

This type of boiler is usually provided with an overheat thermostat

Pressure-jet boiler – options

Boilers are available as wall-hung or floor-standing models. These in turn can be traditional or condensing boilers. Combination boilers are also available (these tend to be floor-standing), as are condensing combination boilers.

As these boilers are fully controllable they can be connected to the same system types as their gas counterparts, so for new installations and replacements they fall under the same requirements for gas as detailed under Part L1 of the Building Regulations.

There are numerous room-sealed and open-flued options, as can be seen from the few examples below. There's the conventional open-flue model, requiring a flue pipe or chimney – you know enough about that so there is no drawing here.

Figure 5.30 shows an open-flue low-level discharge boiler – the same-model is also available as a room-sealed model.

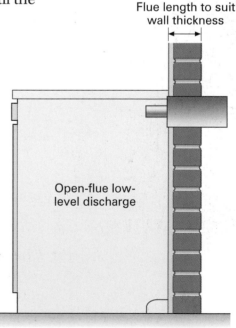

Figure 5.30 Low-level discharge flue – rear outlet

Figure 5.31 shows an open-flue low-level discharge again – but this time with a side outlet connection – also available as a room-sealed model.

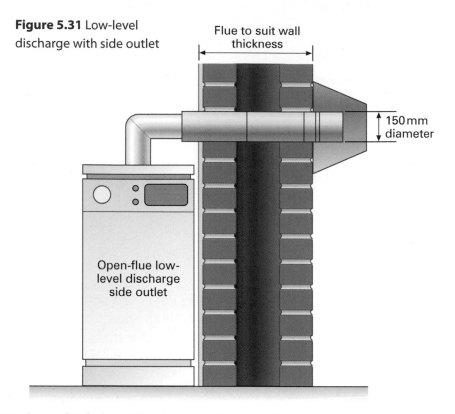

Figure 5.31 Low-level discharge with side outlet

Flue to suit wall thickness

150 mm diameter

Open-flue low-level discharge side outlet

Figure 5.32 shows high-level discharge with open flue, but a room-sealed model is also available.

There is also a room-sealed boiler designed to terminate through a roof, which draws its air supply from the terminal sited on the roof – a vertical discharge flue.

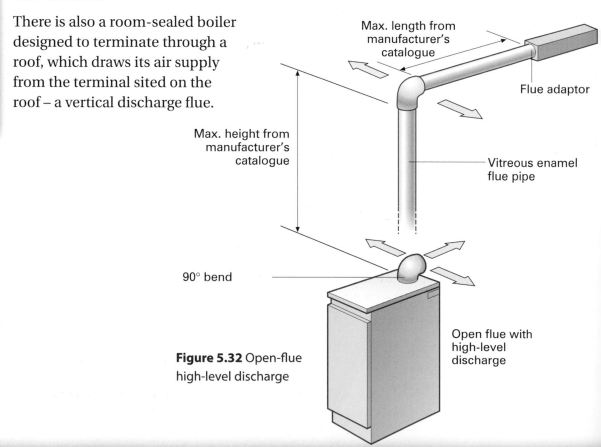

Max. length from manufacturer's catalogue

Flue adaptor

Max. height from manufacturer's catalogue

Vitreous enamel flue pipe

90° bend

Open flue with high-level discharge

Figure 5.32 Open-flue high-level discharge

Oil-boiler commissioning

The correct commissioning of oil boilers is an absolute must. They are extremely sensitive to:

- the right nozzle being fitted
- the correct amount of fuel being delivered at the nozzle
- the correct amount of air being available
- the correct flue pull through the appliance.

To set the boiler up we therefore need to use a number of test instruments to take the following readings:

- oil pressure – measured and adjusted at the oil pump using an oil-pressure gauge
- smoke reading – measures the amount of carbon in the products of combustion. It is used to adjust the amount of air entering the appliance
- flue temperature – using a flue thermometer. This measures the temperature of the flue gases and can be a cross-check to confirm air supply and flue-draught rates
- flue draught – measures the amount of flue pull in the chimney or flue pipe. If the amount is too high, the boiler will be inefficient and may require a draught stabiliser like solid fuel
- carbon dioxide – this is used for fine tuning but also as confirmation of the overall efficiency of the appliance.

Remember

You can't commission an oil-fired appliance without proper equipment. If incorrectly commissioned they tend to operate very inefficiently or the component parts can 'soot up' in a matter of days. When operating correctly, oil pressure-jet boilers are highly efficient

Gas-fired appliances

Here we have the choice of appliances that may be sited:

- traditional floor-standing
- wall-mounted
- fireplace opening.

These in turn (with the exception of fireplaces) may be traditionally operated combination appliances (feeding heating and hot water) or condensing appliances. There are also condensing combination appliances.

First we have the traditional open-flue arrangement, with the appliance connected to a flue pipe or chimney – we've seen this many times so we'll not look at a drawing here; the boiler can be either floor-standing, wall-mounted or in the fireplace opening. Remember: room-sealed models are always preferable to open-flued models as they are safer by design.

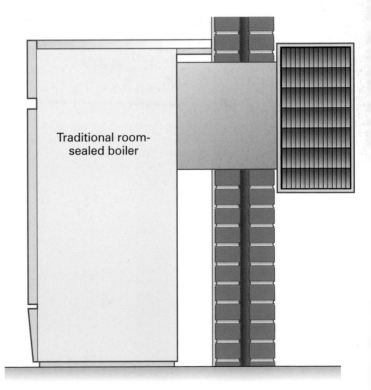

Traditional room-sealed boiler

Figure 5.33 Room-sealed boiler

Figure 5.34 shows a traditional floor-standing boiler but this time with a fanned flue. The flue system discharges from the rear.

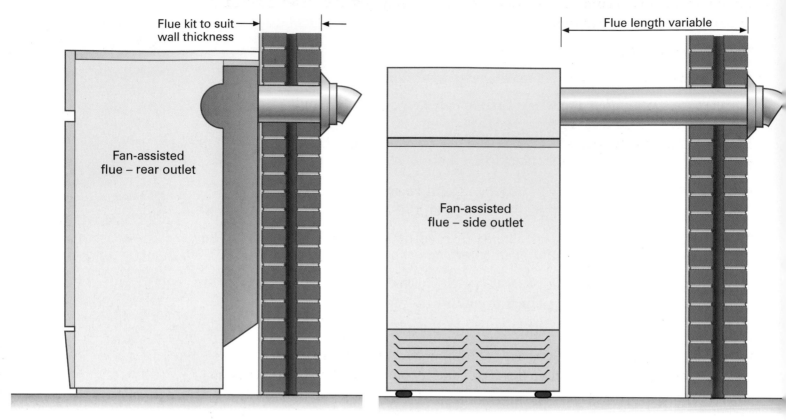

Flue kit to suit wall thickness

Flue length variable

Fan-assisted flue – rear outlet

Fan-assisted flue – side outlet

Figure 5.34 Fan-assisted rear-outlet flue

Figure 5.35 Fan-assisted side-outlet flue

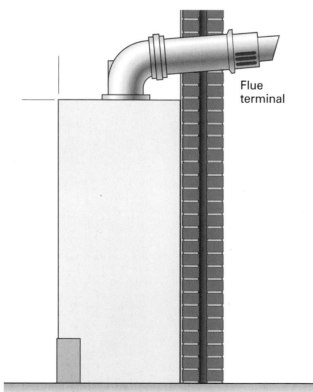

Flue terminal

Figure 5.36 Flue terminal – low level

Figure 5.35 is fanned-flue floor-standing again but discharges from the side. The flue length can be quite considerable to allow more flexibility when siting.

All the various flue types we have seen are available in wall-mounted models. There is also a room-sealed vertical discharge system with the terminal sited on the roof of the property.

Figure 5.36 shows an outline of a condensing boiler. The key feature is the flue falling back to the boiler. This tends to be a key installation point for condensing boilers. It is required to ensure that condensate produced in the flue system falls back into the boiler and can be collected and discharged to waste through a discharge pipe. The discharge pipe should be manufactured from suitable plastic material (copper will corrode); the pipe must have a minimum fall of 2.5° to ensure that it clears the condensate properly. There aren't many restrictions on its length of run but there are restrictions on how it may discharge (this is covered in Chapter 7).

Selection of combination boilers

Combination boilers come in different rates of heat output to the water. They tend to be categorised into two types:

- standard flow rates
- high flow rates.

Those with standard flow rates tend to be suitable for application only in smaller properties where there are fewer outlets or a small number of people to use them. This is because the hot-water flow rate from them tends to be relatively low. So they are unsuitable for:

- properties in which a relatively large family is to bathe one after the other – during winter the heating will be non-operational during the hot-water period

- properties that have baths with a larger than normal capacity. Here the fill time will often be regarded as too long

- where multiple draw-offs are required (in two bathroom properties etc.), the standard combi will not cope with multiple draw-offs.

In these circumstances a high-flow combination boiler may present a suitable alternative; this provides an increased water-flow rate. The manufacturer's instructions will provide detail on flow rates and recommended property types, so that's the main information source. In circumstances where sizeable water-flow rates or hot-water demands are required, a combination boiler will not be the best option; in this case an unvented system may be a better solution.

FAQ

Can I fit any type of boiler?

In most cases, according to the Building Regulations, you should fit condensing boilers. There are some exceptions, but these are limited. Remember, to fit gas boilers you must be CORGI registered.

Wiring central-heating system controls

At the end of this section you should be able to :

- identify the operating principles and wiring requirements of systems with:
  - 2 x 2 port motorised valves
  - mid-position valves
  - combination boilers
- identify the electrical testing requirements of domestic heating-control systems.

You need to understand the basics of the electrical side, the aim being that you will have a thorough understanding of how central-heating system controls work both mechanically and electrically, so that if you don't wire the systems yourself you will at least be able to maintain them properly.

This section therefore provides preparation for the necessary learning that is required to wire heating systems. However, new legislation which came into effect in early 2005 requires you to achieve separate certification to undertake this activity on site.

Remember

Controls manufacturers produce extensive installation instructions, so follow them at all times – here you will learn general principles

Controls systems – electrical function

Programmer

Figure 5.37 shows a terminal strip to which the external wires are connected (marked L across to 8). The internal components of the programmer are connected to this strip – these primarily being a clock or timing device and two simple switches. The purpose of the clock is to control automatically the operation of the switches (independently of each other) when we require them.

By now we should already know that a heating system must have an isolation point – this is usually a double pole switch or a socket outlet and plug. This is also the point at which the over-current protection device (this being a fuse) is located. The fuse is rated at 3 amps for domestic work.

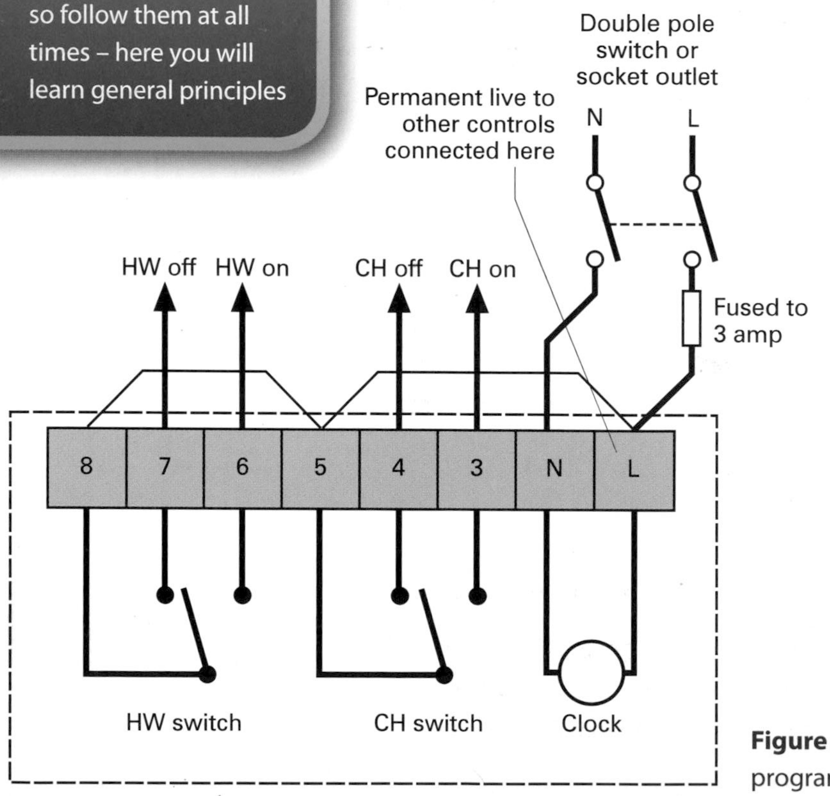

Figure 5.37 Two-channel programmer – wiring

The isolation point is usually located near to the programmer, and a feed wire 3-core (minimum size 1.0 mm²) is taken from the isolation point directly into the programmer (L-N-E).

Some programmers do not have an earth connection. However, other system controls will require an earth, so a connection is made in the programmer using a terminal strip with a further earth wire which connects to the main earth point on the system wiring centre.

So we have coming into the programmer our power supply, connecting to the L, N and E points. With this programmer you are required to provide live link wires across terminals L-5-8. The purpose of these is to provide the power supply to make the switches function correctly.

The switches are classed as what is known as two-way switches. As an example, with the hot-water switch, terminal 7 is only live when the hot-water circuit is off, and terminal 6 is live when the hot-water circuit is on. To operate a circuit we may require power for other components, even when the circuit is off.

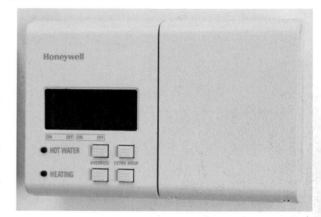

Wires are therefore taken from the required circuits to the main wiring centre to provide the key on/off functions for our components to operate, e.g. room thermostat, motorised valve etc.

We also take a connection from the live or L terminal, known as a permanent live (this remains **permanently live** without being affected by any programmer switching functions, and we need this to get the boiler and pump to work properly – this is for a 2 x 2 port valve system only).

Programmer – time clock

With 2 x 2 port valves

We have a L-N-E into the programmer from the isolation point, and leaving the programmer to go to the main wiring centre we have:

- permanent live (L connection)

- neutral (N connection)

- heating circuit on (switched live connection terminal 3)

- hot-water circuit (switched live connection terminal 6)

- earth (E connection not shown but from isolation point to system main earthing).

The 2 x 2 port valve has 3-core cable into the programmer and 5-core cable to the wiring centre (L-N-E and two switched lives). The wires must be properly colour-coded, neutral colours must not be used for live etc. Uninsulated earth cable cores must never be used for neutral or live connection.

With mid-position valve

Here it's slightly different. We have the same L-N-E into the programmer from the isolation point, and leaving the programmer to go to the wiring centre we have:

- neutral (N connection)
- heating circuit on (switched live connection terminal 3)
- hot-water circuit on (switched live connection terminal 6)
- hot-water circuit off (switched live connection terminal 7)
- earth (E connection not shown but from isolation point to system main earthing).

The mid-position valve has 3-core cables into the programmer and 5-core cable to the wiring centre.

Room thermostat

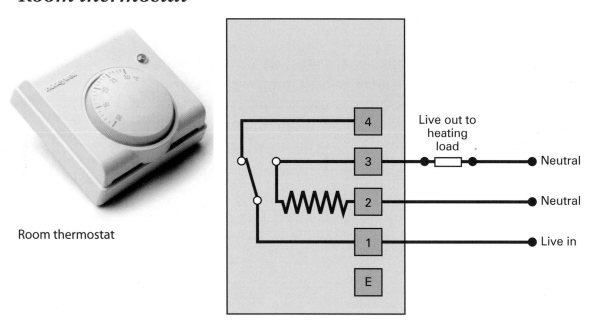

Room thermostat

Figure 5.38 Room thermostat

A room thermostat is a switch, so here we have a live connection. This comes via our wiring centre from the 'on' switch connection for central heating (terminal 3 at the programmer) to the live in terminal 1 on the room thermostat. A live out connection back from terminal 3 on the room thermostat connects via the wiring centre to the heating motorised valve. The neutral and earth connect back to the main neutral and earth connection points at the wiring centre.

So we have a simple switch in the line across terminals 1 and 3. Four-core cable is required to feed the room thermostat from the wiring centre (N-E and two live switch wires).

Cylinder thermostat

With the Honeywell thermostat no earth or neutral connection is required; other manufacturers may have different requirements. Here it's different for the system types.

With 2 x 2 port valves

There is a live in connection via the wiring centre from the 'on' switch connection for hot water (terminal 6 on the programmer) to the 1 terminal in the cylinder thermostat. There is a live out connection from the C terminal which connects via the wiring centre to the hot-water motorised valve.

So we have a simple switch in the line across terminals 1 and C. Two-core cable is required to feed the cylinder thermostat from the wiring centre.

With mid-position valve

This is quite different. It has a live in connection via the wiring centre from the 'off' switch connection (terminal 7 on the programmer) to the 2 terminal on the cylinder thermostat. It has a live in connection via the wiring centre from the 'on' switch connection for hot water (terminal 6 on the programmer) to the C terminal in the cylinder thermostat. A live out connection from the 1 terminal connects via the wiring centre to the mid-position valve.

So here we have two live feeds in and one out (much as we saw at the programmer). Three-core cable is required to feed the cylinder thermostat from the wiring centre.

Boiler and pump

These normally just take a L-N-E from the wiring centre, which is pretty straightforward. The N and E come straight from the main wiring centre terminals, the live feeds to each come from one of the wires on the motorised valve (for both 2 x 2 port and mid-position valve systems).

Two-port motorised valve

Two of these are included on a standard fully pumped system, one controlling heating and one hot water. These valves are really what brings the key circuits together to get the system to function fully.

We normally use spring return 2-port valves for this function – the valve always returns to the closed position by means of a spring when there is no electrical supply to the valve motor.

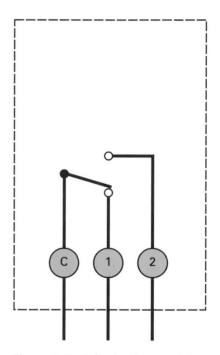

Cylinder thermostat

Figure 5.39 Cylinder thermostat

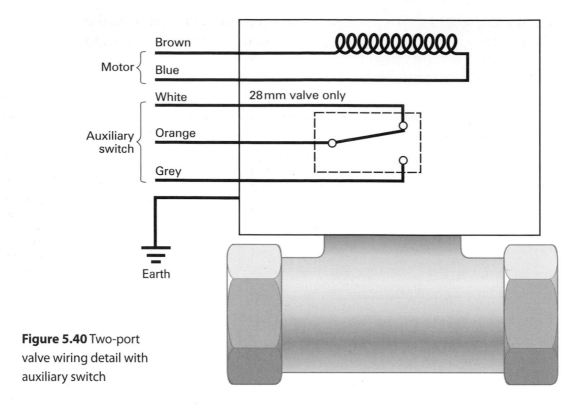

Figure 5.40 Two-port valve wiring detail with auxiliary switch

With regard to the electrics, the valve has two fundamental parts: an electric motor used to drive the valve open so that water can flow through it; and an auxiliary switch (often called a **micro-switch**) which powers the supply to the boiler and pump. The auxiliary switch is there to ensure the effective separation of the electrical supply from each circuit to feed the pump and boiler, and ensures that hot water and central heating can work independently of each other.

Essentially the live out connection from either the cylinder or the room thermostat connect to the brown wire on the motorised valve. This drives the motor open. As the motor moves to its fully open position a mechanical connection is made with the auxiliary switch (it's just like you turning on a light switch). The auxiliary switch receives a permanent live supply (grey wire) from the L connection – permanent live on the programmer via the wiring centre. As the switch is turned on by the valve motor the permanent live supply is allowed to flow through the orange wire from the motorised valve which goes to the live supply (L) to feed both the pump and boiler.

When the programmer turns off or the thermostat reaches temperature, the live supply to the valve motor (brown) is cut and the valve shuts due to the action of the spring return. The auxiliary switch is disconnected and power stops flowing from the permanent live (grey) wire through to the pump and boiler feed (orange wire). If that is the only circuit currently operating it will shut pump and boiler off. If the other circuit is calling for heat, pump and boiler will stay on – so both circuits work independently of each other.

Before we look at the mid-position valve, let's pull all the information together in the form of the full 2 x 2 port valve wiring diagram, and we'll take you through it again to try to fully understand it.

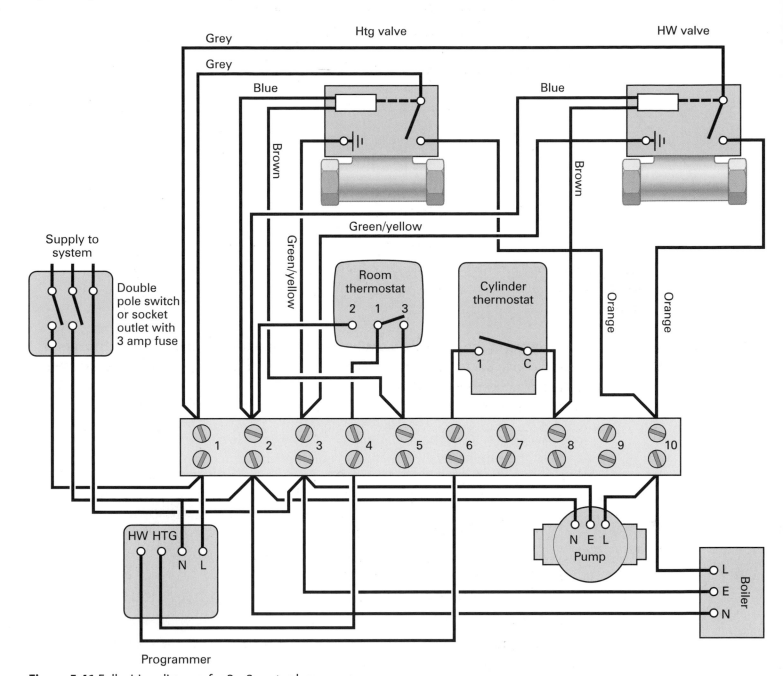

Figure 5.41 Full wiring diagram for 2 x 2 port valves

We'll remove the neutral and earth connections for you, as by now you should know that essentially the components that take a neutral and earth connect together to feed back via the main isolation point to the electrical circuit – so they are not involved in any of the switching arrangements.

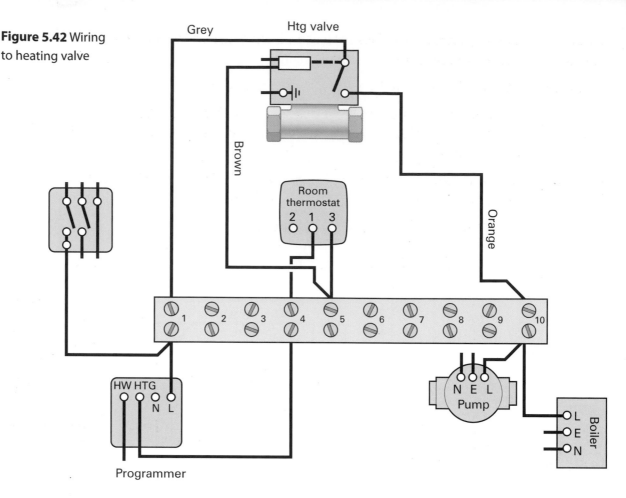

Figure 5.42 Wiring to heating valve

Let's take a look at how each circuit works on its own – first the central heating. In Figure 5.42 you'll see that the main feed to the isolation point is via the wiring centre and not the programmer. Either is acceptable, but remember that with this you'll still need to connect a permanent live neutral and earth back to the programmer. So we have:

- The programmer calls for heating and energises the heating On terminal.

- Power is supplied to wiring centre terminal 4 which in turn feeds terminal 1 on the room thermostat.

- If the thermostat is calling for heat (remember, if it is up to temperature then the circuit will not be made and the rest of the circuit remains dead) then the circuit will be made and terminal 3 will be live, feeding back to terminal 5 in the wiring centre.

- The brown (live motor) wire to the motorised valve is connected to terminal 5 so will be live if the thermostat is calling for heat.

- The brown wire drives the motor open. On fully opening, the motor mechanically makes the contacts on the auxiliary switch.

- Power is then allowed to flow via the grey permanent live wire which connects to terminal 1 in the wiring centre (permanent live direct from isolation point or programmer terminal) through the orange wire to terminal 10 in the wiring centre, which feeds both live connections to the pump and boiler.

- The boiler and pump begin to operate.

When either the programmer reaches the end of its timing period or a thermostat reaches the desired temperature:

- the live feed to the brown wire in the motorised valve is cut

- the valve motor returns to the closed position. During that action the mechanical connection to the auxiliary switch is broken and the live feed to the pump and boiler by the orange wire is cut

- the boiler and pump turn off if it is the last of the circuits to close (if another circuit is calling for heat then they remain operational).

Let's look at the hot-water circuit now – there's not much difference here.

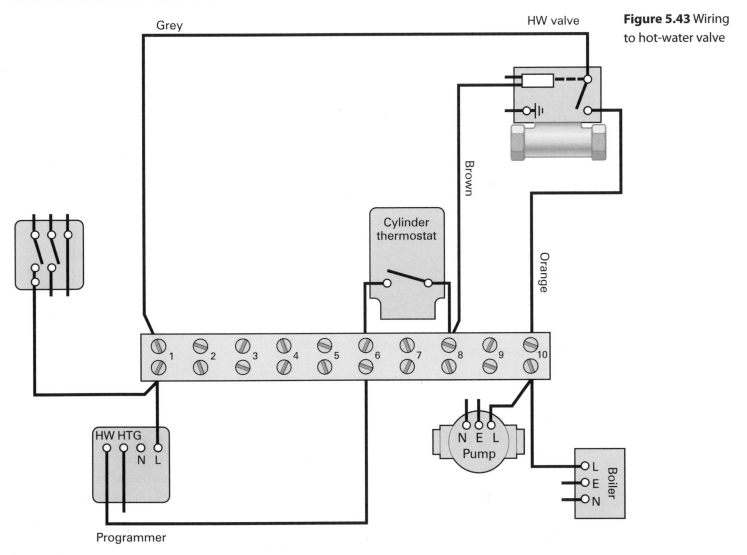

Figure 5.43 Wiring to hot-water valve

So we have:

- The programmer calls for heating and energises the heating On terminal.

- Power is supplied to wiring centre terminal 6 which in turn feeds terminal 1 on the cylinder thermostat.

- If the thermostat is calling for heat (remember, if it is up to temperature then the circuit will not be made and the rest of the circuit remains dead) then the circuit will be made and terminal C will be live, feeding back to terminal 8 in the wiring centre.

- The brown (live motor) wire to the motorised valve is connected to terminal 8 so will be live if the thermostat is calling for heat.

- The brown wire drives the motor open. On fully opening, the motor mechanically makes the contacts on the auxiliary switch.

- Power is then allowed to flow via the grey permanent live wire which connects to terminal 1 in the wiring centre (permanent live direct from isolation point or programmer terminal) through the orange wire to terminal 10 in the wiring centre, which feeds both live connections to the pump and boiler.

- The boiler and pump begin to operate.

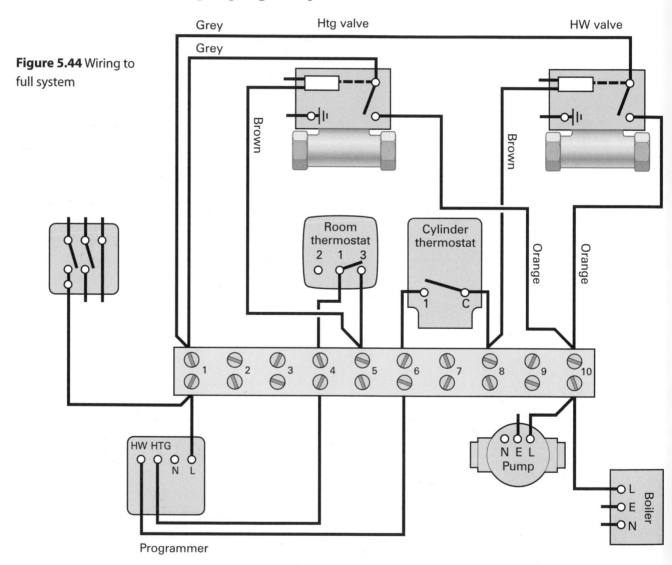

Figure 5.44 Wiring to full system

When either the programmer reaches the end of its timing period or a thermostat reaches the desired temperature:

- the live feed to the brown wire in the motorised valve is cut

- the valve motor returns to the closed position. During that action the mechanical connection to the auxiliary switch is broken and the live feed to the pump and boiler by the orange wire is cut

- the boiler and pump turn off if that is the last of the circuits to close (if another circuit is calling for heat then they remain operational).

Figure 5.44 shows the wiring diagram with both circuits together but with the live and neutral connections removed for the purposes of making it clear to you – now you can see how both circuits can work together and independently of each other.

Mid-position valve

To make Figure 5.45 easier to understand we've removed the live and neutral connections again, but this time we'll have to look at both heating and hot water together to understand it.

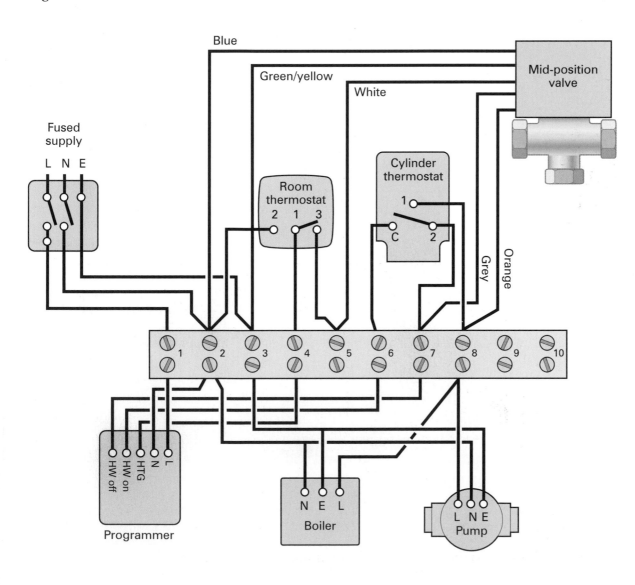

Figure 5.45 Wiring to mid-position valve

Figure 5.46 Wiring to mid-position
– without neutrals shown

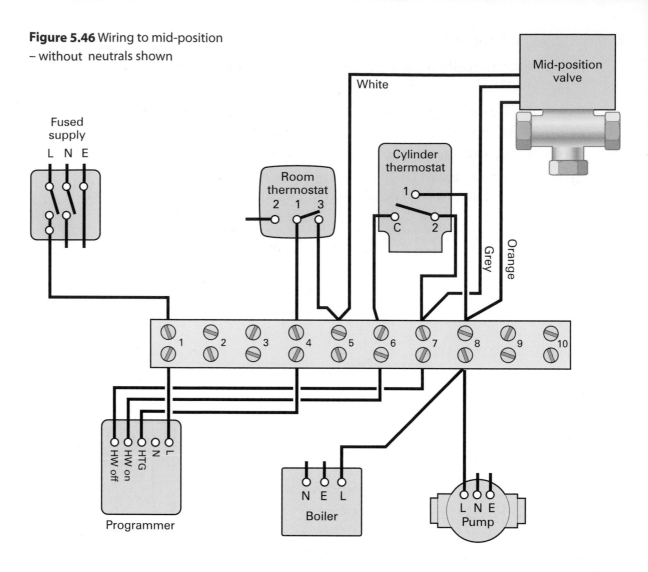

The mid-position valve in its de-energised state is open to the hot-water-only port. So we have:

(a) Hot water only:

- The programmer calls for hot water and energises the hot water On terminal.

- Power is supplied to wiring centre terminal 6 which in turn feeds terminal C on the cylinder thermostat.

- If the thermostat is calling for heat (remember, if it is up to temperature then the circuit will not be made and the rest of the circuit remains dead) then the circuit will be made and terminal 1 will be live, feeding back to terminal 8 in the wiring centre. At this point terminal 2 on the thermostat is not receiving supply from the hot water Off terminal.

- The pump and boiler are directly connected to terminal 8 so they become operational; the connection to the motorised valve orange wire is live but performs no function at this stage. The valve position does not alter as it is normally open to hot water.

(b) Heating and hot water:

- The programmer calls for heat and energises the hot water On and heating On terminals.

- Power is supplied to wiring centre terminal 6 which in turn feeds terminal C on the cylinder thermostat and power is supplied to wiring centre terminal 4 which in turn feeds terminal 1 on the room thermostat.

- Assuming both circuits are calling for heat, the pump and boiler are activated by the live connection from the cylinder thermostat (terminal C).

- The motorised valve is driven to its mid-position (supplying heating and hot water) by the white wire connected to the outlet of the room thermostat terminal 3 (via wiring centre connection 5).

- Pump and boiler supply both heating and hot water.

(c) Heating only:

- With the programmer calling for heating only or the cylinder thermostat up to temperature the grey wire in the valve is activated (either by the hot water Off terminal at the programmer or the 2 terminal at the cylinder thermostat).

- The valve travels fully across to the heating only port where an auxiliary switch connection is made by a mechanical connection. This in turn energises the orange wire which feeds the boiler and pump.

- The boiler and pump are operational.

On this principle the valve can move backwards and forwards through its operating sequence to respond to the demands of whichever circuits are requiring heat. It's a bit more complicated than 2 x 2 port valves and can be a bit more difficult to diagnose faults on – but you should now know the principles.

On the job: Wiring a central-heating system

While working with his plumbing supervisor on the wiring to a central-heating system Todd was faced with a dilemma. The supervisor realised at the last minute that the gas fire/boiler unit in the lounge required a separate permanent live in addition to the switched live from the programmer which is located in the airing cupboard some distance away. Only a two-core plus earth (brown, blue and yellow/green) has been installed and the house is in the final stages of decoration, ready for moving in.

The supervisor tells you he intends to use the earth as the extra live wire needed, by putting tape on it for identification and asks you to assist.

1   Is this the correct thing to do?   3   What would you do now?

2   If not, what should be done?

Frost thermostats

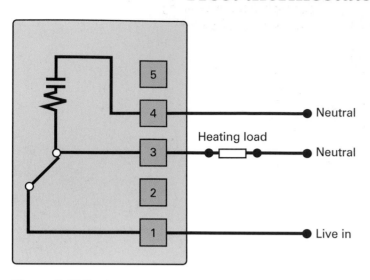

Figure 5.47 Frost thermostat

The frost thermostat is not much different to the room thermostat except that it reads much lower temperature levels and has a tamper-proof cover. It is sited near to the components that are to be guarded against freezing, e.g. in an external boilerhouse, and is usually set to about 4°C, the point at which freezing can begin.

The thermostat is designed to override the time and temperature controls to a particular circuit (usually one of the heating zones, if there are more than one). So a permanent live connection is taken from the wiring centre to the frost thermostat, which in turn feeds the motorised valve (there are differences for mid-position and 2 x 2 port valve systems, as we'll see later).

So if the frost thermostat activates, the pump and boiler are activated via the feed to the motorised valve until the temperature that has been set on the pipe thermostat has been reached. At that point the frost thermostat is still activated but the pipe thermostat is not. The water in the pipework is well above freezing. The temperature in the pipework will begin to fall, and the boiler and pump are reactivated by the pipe thermostat... and so on.

Figure 5.48 shows the wiring layout for these components for a 2 x 2 port valve system. Figure 5.49 shows a mid-position valve system.

> **Remember**
>
> To be energy-efficient the frost thermostat must be used in conjunction with a pipe thermostat, to reduce any energy wastage. The pipe thermostat is just a simple switch rather like a cylinder thermostat and sits between the outlet from the frost thermostat and the connection to the motorised valve. The pipe thermostat is usually set to around 40–50°C

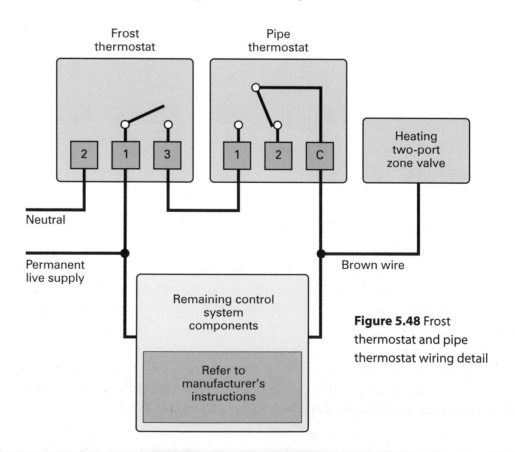

Figure 5.48 Frost thermostat and pipe thermostat wiring detail

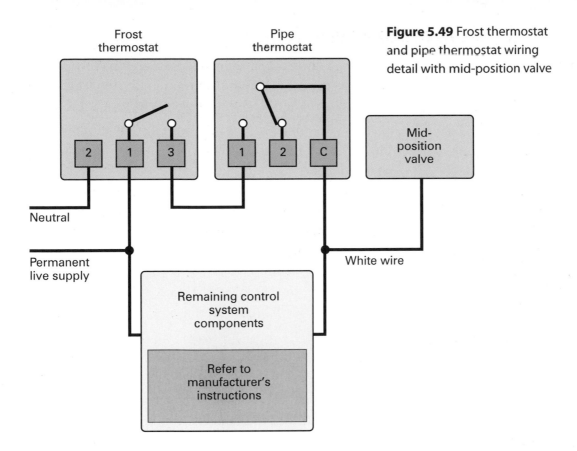

Figure 5.49 Frost thermostat and pipe thermostat wiring detail with mid-position valve

Boilers with pump-overrun thermostats

You probably won't see these much on new boilers but you might come across them on existing installations. They have a slightly different wiring arrangement. The boiler is involved to a much greater extent in the controls function. The differences for a 2 x 2 port valve system are:

- boiler takes a neutral and earth connection as normal

- a permanent live supply is taken direct from the wiring centre to the boiler

- the orange wires from the motorised valves are connected via the wiring centre directly to a terminal on the boiler

- the pump live connection is supplied via the wiring centre by the boiler.

So we need 5-core cable to the boiler (N-E and live from orange wires, permanent live and live to pump).

Essentially when the boiler receives power via the orange wire(s) from the motorised valve(s), both pump and boiler are operational. When the power is cut to the orange wire and the boiler is up to temperature a separate pump-overrun thermostat will have its contacts made (it responds to the temperature in the boiler heat exchanger made at high temperature, not made at low temperature) – power is supplied via the permanent live connection to the pump only. So the pump continues to operate. The boiler is not firing; it continues to operate until the water cools to the temperature setting of the overrun thermostat, then it turns off.

Fire back-boilers with lights

Here we need to ensure that along with the L-N-E connections there is a permanent live provided; this connects directly from the permanent live connection at the wiring centre to the live at the boiler for the fire lights.

We need 4-core cable to the boiler (N-E, live connection to the boiler from the motorised valve orange wires [2 x 2 port valve system] and permanent live to the lights).

Electrical testing of central heating systems

Here the first issue is to ensure that a visual check of the installation is carried out before it is tested. Key checks are:

- correct cable sizes
- correct cable terminations
- correct over-current protection devices
- equipotential bonding of the system – the system will need bonding to earth correctly including any extraneous metal parts that may not be naturally connected to the main fuel or cold-water system.

When the visual check has been carried out key checks and tests can be conducted with no power onto the system.

Earth continuity

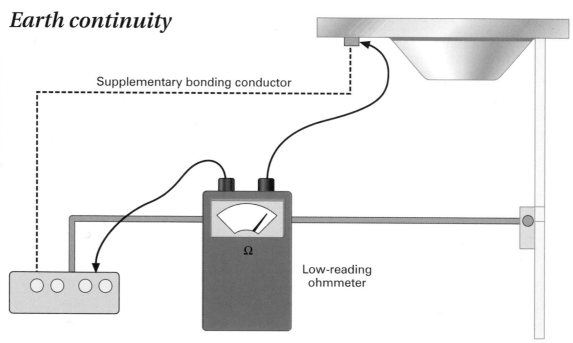

Figure 5.50 Testing for continuity

Remember that earth continuity is making sure that, should there be an electrical fault, all exposed metalwork in a building is bonded together and connected to the earthing block in the consumer unit, leaking the current to earth and automatically disconnecting the supply. The earth continuity test will verify that.

The ohmmeter leads are connected between the points being tested, between simultaneously accessible extraneous conductive parts, i.e. pipework, sinks etc. or between simultaneously accessible extraneous conductive parts and exposed conductive parts (metal parts of the installation). This test will verify that the conductor is sound. To check this, move the probe to the metalwork to be protected. This method is also used to test the main equipotential bonding conductors. There should be a low resistance reading on the ohmmeter.

Testing for continuity

Polarity

Testing for polarity is to make sure that phased conductors are not crossed somewhere, i.e. neutral from mains connected to live and vice versa (reversed polarity). This situation would mean that the system might still function as expected; however, when isolated from a switch the system would be in dangerous mode.

Testing across the conductors is carried out to make sure that no wires have been crossed.

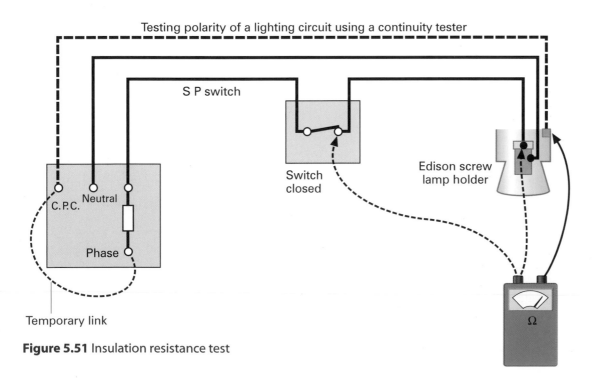

Testing polarity of a lighting circuit using a continuity tester

Figure 5.51 Insulation resistance test

Insulation resistance

Insulation-resistance tests are to make sure that the insulation of conductors, electrical appliances and components is satisfactory and that electrical conductors and protective conductors are not short-circuited, or do not show a low insulation resistance (which would indicate defective insulation).

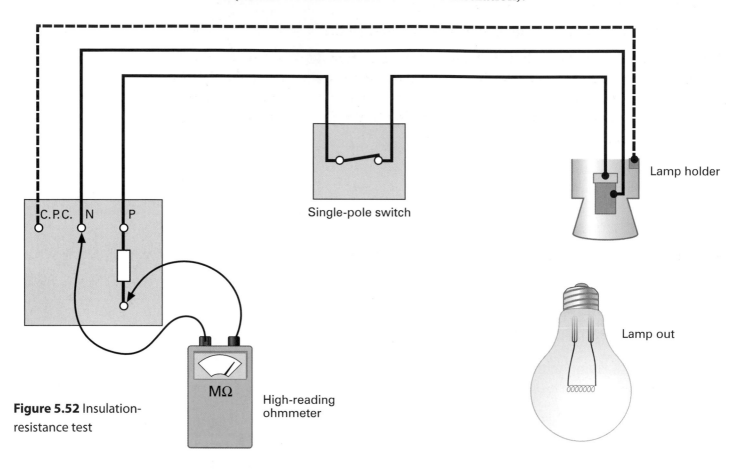

C.P.C. N P

Single-pole switch

Lamp holder

MΩ

High-reading ohmmeter

Lamp out

Figure 5.52 Insulation-resistance test

Before testing, ensure that:

(a) pilot or indicator lamps and capacitors are disconnected from circuits to avoid an inaccurate test value being obtained

(b) voltage-sensitive electronic equipment such as dimmer switches, delay timers, power controllers, electronic starters for fluorescent lamps, emergency lighting, residual current devices etc. are disconnected so that they are not subjected to the test voltage

(c) there is no electrical connection between any phase or neutral conductor (e.g. lamps left in).

An insulation resistance of no less than 0.5 ohms should be achieved.

Remember

Only when the system has been properly tested can it be put into service

Commissioning central-heating systems

At the end of this section you should be able to:

- state the requirements for flushing, cleansing and corrosion protection within central-heating systems
- identify correct commissioning and handover procedures.

Now we've got to the point when we have the whole system installed and we can think about putting it into service. This is probably the most important phase of the installation, as our aim should be to leave the site with a correctly operating system that meets the design specification. But it's not as simple as just turning it on, both from a safety and an operational perspective.

The visual inspection

Just as with hot- and cold-water pipework, this includes thoroughly inspecting all pipework and fittings to ensure:

- that they're fully supported, including F&E cisterns, hot-water storage cylinders, expansion vessels etc.
- that they're free from jointing compound and flux
- that all connections are tight
- that in-line valves and radiator valves are closed to allow stage filling
- that the inside of the F&E cistern is clean
- that all the air vents are closed
- that before filling the pump is removed and replaced with a section of pipe. This will prevent any system debris entering the pump's workings. It is also good practice to take the expansion vessel out of the system at this stage if possible, or to leave it out until after testing for leaks has been completed, to prevent any dirt or residue getting to the membrane. Remember to plug its connection for the test, though
- that the customer or other site workers are advised that soundness testing is about to commence
- that the flue system is correct as well as any ventilation that is required for combustion.

Testing for leaks

- Turn on the stop tap if it's a complete cold-water, domestic hot-water and central-heating system installation, or the service valve to the F&E cistern if it's just the central-heating circuit. Allow the system to fill.
- Turn on the radiator valves, manually open the motorised valve and open any bypass valves fully, and bleed each radiator.
- Visually check all the joints for signs of leaks and repair as necessary.

Pressure testing

The Water Regulations don't require closed systems to be tested. However, it's good practice to test equipment and procedures as outlined in Section 3 under the Water Regulations. On larger jobs it could also be part of the contract specification. You'd usually expect to test the system to 1½ times the normal working pressure; the test pressure would be achieved by using hydraulic test equipment.

Hydraulic test equipment

Remember to seal any open ends prior to testing. Test to 1½ times the working pressure and leave to stand for one hour, and again repair any leaks as necessary.

The cold flush

Once you've checked and tested for leaks it's time for the cold flush. Here you need to drain the system from the drain-off point(s) as quickly as possible, so the first job is to turn off the water supply and drain off with a hose pipe, letting air into the system at the various bleed points.

Once emptied you can close all bleed valves and reinstall the pump/expansion vessel ready for the next phase. Be sure to check the pressure in the expansion vessel before it's refitted to see if it is correct for the system – tyre-pressure-type valve under the base cap.

Pressure tester – hydraulic

System cleaning

BS 7593 details the requirements for the treatment of water in domestic heating systems, and identifies that all systems, as part of the commissioning process, should be treated with a proprietary system cleaner to ensure the removal of foreign matter and bodies that may cause corrosion, such as flux residues. This is done by introducing a system cleaner into the system:

- open-vented via the F&E cistern as it is refilling, not just dumped into the cistern after it has refilled, or else it won't work

- sealed – remove a radiator plug or automatic air vent and fill with a funnel (connected with a pipework rig that you may have to make up).

The amount of cleaner required will be as detailed by the manufacturer, based on the system volume in litres. Once the cleaner has been introduced the system can be refilled and vented.

At this point the electrical controls are turned onto the system (boiler isolated) and the pump is allowed to circulate water round the system for approximately 10 minutes, during which time you can check for correct initial operation of the controls (whether they work in the correct sequence, whether the right thermostat is working the right valve etc.).

After the 10-minute period you can try venting any air again from the radiators. Some more is likely to have accumulated. Now's the point to check the operation of the pressure-relief valve on a sealed system by twisting the cap to let it discharge water for 30 seconds and then making sure it reseats.

Boiler commissioning

You're now ready to commission the boiler. Assuming that you have tested the fuel supply already and it's sound, you can commence the process. For solid fuel and oil we have already given a brief overview of the requirements, and for gas you'll look at the requirements in greater detail in Chapters 7 and 8.

Essentially we are looking to establish that the rate of fuel supply is correctly set, together with the air supply, and that the flue system is operating correctly.

Once all these points have been correctly checked and set, it's time to fully run the boiler. So, with the cleaner in the system, the boiler should be run. The first check is to run it up to temperature to see if the boiler thermostat is working. Set the boiler to its correct operating temperature (about 80°C) and connect a digital pipe thermostat on the flow pipe next to the boiler. It should turn off when the reading is about the right temperature and begin to cycle on and off.

Next, run the water up to temperature and establish that the water supplied is at a higher temperature than the cylinder thermostat setting of 60–65°C. It should be higher, as the temperature of the water at the top of the cylinder will be higher than the point at which the cylinder thermostat is positioned.

Now check the room thermostat operation, again using the digital thermometer. Set the room thermostat to just above the temperature currently in the room and run the heat up till it shuts down – the air temperature near to the thermostat should be very similar to the point at which the thermostat has turned off.

System balancing

This is probably the most crucial phase, as you will need to ensure that all the radiators are working uniformly and the components are receiving the correct flow rate from the pump.

To understand this fully you need to know a bit about system design. So an overview is required here.

When designing a system we include a pump that will provide a certain flow rate, to get the required amount of water from the boiler to the component parts of the system's cylinder and radiators. In delivering the required flow to all points of the system we have to overcome pressure created by the frictional resistance in the pipework etc. However, the pressure that we have to overcome is in relation to the frictional resistance of (usually) just one of the radiator circuits – this is known as the **index circuit**. In any system this will be the circuit that has a combination of the highest flow rate and highest resistance. So the pump is sized to meet the resistance of just that index circuit. Any other radiator, or indeed cylinder circuit, on the system will have less resistance and so the pump will be able to deliver the flow rate required to those circuits much more easily.

To ensure that all the circuits work uniformly we need to balance the radiators that are not in the index circuit. This balancing ensures that there is the same frictional resistance at the other circuits as there is in the index circuit, and is achieved by adjusting the lockshield radiator valve, so placing a restriction on the other circuit.

The starting point is to put the pump on its correct setting. Now with only the index radiator working (and the valves fully open) and using the digital thermometer (which should be capable of taking readings on both the flow and return pipes), you should have the required temperature drop across the flow and return pipes near the boiler of 80°C and 70°C respectively.

If the boiler requires a bypass, or the pump is sized much higher than the system flow rate usually requires (e.g. Combi boiler with bypass), then you should start taking the readings at the flow and return pipes with the bypass fully closed. This will have the effect of raising the return temperature at the boiler. The bypass should be eased open gradually, turn by turn of the valve, until the return temperature achieves the required temperature drop across the boiler – 80°C and 70°C respectively. The bypass should then be left at this setting.

The remaining radiator circuits should now be balanced. This means going round the radiators and opening them gradually at the lockshield valve until a point occurs at which the difference in the flow and return pipes at the inlet and outlet to each radiator is 10°C. You might have to go round the system a couple of times to get this right, as opening one valve tends to have an effect on other parts of the system already running.

So once you've go them all running, with the required temperature drop across them, do a final check on the boiler flow and return, which should be 80°C and 70°C, with the 10° drop, and your system is balanced.

The final job is to check to see if there is any discharge from the pressure-relief valve (if it's a sealed system) and that the pressure gauge does not show an excessively high pressure in the system. If it does, and the expansion vessel has been properly charged, it's too small.

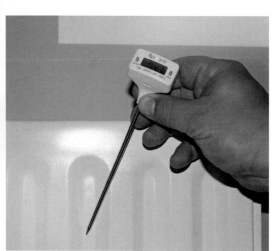

Using a radiator thermometer

The hot flush

The system cleaner should now be emptied out of the system. Most manufacturers recommend a two-hour period for the cleaning operation. The balancing procedure etc. that we have just shown will usually take this period of time. Turn off the system electrics so that the controls are isolated, open the drain valves full bore and empty the system contents fully, opening any bleed valves and motorised valves. Once the system contents have been drained, close all valves and introduce the required amount of corrosion inhibitor – this is done prior to refilling at the same fill point as you introduced the cleaner. Refill the system and vent the air. Run the system back up to temperature, venting air as required and ensuring all circuits are operating correctly.

Fit any thermostatic radiator valve heads and set the temperature controls to the correct levels. You're now ready for handover.

Handover

Hand over the system to the customer with a full demonstration of how it works, showing where key valves and isolation points are located. Provide the customer with:

- all manufacturers' instructions for safe-keeping
- commissioning certificate, signed and dated, with full details of the system (an example is the Benchmark Logbook – ask your lecturer if you can look at a copy).

Just tidy up now and the job is complete. There may be a bit more to commissioning than you thought there was, but if you don't do it properly the system won't work correctly.

Fault diagnosis

At the end of this section you should be able to:

- diagnose a range of common faults on central-heating systems.

Here we're going to look at a range of common system faults. Boiler faults will be looked at in Chapter 7 Gas appliances.

System faults

Poorly designed and installed systems account for a great number of call-backs with central-heating systems. If it's a system that you haven't installed, then a visual inspection is required for correct compliance with installation standards – this may reveal a lot.

The system works but there's noise from the pipework

Key questions are:

- Has it just started?
- Has it been there all the time?

If the noise has just started then do the preliminary checks:

- Has the system got water in it?
- Is it topped up?
- Has the float-operated valve stuck?

Then run it up to temperature:

- Is the pump on the right setting?
- If this checks OK, then are all the valves open properly?
- Is the bypass set correctly?

You are beginning to work out the problem, based on what you see. If it's been there a long time you might be checking on such issues as whether it needs a bypass. Following these checks, one of your findings might be that key pipe sizes in the system are incorrect, particularly if it's been a long-term problem. Here you might have to re-pipe some aspect of the system.

Remember

Expect the unexpected when working out the problem, but always look for the simple points first, as they are normally to blame

Radiators are not getting hot in some parts of the system

The first thing to identify is which radiators are not heating up:

- Upstairs only and not as hot as they usually are: it's a possibility that the pump may have stuck or failed.

- Downstairs only: the system could have air in the upstairs radiators and/or the system is running dry and may need topping up, or the float-operated valve in the cistern has stuck.

- Individual radiators not as hot as they used to be: have the radiators been off for decorating? Does the system need re-balancing? Is the system getting 'sludged up'? If it is, then there will be reasons for it and these will need investigating.

- Individual radiators not working at all: are the radiator valves open? On thermostatic valves, check the pin that operates the function of the valve: they have a tendency to stick closed in old age and may need greasing or replacing.

- No radiators at all: it could be a component fault or an electrical fault (you can discount the boiler if hot water is available):
  - First check the major components to ensure that the fuse has not blown.
  - If power is available then check the operation of the following:
    - programmer, to
    - thermostat, to
    - motorised valve (here it could be due to a defective motor or a defective auxiliary switch), to
    - pump.

With a combi you'll need to follow the manufacturer's guidance on checking its components for correct operation.

No hot water

What type of system is it? If it is fully pumped, you are looking at establishing that the system's got water in it. Assuming that the radiators work, it's probably down to looking at component faults in a similar way as for heating-circuit faults:

- First check the major components to ensure that the fuse has not blown.

- If power is available then check the operation of the following:
  - programmer, to
  - thermostat, to
  - motorised valve (here it could be due to a defective motor or a defective auxiliary switch), to
  - pump.

With a combi you'll need to follow the manufacturer's guidance on checking its components for correct operation.

With gravity systems, check whether the circulators are air-locked. If they are, investigate the pipework to make sure it's run properly.

Dirty-coloured hot water

With single-feed indirect cylinders this is usually a sign that the bubble has been lost. This could need investigating further – to address such issues as whether the boiler thermostat may have failed, overheated and removed the bubble, or whether the system may have been extended and the cylinder requires replacement.

Noise at the boiler

The key question here is: what is the noise like and how long has it been going on?

If the noise has been short-term and sounds like a boiling noise, the cause is likely to be a component failure on the boiler.

If it's more long-term and it's like a kettle heating up until it shuts off under temperature, then starts again, you're more likely to find that it's a boiler circulation problem. This could be due to:

- no bypass fitted – the boiler heat exchanger could already be damaged
- pump at the wrong setting (again the boiler heat exchanger could already be damaged)
- sludge in the boiler causing poor circulation (cleaning may be an option but an investigation into how the sludge has collected is required).

No power to the system

First check the fuse; try replacing it. If it blows again then it is either a component or wiring fault. With the electricity off it's time for a check. The first thing is probably the pump – if it sticks then there's a good chance that it will blow a 3 amp fuse. Then it's down to checking such items as:

- Has water got into any of the electrical components? (Badly positioned motorised valves suffer from this problem, quite often with drips from valves entering the electrics and the motor.)
- Have any flexes connecting to the boiler strayed too close to the boiler? (Badly installed flex to fireback boilers can be a problem, even if it's heat-resistant.)
- Have any badly installed cables strayed too close to heating pipes, or has somebody recently been working in the property and damaged a cable? You'll see everything if you do enough maintenance on heating systems...

System keeps filling up with air

This signifies a major problem of some description. The question is how often do you have to let the air out?

If it's a sealed system then it's probably going to signify a leak, and a big clue is that the system needs ongoing topping-up (normally it should require it only very rarely).

With an open-vented system it could be a leak again or a sign of more serious problems such as pumping over and sucking down at the F&E cistern – it's surprising how long these faults can remain unreported after installation – in which case it's going to be probable surgery to the system.

In addition to the faults highlighted above you will also find some really unusual ones that you can't plan for but which experience will help you to deal with. For example, there was the system fitted with thermostatic valves, with no radiators working, in which the insulation jacket had collapsed into the F&E cistern (because it had no lid) and had blocked all the thermostatic valve ports as well as other parts of the system – what a job that was to clear up!

Knowledge check

1   When dealing with gravity systems, what is meant by 'circulating head'?

2   What other component should a frost thermostat be connected to in order to save energy when providing frost protection?

3   What is the normal setting on a pressure-relief valve at which it will begin to discharge the system contents?

4   As a rule of thumb, what maximum heat load, in kW, will a 15 mm pipe normally carry?

5   What additional safety control measure can be applied to a solid-fuel heating system that has a heat leak fitted to prevent overheating?

6   What is the minimum fall on a condensate pipe from a condensing boiler?

7   What generates the spark in an oil-fired pressure-jet boiler?

8   Which fuel is exempt from some of the key requirements of Part L1?

9   Part L1 refers to insulating the pipework adjacent to the cylinder on new and existing installations. What are the requirements?

10 What are the special requirements for heating circuits with floor areas over 150 m²?

11 What is the recommended system soundness test pressure?

12 In order to balance a heating system successfully, you would need to know which radiator formed the 'index cicuit'. What is the index circuit?

13 What requirements does Regulation L1 lay down for the commissioning of systems?

chapter 6

Gas safety and supply

OVERVIEW

Domestic gas is used as a fuel for cooking, hot water and heating systems. The two main gases used are natural gas (NG)(methane) and liquefied petroleum gas (LPG) (propane and butane). All work is covered by the Gas Safety Regulations. It is now law that any businesses working with natural gas or LPG must be registered with CORGI. In order to prove competence all Operatives who work on gas must sit a test every five years; this test is known as the Accredited Certification Scheme for Individual Gas Fitting Operatives (ACS). This section deals with general gas safety relating to the CCN1 assessment requirements of a separate ACS, on which the plumbing technical certificate unit that you are studying is based. Chapter 7 covers gas appliances.

What you will learn in this unit:

- **Gas safety and legislation**
- **Installation of gas pipework**
- **Pressure testing domestic systems**
- **Purging and re-establishing gas supplies**
- **Checking pressures and gas rates**
- **Combustion**
- **Ventilation**

- **Open-flued systems**
- **Room-sealed flue systems**
- **Flue inspection and testing**
- **Unsafe situations**
- **Gas controls and devices**
- **Decommissioning gas systems**
- **Liquefied petroleum gas (LPG)**

Gas safety and legislation

At the end of this section you should be able to:

- recognise the statutory requirements
- define 'work' in relation to gas fittings
- state the function of CORGI
- make reference to the Gas Safety Regulations

The purpose of the following notes is to give you the skills and knowledge you need to work safely. You will be required to pass a practical and written test to complete this section of the technical certificate. This proof of competence, together with records of actual work experience on a range of gas installations working with a CORGI-registered installer, will allow you to make an application to sit the ACS test and (if successful) allow you to join CORGI, if your application to them is satisfactory.

Although the areas in this section have been separated for study purposes, it is not possible to separate the sections in terms of overall knowledge and skills required. It is necessary to understand *all* sections, as they combine to give an overall picture of gas installation requirements.

Responsibilities

It is the responsibility of *any* person who knows or suspects that gas or fumes are escaping into a property to take reasonable steps immediately to turn off the gas supply and ventilate the property.

Reported gas escape – advice to user

When a gas operative is advised of a gas escape but is not able to inspect the job, the gas user should be instructed as follows:

- Turn off the gas supply immediately at the emergency/meter control valve.
- Extinguish all sources of ignition.
- Do not smoke.
- Do not operate electrical light or power switches (on or off).
- Ventilate the building(s) by opening doors and windows.
- Ensure access to the premises can be gained.

Safety tip

The gas supply should not be used until remedial action has been taken to correct the defect and the installation has been recommissioned by a competent person

The gas user should be instructed to report the gas escape to the appropriate gas emergency service call centre listed below:

- in England (except the Isle of Man), Scotland and Wales, contact the National Gas Emergency Service Call Centre, on 0800 111999

- in Northern Ireland contact Phoenix Natural Gas on 0800 002001

- in the Isle of Man contact Manx Gas on 01624 644444.

There are a number of statutory requirements which have to be complied with when installing, servicing, maintaining or repairing gas appliances and fittings. These requirements apply to all of Great Britain and cover all aspects of work including gas, electricity and related building work.

It is important to ensure that all 'work' is in compliance with legislation and in accordance with relevant standards and manufacturers' installation instructions.

CORGI (Council for Registered Gas Installers)

The HSE (Health and Safety Executive) has approved CORGI to maintain a register of competent gas installers and businesses. These businesses are defined as a 'class of persons approved by the HSE to undertake gas work'. The rules for registration are strict and require individual operatives to be competent in gas work and to hold valid certificates. CORGI has a team of inspectors who check on the quality of work carried out by gas businesses, and they investigate any complaints of unsatisfactory work. Any business found not to comply with the requirements, when inspected, and if a substantial breach of Regulations is found, can be subsequently informed in writing of their removal from the register and no longer be able to work on gas. In serious cases they may face criminal action. Furthermore, it is an offence if they then continue to work or falsely claim to be a member.

The legislation, although very important, can be a little daunting to learn as there is a lot of legal jargon and there are many regulations to be conversant with. It is *not* the intention to reproduce the legislation in this section. The subject is covered in a series of questions for you to work through, by referring to the Gas Safety (Installation and Use) Regulations. You can get these from your library or the college that you are attending.

This will give you the essential practice that you need to become familiar with the Regulations. When you attend a college your tutor will work with you and will be able to confirm that your response/answers are correct.

Take your time and work through the questions methodically. You will become much more familiar with the Regulations and have the confidence to refer to them in the future – as all good installers have to from time to time.

When you sit the ACS Gas Safety Test at Level 3 you will be expected to do something very similar to this.

Did you know?

'Work' is defined as activities in relation to a gas fitting by any person, whether an employee or not. Activities include:

- installation
- maintenance
- servicing
- repair
- renewing
- replacing
- purging

Questions on gas-safety legislation

1. Do the Gas Safety Regulations apply to gas fittings connected to an LPG installation?

2. Can a self-employed person carry out gas work if he does not have a current certificate of competence?

3. Who is responsible for ensuring that an installer who carries out gas fitting work complies with the Regulations?

4. What device must be fitted to a meter to prevent 'foreign matter' from entering the system?

5. What is the pipework carrying gas from the meter to the gas appliances called?

6. What is the name and number of the standard that applies to all pipework used for gas installations?

7. In what circumstances do you need to apply corrosion protection to copper tube used in gas?

8. State the overall requirements for the positioning of an emergency control.

9. State the information that a label to an emergency control should contain.

10. What safety device must be fitted to any primary meter?

11. The regulator on a primary meter is sealed to prevent unauthorised access. Who is allowed to break the seal?

12. Give some examples of considerations when installing pipework.

13. Only in a special situation are you allowed to run a pipe in a cavity wall. What situation is this?

14. Special precautions must be taken when pipes are installed under the foundations of a building. What are they?

15. What is the requirement for purpose-made ducts when gas pipework is installed within them?

16. What considerations should the installer have for the building itself when installing gas pipework?

17. What must an installer do to the installation and appliances when the gas service/meter has just been installed and gas is available for the first time?

18. Pipes conveying gas in commercial premises are required to be colour-coded. What is the colour? Do pipes in domestic installations require colour-coding?

19. What safety requirements must gas installers comply with when installing appliances?

20. Do the Regulations apply to disconnection or purging through an appliance such as a cooker?

21. When a gas appliance has been installed how should it be left?

22. What do the Regulations state regarding a separate shut-off valve for each appliance?

23. Is it permissible to modify/alter an appliance to suit the customer's requirements or a particular installation?

24. When a flue pipe is installed into a brick or masonry chimney, the pipe can be boxed in but special requirements must be met. What are these?

25. Can an open-flued gas fire be installed in a bedroom? If so, are there any requirements.

26. Can an open-flued water heater be installed in a bedroom? If so, are there any requirements?

27. If manually operated dampers are found in a flue, what action should be taken?

28. What checks must an installer carry out to a completed installation that is to be commissioned for the first time?

29. Who is regarded as being responsible for permitting the use of a gas appliance in a property?

30. How often should a gas-safety check be carried out?

31. Who is permitted to complete a landlord safety check?

32. Within how many days must a copy of the landlord's safety record be made available to the tenant?

33. What items should be included in a safety check?

34. What is the response time in which a supplier must respond to a gas escape?

35. What is your responsibility if a smell of gas persists even after you have carried out a tightness test that proved there was no leak?

36. You have been asked to service a boiler, but on arrival you find that the gas has been turned off; the customer informs you that he has carried out a repair to a gas pipe which was leaking. What actions do you take?

37. A supplier has a duty to respond to gas escapes. Does the supplier have a similar responsibility to respond to a report of 'fumes'?

Once you have completed this section, move on to the installation of pipework. You will need to return to the Regulations throughout this topic many times, however, to check that installations comply.

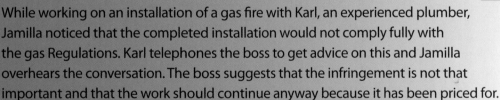

On the job: Installing a gas fire

While working on an installation of a gas fire with Karl, an experienced plumber, Jamilla noticed that the completed installation would not comply fully with the gas Regulations. Karl telephones the boss to get advice on this and Jamilla overhears the conversation. The boss suggests that the infringement is not that important and that the work should continue anyway because it has been priced for.

Karl is not totally happy with the outcome but carries on with the work as instructed. A further check with the Regulations confirms that Karl is indeed correct: the installation will not comply fully and will infringe the Regulations.

1 What should Karl do now?

2 What would you do in these circumstances?

3 What are the possible implications of doing a job incorrectly?

4 What are the dangers to the end users?

Installation of gas pipework

At the end of this section you should be able to:

- identify the various materials for:
  - interior pipework
  - exterior pipework
  - jointing methods
- state the installation requirements for:
  - buried pipework
  - pipework in shafts/stairways
  - pipework in proximity to other services
  - cross-bonding
  - protection of pipework.

So far, in terms of the installation side of plumbing, we've looked at a variety of materials and jointing and fixing methods. The general principles are exactly the same for gas but there are some special installation requirements which must be followed.

There is no need to try to remember the BS numbers, but you should be aware of the general content and be able to refer to anything out of the ordinary that may arise during your work.

It is also worth stressing that these training notes are a much simplified version of a rather complicated and strict set of rules/regulations. It is essential to refer to the actual Regulations and Standards to confirm exact installation practice, and you'll also need these to complete this gas section.

Materials

Steel

- Steel pipe and fittings shall conform to BS 1387 (medium or heavy grade), BS 3601 and BS 3604.
- Rigid stainless steel must conform to BS 3605 or BS 4127.

Tracpipe joint

Corrugated stainless steel to BS 7838

This type of pipe is now being used for gas installations and was not covered at Level 2, so the following extracts from the 'Tracpipe' brochure should help you to understand correct methods of jointing.

'CUT-TO-LENGTH: Determine proper length. Cut through plastic cover and stainless steel pipe using a tube cutter with a sharp wheel. Cut must be centred between two corrugations. Use full circular strokes in one direction and tighten roller pressure slightly (a quarter turn) after each revolution. DO NOT OVERTIGHTEN ROLLER, which may flatten pipe.

STRIP COVER: Using a utility knife, strip back the plastic cover about 25 mm from the cut end to allow assembly of fittings.

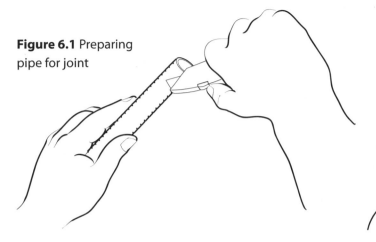

Figure 6.1 Preparing pipe for joint

INSTALL BACK NUT: Slide the back nut over the cut end; place the two split rings into the first corrugation next to the pipe cut. Slide the back nut forward to trap the rings.

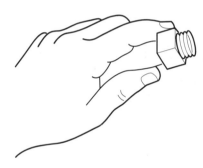

Figure 6.2 Joint preparation

FIT AutoFlare® FITTING: Place the AutoFlare® fitting into the back nut and engage threads. Note that the AutoFlare® fitting is designed to form a leak-tight seal on the stainless piping as you tighten the fitting. (The piloting feature of the insert will not always enter the bore of the piping before the tightening operation, but will centre the fitting when tightened.)'

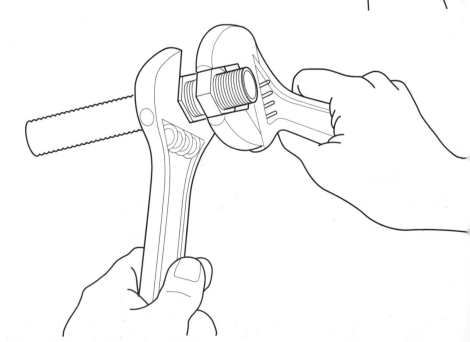

Figure 6.3 Tightening Tracpipe joint

Malleable iron

Malleable iron fittings shall conform to BS 143 and BS 1256.

Copper

- Copper tube shall conform to BS EN 1057.
- Capillary and compression fittings shall conform to BS 864 Part 2.
- Brazed joints are rarely used and must conform to BS 1723.

Plastics

Polyethylene pipes and fittings are used for exterior work only. This type of installation is a specialist area and should not be attempted by anyone without the expertise. Specialised courses and qualifications are available for this area of work.

Jointing

We will now consider some special pipework jointing requirements for gas:

- Any flux must remain active during the heating process only as it cannot be flushed out of the system.
- No flux should be allowed to come into contact with stainless steel.
- Compression fittings shall only be used where they are readily accessible, not under floors, in ducts etc.
- Push-fit and quick-release fittings are not normally used for gas installations.
- Union joints for steel pipe shall be sited in accessible locations.
- Hemp shall not be used on threaded joints.
- Jointing pastes shall not be used with PTFE tape.

Pipework installation

When we refer to installation pipework in a gas system we are referring to the pipework downstream of the meter; that which is upstream of the meter is the responsibility of the gas supplier (see Figure 6.4).

Figure 6.4 Pipework installation

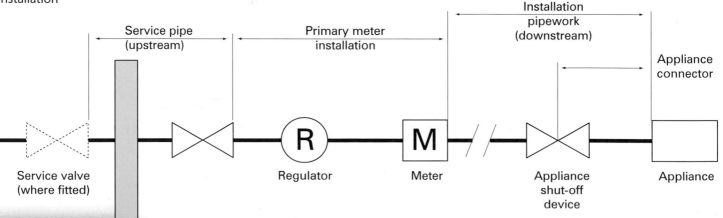

Generally, any installation pipework shall be physically protected or positioned where it cannot be damaged. Damaged water pipework will cause a problem, but damaged gas pipes will probably cause an explosion!

When work is in progress take care to keep any dirt/water from entering gas pipes, and never leave an open end unattended (even for a tea break); it must be capped off. If work is being carried out on an existing installation then the meter must be removed and capped to prevent any risk of flames/sparks causing an explosion and of flames entering existing pipework that you are working on.

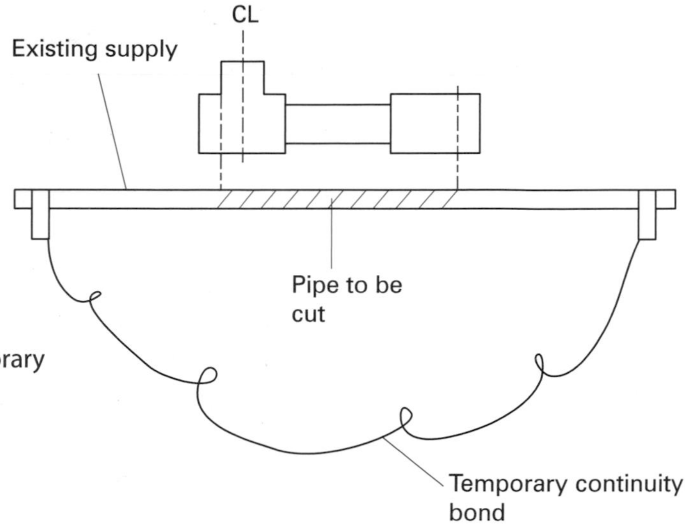

Figure 6.5 Temporary continuity bond

Special fittings such as 'micropoints' may be used when working on gas. These allow a neat finish to the ends of a gas carcass installation; some even have their own shut-off valve. Small-diameter rigid pipes are then used to connect to appliances, although flexible hoses can be used for cookers or refrigerators.

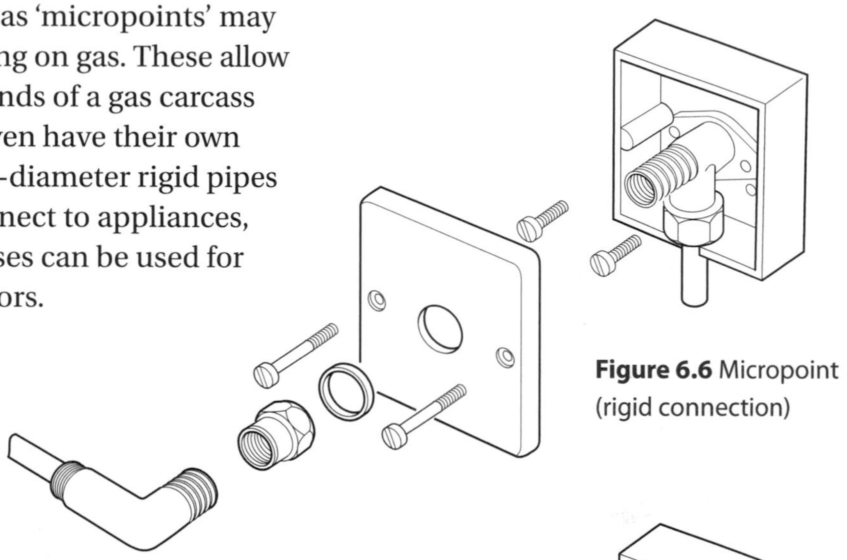

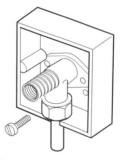

Figure 6.6 Micropoint (rigid connection)

Safety tip

If any section of pipework is going to be removed then a temporary continuity bond must be attached to prevent the risk of a spark or a shock causing a hazard. This bond should be removed only when work is complete

When any installation pipe is no longer required, the pipe(s) should be disconnected as close to the supply point as possible and capped off.

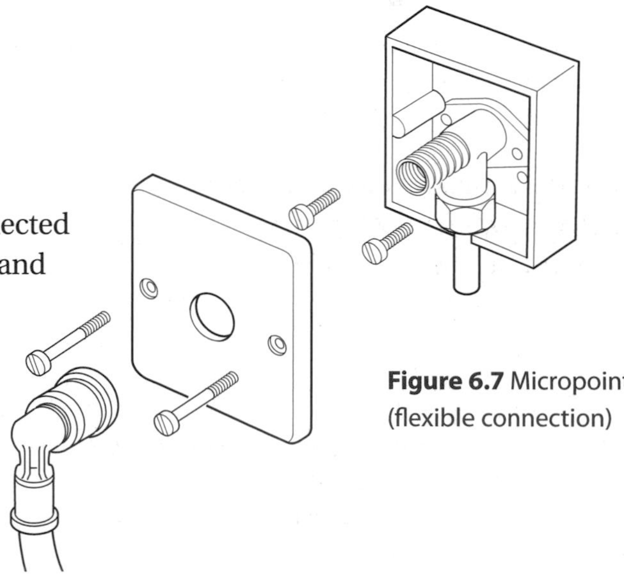

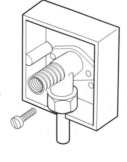

Figure 6.7 Micropoint (flexible connection)

Pipes laid in wooden floors

The following points must be followed when placing pipework underneath wooden flooring surfaces:

- Pipes must be supported.

- Purpose-made notches or holes should be provided (these should be in accordance with Building Regulation requirements as shown in Figures 6.8 and 6.9 for joist notching and drilling holes through joists).

- Flooring should be marked to indicate that there is a gas pipe below.

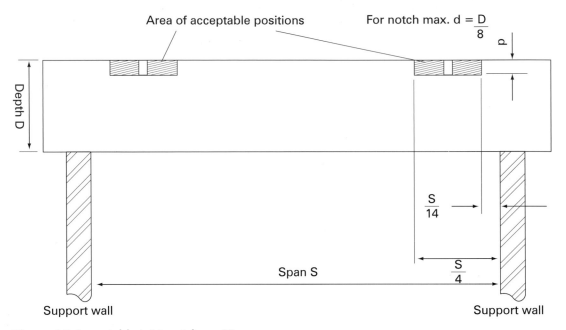

Figure 6.8 Acceptable joist notch positions

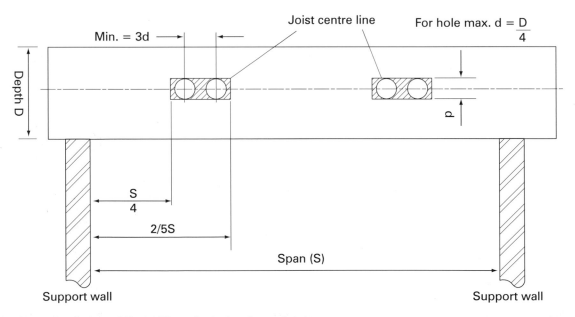

Figure 6.9 Acceptable positions for holes through joist

On the job: Laying pipes below a wooden floor

Jason, a final-year apprentice, has lifted a couple of floorboards at each end of the room and has guided a 22 mm copper pipe under the floorboards from one side of the room to the other. He suspects that the copper gas pipe may be touching an electric cable, although it cannot be reached as it is halfway along and out of sight.

Jason thinks, 'Don't worry, it's OK, and I'm on a bonus anyway!'

1   What should Jason actually have done?

2   What are the risks of leaving things as they are?

Pipes laid in concrete floors

Pipes must be protected against damage and corrosion to meet the requirements of BS 6891. Figures 6.10, 6.11, 6.12 and 6.13 show acceptable methods of locating pipework in various types of solid floor.

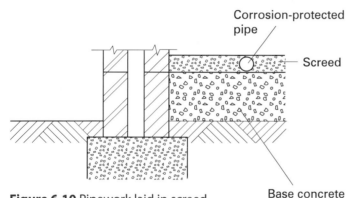

Figure 6.10 Pipework laid in screed

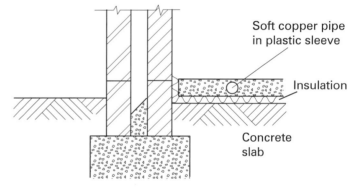

Figure 6.11 Pipework laid in screed with insulating material

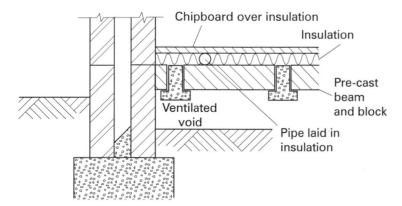

Figure 6.12 Pipework laid in chipboard with pre-cast block and beam

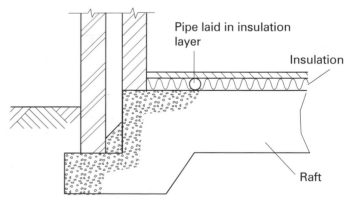

Figure 6.13 Pipework laid in insulation material with raft construction

When laying pipes in solid floors the following must be taken into account:

- Joints must be kept to an absolute minimum.
- Compression fittings must NOT be used.
  - Acceptable pipework protection methods are:
  - factory-sheathed soft copper passed through a larger-diameter plastic sleeving previously set into the concrete. No joints are to be located in the plastic sleeving
  - pipes laid on top of the base concrete and in the screed must be factory-sheathed (plastic coated) or protected on-site using appropriate corrosion-resistant wrapping material
  - pipe laid in pre-formed ducts with protective covers
  - pipe fitted with additional soft covering material at least 5 mm thick and resistant to the ingress of corrosion materials such as concrete.

Protection against corrosion

The following principles need to be applied:

- All pipes should be protected from corrosion.
- Factory-finished protection is preferred, e.g. plastic coated where pipework is to be routed through corrosive environments.
- Wrapping with protective tape is acceptable.
- Soot is very corrosive: protect pipes in fireplace openings.
- Test pipework BEFORE wrapping.
- Use stand-off clips to avoid contact with wall surfaces.

Relation to other services

- Installation pipes need to be located or insulated to prevent contact with metal fitments which may cause electrolytic corrosion.
- Installation pipes *must* be 150 mm from electrical meters and 25 mm from any electricity supply cables or metallic services.

Pipes in walls

The following must be applied when siting pipework within wall surfaces:

- Keep pipes to be covered in plaster vertical (see following drawings).
- Provide ducts/access wherever possible.
- Pipes are not allowed to run in the cavity.
- Pipes passing through a cavity must be via the shortest route and be sleeved.

- Pipes behind dry lining should be encased by building material.
- The number of joints must be kept to a minimum.
- Pipes in timber studding must be secure.
- All pipes are to be protected from mechanical damage and corrosion.

Figure 6.14 shows an example of typical pipe runs within wall surfaces.

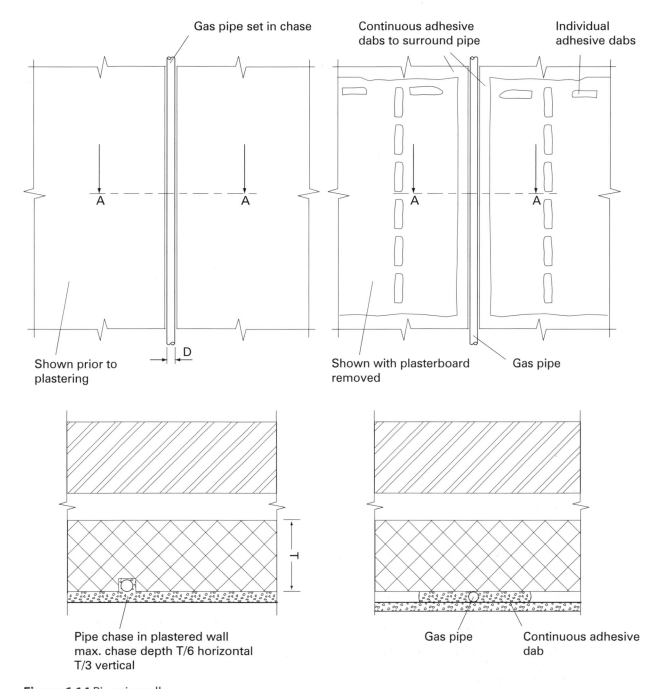

Figure 6.14 Pipes in walls

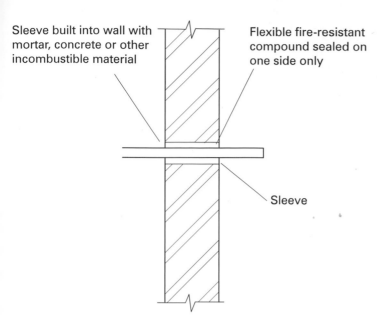

Sleeve built into wall with mortar, concrete or other incombustible material

Flexible fire-resistant compound sealed on one side only

Sleeve

Figure 6.15 Pipe sleeve

Pipes passing through walls (sleeves)

A pipe sleeve must be provided where it passes through a solid wall to protect the pipework from movement in the wall and from the corrosive effects of the materials used in the wall construction. Key features of the sleeve are:

- It must be made of a material capable of containing or distributing gas, e.g. copper or PVC.

- Copper pipework should not be sleeved with iron or steel due to the possibility of electrolytic corrosion.

- The space between the sleeve and the pipe must be capable of being sealed with an appropriate fire-resistant non-setting material; the seal should be made at one end only.

- It should span the full width of the wall and be continuous without any splits or cracks.

- The sleeve should vent to outside air, i.e. the seal should occur on the inside wall (other than in the case of a pipe entry to a gas meter box, which must include a seal inside the box itself).

- There must be no joints inside the sleeve.

- It must be sealed on its outside at each end of the structure with a building material, e.g. mortar.

Ducts and protected shafts

Ducts carrying pipework must be ventilated to prevent any minor gas leak causing a problem, and minimum openings are required as shown in Table 6.1.

Cross-sectional area of duct m²	Minimum free area of each opening m²
Not exceeding 0.01	Nil
0.01 but not exceeding 0.05	Cross-sectional area of duct
0.05 but not exceeding 7.5	0.05
Exceeding 7.5	1/150 of the cross-sectional area of the duct

Table 6.1 Free area of ventilation openings for ducts

It is also important to consider fire stopping. As with any pipework passing from one floor to another in flats/maisonettes, the alternative is to have a protected shaft which is ventilated at top and bottom but sealed where installation pipes enter each flat.

Any gas pipe installed in a protected shaft/stairwell/fire escape should be of screwed or welded steel; copper is not allowed.

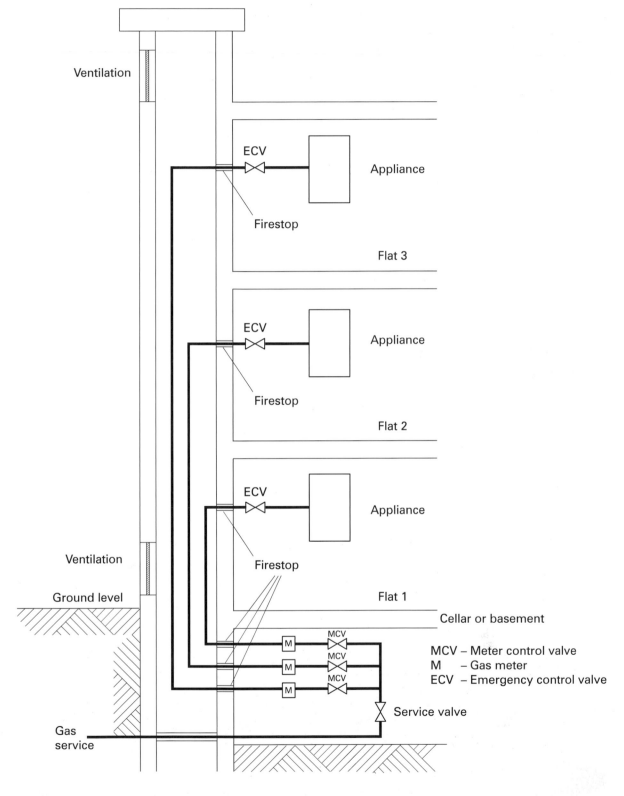

Figure 6.16 Pipework in a multi-storey dwelling with ventilated duct

Figures 6.16 and 6.17 show examples of possible pipework layouts and the necessary protection measures.

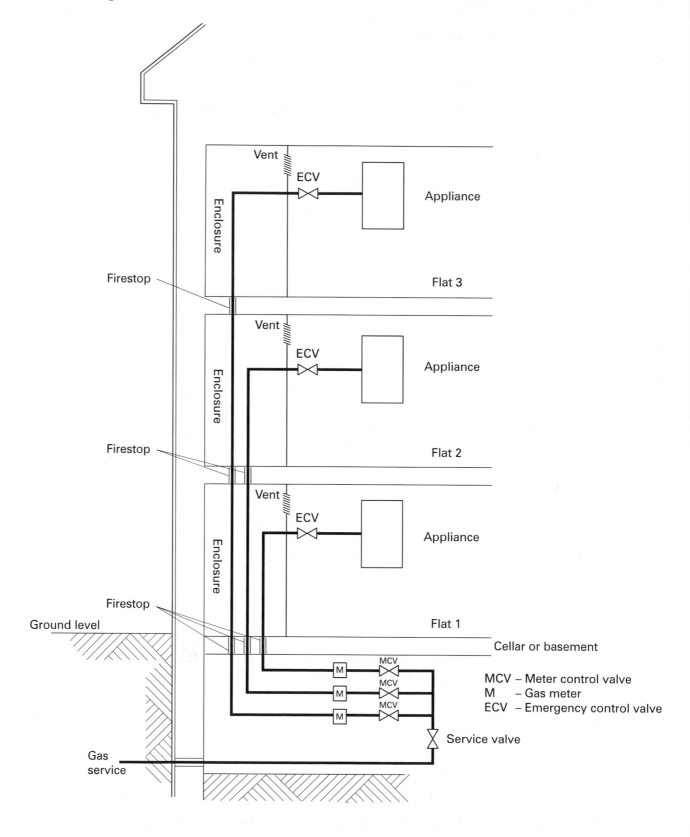

Figure 6.17 Pipework in a multi-storey dwelling with fire-stopped pipework and in a ventilated, enclosed area

Valves

The emergency control valve may be fitted:

- to the inlet of the primary meter; or
- to the installation pipe where it enters the building, where the meter is sited 6 metres or further away from the building; or
- inside individual flats served by a large single meter or a multiple meter installation located in a remote or communal area.

An emergency control valve (ECV) shall:

- always be fitted and labelled to show open and closed position
- be fitted in an accessible position and be easy to operate with a suitably fixed handle which falls safely downwards to an 'off' position.

Pipe supports

Pipes shall be adequately supported as per Table 6.2. Pipe clips or supports must be of a type not likely to cause corrosion.

Material	Nominal size	Interval for vertical support	Interval for horizontal support
Mild steel	Up to DN 15 DN 20 DN25	2.5 m 3.0 m 3.0 m	2.0 m 2.5 m 2.5 m
Copper tube	Up to 15 mm 22 mm 28 mm	2.0 m 2.5 m 2.5 m	1.5 m 2.0 m 2.0 m
Corrugated stainless steel	DN 10 DN 12 DN 15 DN 22 DN 28	0.6 m	0.5 m

Table 6.2 Maximum interval between pipe supports

Exterior pipework

- The use of fittings should be kept to a minimum.
- An external control valve needs to be fitted where the gas supply leaves the dwelling if connecting to an external appliance such as that shown in Figure 6.18.
- Buried pipework in soil or vehicular driveways must have at least 375 mm of cover.
- Buried pipework under concrete paths for pedestrians need have only 40 mm of cover.

- Compression fittings are not allowed below ground.
- Where pipework is run externally above ground level it is preferable for it to be protected against corrosion with factory-applied sheathing. However, if stand-off clips are used then it is permissible to install bare copper pipes. Pipework below ground level MUST have additional protection against corrosion.

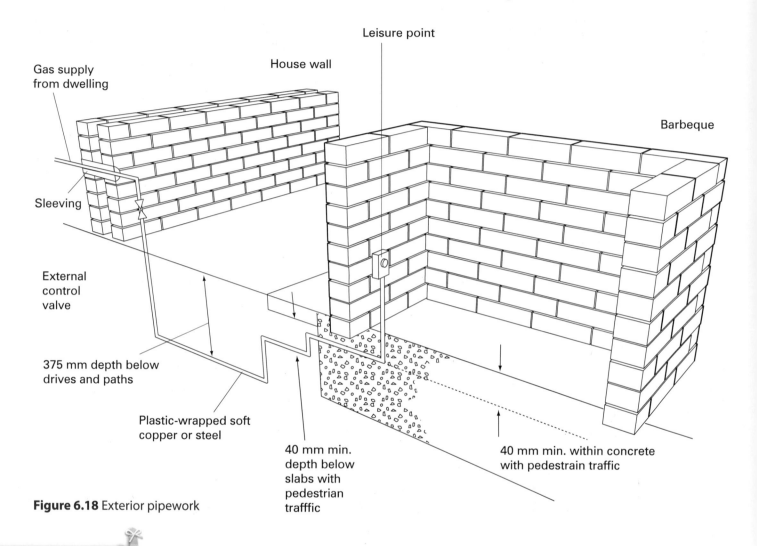

Figure 6.18 Exterior pipework

Main equipotential bonding (cross-bonding)

- All domestic gas installations shall have main bonding of the pipework on the customer's side of the supply as *near as possible and in any case within 600 mm of the meter and before any branch is taken off*.
- The bond should be mechanically sound, visually observed, and labelled correctly: 'Safety electrical connection, DO NOT remove'.

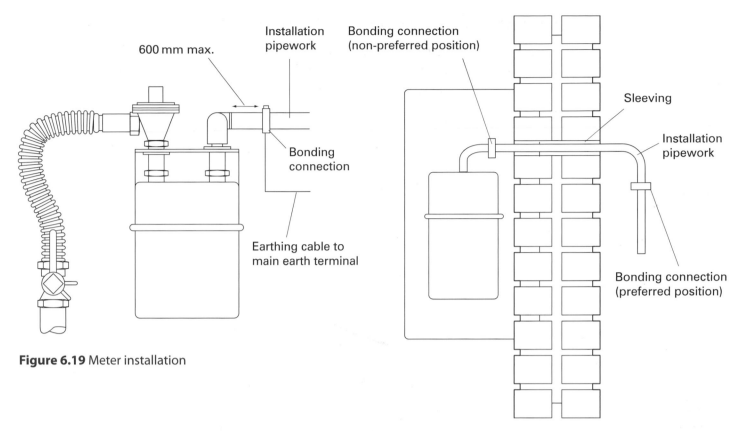

Figure 6.19 Meter installation

Internal gas meter

External gas meter/meter box

Domestic gas pipe sizing

Before we begin to install pipework for gas systems we need to consider the correct size to be used. Factors that affect the size are:

- gas rate of appliance
- length of pipe run
- permissible pressure drop.

We should also remember that the use of elbows and kinks or tube-cutter burrs in the pipe will have an effect on the amount of gas reaching the appliance.

It is recommended practice to allow only a **maximum pressure loss of 1 mbar** between the meter outlet and the appliance connection point.

This can only be tested when the installation is complete, so you can see that it is essential to make sure the pipe size is correct BEFORE we start installing. If the pressure loss was greater than 1 mbar, we would have to start installing all over again with a bigger pipe size.

At first, the sizing of pipe for gas installations looks complicated, but it is really not difficult. The secret is to do it in easy stages and to use the simple method that we are going to look at here. You can follow the example and change the system layout/lengths/appliances as you wish to size the pipework for any job that you will be working on.

Sizing gas supplies

The size of the pipe selected should be of sufficient diameter to supply all the appliances on the installation when they are used at the maximum gas rate.

Let us take a very simple example of a boiler that requires a gas rate of 2.5 m³/hr. The pipe from the meter to the boiler is 8 metres long, with four elbows (let's call it section A–B).

The elbows are converted into an 'equivalent straight length' in order to do the calculation. The first example shows that four elbows at 0.5 metres each becomes 2 metres, added to the pipe length of 8 metres, which equals 10 metres. Bends offer less resistance than elbows, so bends are calculated at 0.3 metres each.

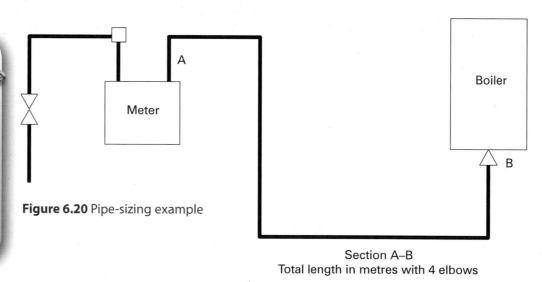

Figure 6.20 Pipe-sizing example

Section A–B
Total length in metres with 4 elbows

Use Table 6.3 to produce your own sizing results, as follows. It can be seen that a 15 mm pipe will carry only 1.3 m³ of gas, so therefore a 22 mm pipe is required.

Size of tube mm	Length of pipe (add 0.5 m for each elbow or tee and 0.3 m for each bend)							
	3 m	6 m	9 m	12 m	15 m	20 m	25 m	30 m
	Discharge in cubic metres per hour (m³/h)							
10	0.86	0.57	0.50	0.37	0.30	0.22	0.18	0.15
12	1.5	1.0	0.85	0.82	0.69	0.52	0.41	0.34
15	2.9	1.9	1.5	1.3	1.1	0.95	0.92	0.88
22	8.7	5.8	4.6	3.9	3.4	2.9	2.5	2.3
28	18.0	12.0	9.4	8.0	7.0	5.9	5.2	4.7

Table 6.3 Discharge of gas in m³/hour with only a 1.0 mbar drop between the ends (from meter to appliance)

Table 6.4 summarises the results of our calculation.

Pipe section	Gas rate m³/hour	Pipe length	Equivalent length		Pipe diameter
			Type	Total length	
A–B	2.5	8 m	4 elbows @ 0.5 = 2.0 m	10 m	22 mm

Table 6.4 Summary of sizing results

Figure 6.21 gives an example of a typical copper-tube install-ation, showing the lengths of pipes and the gas rates of the appliances. The pipes have been sized using the table of discharges, and the results are shown in Table 6.5.

General appliance types	Typical gas rate m³/hour
Warm-air heater	1.0
Multipoint water heater	2.5
Cooker	1.0
Gas fire	0.5
Central-heating boiler	1.5
Combination boiler	2.5

Table 6.5 Gas rates of appliances

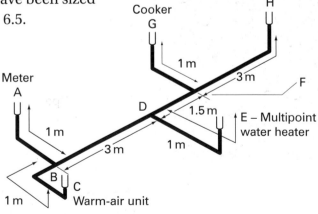

Figure 6.21 Pipe sizing to four appliances

When sizing pipes, it is essential that consideration be given to the permissible pressure loss in each section of the installation. For example, the pressure loss between A and H in the drawing should not exceed 1 mbar. A to H is made up of four separate sections of pipe:

- section A–B
- section B–D
- section D–F
- section F–H.

Each section carries a different gas rate and needs to be sized separately.

Remember that we said that A–H is to have a pressure loss of not more than 1 mbar, so then the pressure losses in **each of the four sections** should be approximately 0.25 mbar.

So A–B, B–D, D–F and F–H should each be sized to give a pressure loss of approximately 0.25 mbar.

The discharge in a straight horizontal pipe given in the table of discharge only allows for pressure losses of 1 mbar. However, pressure loss is proportional to length, so if the pipe size selected from the table of discharges is then calculated at **four times longer** than required, the pressure loss on the actual length will be 0.25 mbar (four times 0.25 equals 1.0 mbar).

Example

Considering length D–F as given in Figure 6.21: D–F has a length of 1.5 m and is to carry a gas rate of 1.5 m³/hr. This is made up of 1.0 for the cooker and 0.5 for the gas fire; it should have a pressure loss of 0.25 mbar maximum. However, a pressure loss of 0.25 mbar in a length of 1.5 m equals:

$(4 \times 0.25) = 1$ mbar in $(4 \times 1.5\,m) = 6\,m$

In the discharge rates table, look up the column under 6 m for a required discharge of 1.5 m³/hr and find:

12 mm = 1.0 m³/hr

15 mm = 1.9 m³/hr

The first size, 12 mm, would give a lower flow rate than is required. The larger size, 15 mm, would carry the 1.5 m³/hr of gas with little pressure loss and could allow for appliances to be added to the installation at a later date, if required. This is the size to be used.

Also remember that a change of direction caused by a tee is similar to an elbow and an equivalent length of pipe of 0.5 m must be added.

Table 6.6 shows the results of our calculations for all the sections of pipe. Check through it yourself to understand how it's done.

Pipe section	Gas rate m³/hour	Pipe length	Equivalent length		Pipe diameter
			Type	Total length	
A–B	9	1	Elbow @ 0.5 m Tee @ 0.5 m	2 m	28
B–C	1	1	Two elbows @ 0.5 m	2 m	12
B–D	8	3	–	3 m	28
D–E	2.5	1	Tee @ 0.5 m Elbow @ 0.5 m	2 m	22
D–F	1.5	1.5	–	1.5 m	15
F–G	1	1	Tee @ 0.5 m Elbow @ 0.5 m	2 m	12
F–H	0.5	0.5	Elbow @ 0.5 m	1 m	12

Table 6.6 Sizing results

If you don't understand it at first then talk it through with your lecturer or a colleague and, when ready, try some calculations for yourself.

Remember

The same calculations can be used for sizing supplies for liquefied petroleum gas (LPG), as this requires smaller diameter pipes and operates at a higher pressure. The property could be converted to natural gas in the future and no problems would be found with the pipe sizes

FAQ

Will I ever remember all these calculations and measurements?

You may not always remember them, but you will be aware of them and will know how and where to find the information when you need it.

Pressure testing domestic systems

At the end of this section you should be able to:

- know when a test is required
- recognise various test instruments
- carry out a 'let-by' test
- carry out tightness testing on new/existing work
- understand regulator 'lock up'.

A very important part of an operative's work on gas installations is the carrying out of a tightness test (this used to be called a soundness test).

A test for gas tightness needs to be carried out:

- whenever a smell of gas is suspected or reported
- on newly installed pipework
- whenever work is carried out on a gas fitting that might affect its gas tightness (including pipework, appliances, meters and connections)
- before restoring the gas supply, after work on an installation
- prior to the fitting of a gas meter on new or existing pipework installation
- on the original installation prior to connecting any extension.

The procedure is very precise, and a suitable pressure gauge with flexible rubber tubing is required, together with a stopwatch/timing device.

Pressure-test gauges

Pressure gauges may be of two types: water or electronic.

Water gauge

- Readings are taken in millibars (mbar).
- Capable of being read to an accuracy of 0.5 mbar.

Gauges are available in a range of sizes, although the most common is the 300 mm, as used on natural gas. Larger gauges of up to 1 m, for example, are used for LPG testing where the pressures are higher. Before use, ensure the gauge is vertical; water levels should be set at zero.

Electronic gauge

- Capable of being read to an accuracy of 0.3 mbar.

Electronic gauges should be calibrated and certificated annually.

As it tends to be more common we'll focus on the use of the 'U' gauge (manometer) here. They are available in several sizes and also in metric or imperial forms; some even have a dual scale showing inches and mbar on the same side but either above/ below the zero line, as shown in Figure 6.22.

A 300 mm gauge will measure up to 30 mbar. One leg of the gauge is connected to the installation and the other is open to the atmosphere after first being set to zero.

Although the gauge should be set at zero to begin, care should be taken that the readings on each leg are the same as in the drawing, e.g. 0 mbar and 0 mbar, 2 mbar and 2 mbar.

If the readings were different, e.g. 8 mbar and 12 mbar, then the correct pressure would be 10 mbar (this is found by adding both legs and dividing by 2, e.g. 8 plus 12 is 20; therefore 20 divided by 2 is 10 mbar).

Note that readings should always be taken from the lowest point of the 'meniscus' marked by a black band in Figure 6.22.

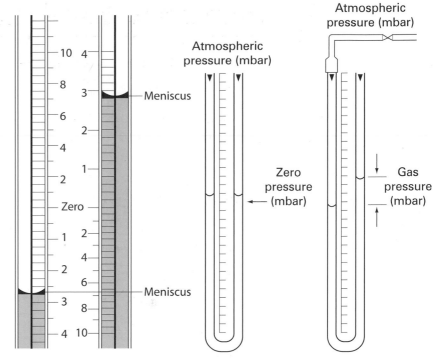

Figure 6.22 Manometer   **Figure 6.23** Pressure readings

On the job: Carrying out a tightness test

Angus is working on an existing installation and is asked to carry out a tightness test BEFORE he commences work. This of course is the correct procedure. On testing, though, there is a 3 mbar pressure drop within the two minutes.

1   Is it acceptable within the Regulations for Angus to commence work?

2   Should he tell his supervisor or take a decision himself about his course of action?

3   What would you do in these circumstances?

There may be occasions while testing all or part of a new installation when you have to use air introduced into the system via a test tee and suitable pump, as shown in Figure 6.24.

Before testing for tightness it is important to identify the type and size of meter that is installed. This will determine the permissible pressure loss allowed on an existing installation.

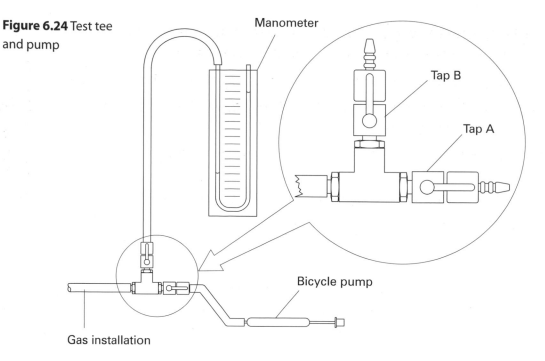

Figure 6.24 Test tee and pump

Manometer

Tap B

Tap A

Bicycle pump

Gas installation

Types of meter

For an existing installation with a U6/G4 meter fitted a loss of up to 4 mbar over a two-minute test is considered acceptable, provided there is no smell of gas.

Figure 6.25 Types of meter

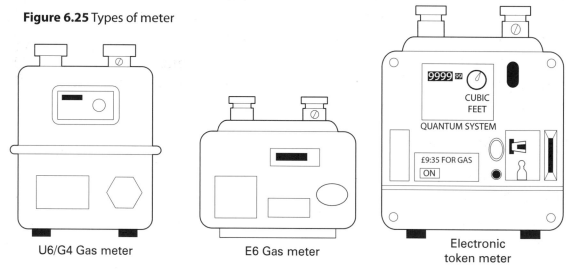

U6/G4 Gas meter

E6 Gas meter

Electronic token meter

For an existing installation with an E6 meter fitted, a loss of up to 8 mbar is considered acceptable, provided there is no smell of gas, and the escape is not in the pipework.

It is very important to remember that the Gas Safety (Installation and Use) Regulations require that joints should be tested BEFORE being painted or otherwise protected against corrosion. It is also recommended that pipework buried in plaster, underground etc. is tested BEFORE being covered.

When a new gas installation is completed it shall be tested according to the following procedure. Where the supply is connected to a meter and gas is available it is recommended that gas is used as a test medium.

- Check the completed installation and ensure all sections to be tested are connected.

- Make a visual inspection of all pipework and joints to ensure they have been correctly made. Ensure all open ends have been suitably sealed.

- Make sure that any appliances fitted are isolated, i.e. turned off at the isolation valve.

It is important that *pipework only* is tested as there is *no allowable drop*. This is whether the pipework is new or existing.

Connect a correctly zeroed pressure gauge to the system at the test point on the meter.

Remember that you are required to apply LDF on pipework and components between the ECV and governor.

'Let by' test

This is to ensure that the main emergency control is not 'letting by' even a very small amount of gas, as this would give a false reading to our actual tightness test. It could be that a leak on the installation is not detected because the 'let by' keeps the gauge up to pressure.

For installations where the gas is connected, slowly raise the pressure with the control valve to approximately 10 mbar and isolate the supply.

If the pressure rises by more than 0.5 mbar over the next one-minute period this will indicate that the meter-control valve is passing gas in the closed position. This can be confirmed by using leak-detection fluid on an open end of the installation if required.

If the valve is 'letting by' then the operative must make safe and contact the gas emergency service provider immediately – the installation must not be put into use until the fault has been rectified.

Manometer at 10 mbar

Pipework only test

All the appliances are off at their isolating valve, the cooker bayonet is disconnected etc.

- If the let-by test is satisfactory, now slowly raise the pressure to 20 mbar and turn off the supply.

- Do not raise to pressures higher than 20 mbar as this may cause the meter regulator to lock up: this means that the regulator is closed and can cause incorrect test results.

- Allow one minute for temperature stabilisation.

- Record any pressure loss in the next two minutes.

- If there is no *perceptible* pressure loss (i.e. below 0.5 mbar for a water gauge and 0.3 for a digital gauge) and there is no smell of gas, then the installation will have passed the test.

- If the test has failed then the leak must be found or the installation sealed and made safe.

Manometer at 20 mbar

Complete installation

Once it has been established that the pipework is not leaking then we must turn on the isolating valves to all appliances and test again to ensure that the complete system is safe.

- If the pipework-only test was satisfactory and all appliance-isolating valves are now open, slowly raise the pressure to 20 mbar and turn off the supply.

- Again, do not raise to higher pressures than 20 mbar as this may cause the meter regulator to lock up.

- Allow one minute for temperature stabilisation.

- Record any pressure loss in the next two minutes.

With new installations there is *no* allowable pressure drop for new appliances.

For existing installations:

- Installation with a U6/G4 meter – loss of up to 4 mbar acceptable, provided there is no smell of gas.

- Installation with an E6 meter – loss of up to 8 mbar acceptable, provided there is no smell of gas.

- Where no meter is fitted in the dwelling, such as a flat supplied by a communal meter, then up to 8 mbar loss is allowed.

If the test has failed then the leak must be found or the installation sealed and made safe.

It is worth mentioning here that many cookers have a fold-down lid which closes the gas off to the burners and the control taps. If the cooker lid was DOWN during the test then we would not have tested the system properly: there could be a leak on the cooker that has not been detected.

This is an example of how the ACS test can be failed; each part of the test is critical as we have to be certain that the installation is safe to use.

Medium pressure regulator/release mechanism

Medium pressure-fed installations

There may be occasions when operatives will come across a medium pressure-fed installation. The configuration for gas control and metering connections is shown in Figure 6.26. If the installation has a meter test valve fitted then the test procedure is exactly the same as that already covered, but if there is no test valve included then we need to adopt a different procedure.

The procedure for carrying out a tightness test differs a little from a normal low-pressure test as we need to check both the ECV (emergency control valve) and the medium- to low-pressure regulator. This is best done in three stages as follows:

Stage 1: Test ECV

- Turn off gas at the ECV and connect gauge.

- Carry out let by test on ECV.

- Release pressure from installation and *hold open* release mechanism lever on side of regulator.

- No more than 0.5 mbar rise should take place over the next one minute; if more than 0.5 mbar then ECV is letting by: contact emergency service provider.

Stage 2: Test regulator

- Allow release mechanism lever on regulator to return to off position.

- Open ECV.

- No more than 0.5 mbar rise should take place over the next one minute; if more than 0.5 mbar then the regulator is letting by: contact the emergency service provider.

Stage 3: Tightness test

- Following a successful let by test on both the ECV and the regulator, release all pressure from installation and switch off.

- Very slowly raise pressure to 19 mbar (no higher to prevent regulator lock-up).

- Allow one minute for stabilisation.

- Check pressure loss over next two minutes.

- The following applies:

  - pipework only: no drop allowed

  - new appliances: no drop allowed

  - existing appliances fitted:

    (i) U6/G4 meter: 4 mbar drop with no smell of gas

    (ii) E6 meter: 8 mbar drop with no smell of gas.

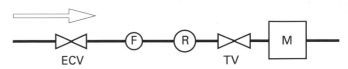

Upstream   Downstream

ECV

Low pressure-fed gas meter installation

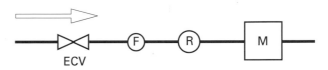

ECV   TV

Medium pressure-fed meter installation with a test valve fitted

ECV

Medium pressure-fed meter installation without a test valve fitted

ECV – Emergency control valve
R   – Regulator
M   – Meter
F   – Filter
TV   – Test valve

Figure 6.26 Low- and medium-pressure fed installations

Making connections to existing systems

Before connecting to an existing system you need to establish that the installation is properly gas tight, so undertake a tightness test prior to making the final connection:

(a) Establish that the system is gas tight to within permissible pressure-loss tolerances. If there are any pressure losses (assuming they are within these tolerances) then record the detail. If the test proves that there is an unacceptable pressure loss then it is usual practice to liaise with the customer and rectify the leak source prior to making the new connection.

(b) On completion of the new connection the system needs to be re-tested.

(c) No pressure loss in the system should exceed acceptable levels.

(d) There should be no smell of gas.

Tracing gas leaks

In the event that you establish that a gas leak is present or the customer reports a smell of gas, the usual first stage is to complete a tightness test. This will usually confirm that there is some form of leakage. The next step is to identify where the leak is coming from, which can be quite a process. Usual methods are:

- detection using a portable electronic gas detector to trace the source of the leakage – key checking points are control taps, appliance connection points, fittings and unprotected pipework in contact with corrosive materials

- testing the pipework system in smaller sections using leak-detection fluid and key potential weep points (as listed above) using a process of elimination until the leak source is identified.

FAQ

All this testing takes so long. Can't I just test at the end of the job?

The testing takes only a few minutes before you start the work and, if you have no leaks, a few minutes at the end of the job. If you don't test at the start, it could mean several extra hours' work at the end if there is an existing leak in the system.

Purging and re-establishing gas supplies

At the end of this section you should be able to:

- understand explosive mixtures
- recognise various meter types
- purge systems
- re-establish supplies.

Now that we have correctly sized, installed and tested the gas pipework it is time to put gas into the system. All the air in the pipework has to be pushed out and replaced with gas. This process is called **purging**.

While we do this some gas and air mixture is let into the building. So before we start this section we need to understand the real dangers of purging. General science principles have been covered at Level 2 but we now need to appreciate the problems of gas and air mixtures.

The problem with purging is that we may be allowing a quantity of gas to accumulate in a room, and there is a very real danger that explosive mixtures can be created. You will know now from your experience with blowlamps that gas will only burn if it is mixed with air; the percentage of gas to air required depends on the type of gas. The percentage of gas to air for natural gas is 5–15 per cent. That is to say that when there is approximately 5–15 per cent of gas mixed with air then we can ignite the mixture and allow it to burn.

For LPG it is 2–10 per cent for propane and 2–9 per cent for butane. It is at these levels that gas will burn if it is ignited. If the gas is not ignited and burned, however, then the build-up of gas is likely to cause an explosion if it is ignited at these levels.

The lowest figure in each case is referred to as the **lower flammability level** (LFL); the higher figure is referred to as the **upper flammability level** (UFL).

It is important not to allow gas to build up in any confined space as an ignition source could cause an explosion. It is essential to have good ventilation: open doors and windows, prohibit any smoking or naked lights and warn people close by of the danger.

So how do we purge? Well, we simply disconnect and open the end of the line and turn on the gas until it fills all of the pipe. Remember we need to get all of the air out. There is a minimum volume that must be put through the meter to comply with BS 6891, and this states the volume shall be 'not less than five times the badge capacity per revolution of the meter mechanism'.

Meters and dials

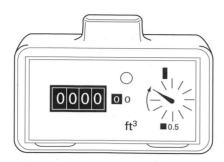
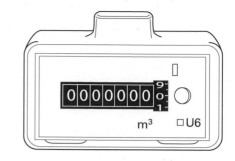

Figure 6.27 Meters and dials

The internal workings of a diaphragm meter also show a considerable content which needs to be purged – as shown in Figure 6.28.

1. Front outlet chamber is full while the front inner chamber is empty and the valve, moving to the right, has reached the mid-point. Rear inner chamber is half full and filling while the outer chamber is half empty and emptying. The control valve has moved fully to the left and will start its return motion.

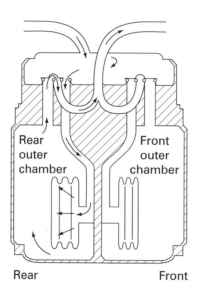

2. Front outer chamber is half empty and emptying while the front inner chamber is half full and filling. The valve has moved fully to the right and will start to return. Rear inner chamber is full while the outer chamber is empty. The valve is moving to the right and has reached mid-point.

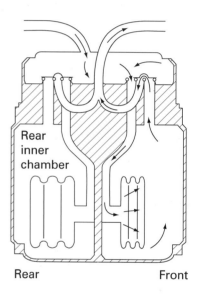

3. Front outer chamber is empty with the inner chamber full. The valve is moving to the left and has reached mid-point. Rear inner chamber is half empty and emptying while the outer chamber is half full and filling. The valve has moved fully to the right and will start to return.

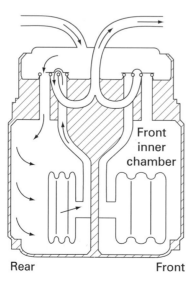

4. Front chamber is half full and filling while the inner chamber is half empty and emptying. The valve has moved fully to the left and will start to return. Rear inner chamber is empty while the outer chamber is full. The valve is moving to the left and has reached mid-point.

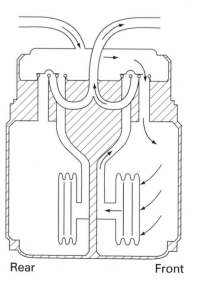

Figure 6.28 Internal workings of diaphragm meter

The data badge on the meter will usually show the meter capacity. The typical badge capacity given for an imperial U6 meter is $0.071ft^3$.

Since we must purge five times the badge capacity, we must calculate this as follows:

$5 \times 0.07 = 0.35ft^3$

Therefore we must purge $0.35ft^3$ as a minimum.

E6 meters (electronic) shall be purged by passing at least $0.010m^3$ of gas.

It is normal in domestic installations for the purging to be done into a well-ventilated building from an appliance or a disconnected appliance at the furthest point from the meter.

Please note that there may be quite a substantial amount of gas removed from the system to ensure that all air is eliminated before we attempt to light appliances. All installation pipes need to be purged. Pay particular attention to ensuring adequate ventilation, no sparks from electric switches etc.

For larger installations more information should be obtained from the Institution of Gas Engineers publication IGE/UP/1/(7).

Remember

Always check the reconnection with leak-detection fluid on completion of work

Re-establishing gas supplies

We now come to another important stage in the work. We have installed, tested and purged each part of the system. We must now put appliances into use, and once again it is essential that the correct procedure is followed. It may be that we are working on an existing installation carried out by someone else, perhaps a long time ago. The rules are the same: we must make several critical checks *before* we leave the installation as safe for use.

Note that there are separate sections dealing with individual appliances. This section only gives an overview of putting an installation into use:

- Make a visual check of the appliances.
- Check there is ventilation available where required.
- Light each appliance in turn, checking that all user controls are working.
- Check all open-flued appliances for spillage.

The checks are carried out to ensure that there is no danger to the user. Any appliance must either:

- be commissioned fully, or
- be disconnected from the installation.

If an appliance proves not to be operating satisfactorily and there is a risk or danger to the user, action should be taken to shut off or disconnect.

Checking pressures and gas rates

At the end of this section you should be able to:

- check meter regulators
- check and set appliance pressures
- check gas rates.

This topic is best tackled by separating it into three distinct areas:

1 Checking meter regulators.

2 Checking/setting appliance pressures.

3 Checking gas rates.

Regulator and seal

1. Checking meter regulators

Natural gas is transported across country at high pressures in excess of 7 bar. It is then reduced locally and distributed to domestic premises at pressures not exceeding 2 bar.

The normal working pressure of a domestic installation is only 20 mbar. It is important that the pressure is kept constant, irrespective of the fluctuating demands on the gas supply. In order to keep this constant pressure in installations of 20 mbar, a pressure regulator is fitted to the gas meter inlet connection, as shown in Figure 6.29 and the photograph.

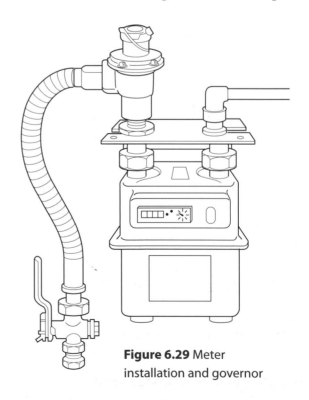

Figure 6.29 Meter installation and governor

Most domestic gas appliances use atmospheric burners, like a Bunsen burner, which deliver gas from an injector mixed with air in the mixing tube to be ignited and burned at the burner head. To make sure we have satisfactory combustion (a good flame) it is essential for the pressure to be kept constant. It is also important to have enough air.

The pressure at the inlet of any appliance is required to be 20 mbar in order to 'entrain' the right amount of primary air for correct combustion. Some appliances, however, are designed to operate at different pressures: boilers, for example, may be able to operate at a range of pressures. These appliances are fitted with an individual appliance regulator which is sometimes built into the **multi-functional control valve**.

When you put an appliance into use you will have to check and adjust the appliance regulator. It is the responsibility of the gas operative to ensure that the meter regulator is set correctly, but only the gas supplier is allowed to adjust it if it is wrong.

So how do we check it? We need to connect a 'U gauge' to the meter test point and put any *one* appliance on full, or the cooker on with three burners on full. Do not make the mistake of thinking that all appliances need to be on: only *one* is required to check the working pressure.

The pressure recorded should be **21 mbar plus or minus 1 mbar**. That is to say, if the working pressure is 20 mbar to 22 mbar, then the meter regulator is working satisfactorily. If the recorded pressure is outside these limits, switch off the supply and contact the Gas Supplier on 0800 002001 (England and Wales).

> ### Remember
>
> Standing pressure is the term given to an installation with *no* appliances working. Working pressure is the pressure with *one* appliance working on full. However, somewhat strangely perhaps, a cooker must have only three burners on

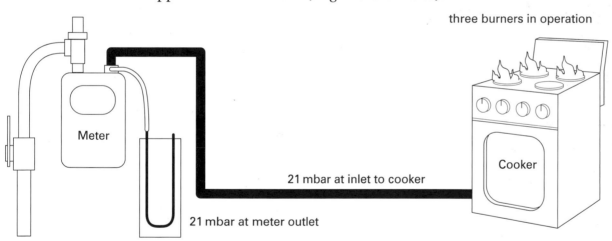

three burners in operation

21 mbar at inlet to cooker

Meter

Cooker

21 mbar at meter outlet

Figure 6.30
Checking working pressure

As an additional check it is now worth trying the standing pressure. All you need to do is turn all appliances off and record the pressure. It is likely to be approximately 21 mbar to 25 mbar – if it is vastly different to these figures then attention is likely to be required by the gas supplier. Never break the regulator seal yourself.

2. Checking/setting appliance pressures

Now that we have confirmed that the meter regulator is working correctly, we need to check and adjust appliance pressures. The correct burner pressure for each appliance is given:

- on the data plate (fitted to all appliances)
- in the manufacturer's instructions (always left with the householder).

If this information is not available then it is not possible to carry out the work and replacement instructions are required.

You need to look at the instructions with an appliance to determine where to test for burner pressures – before and after the regulator or on the multifunctional valve. The pressures to any appliance (before the regulator) should never be more than 1 mbar below the correct working pressure at the meter regulator. Remember that this will usually be between 20 and 22 mbar.

If it is less than this then the pipework may be undersized/partially blocked or kinked, the installation does not meet the Regulations and corrective work is necessary or the installation must be shut down and made safe.

Once we have determined that the inlet pressure is satisfactory then we must connect the gauge to the outlet side in order to read the actual burner pressure. Attaching the gauge must be done with the switch in the off position, as gas will escape from the test point. Now turn on the appliance for 10 minutes and check the actual burner pressure as stated in the manufacturer's instructions.

It may be necessary to adjust the manufacturer's burner pressure adjuster on a range-rated appliance (as shown in Figure 6.31) to give the stated pressure and/or adjust the pressure within the range given – for example, for a boiler of varying outputs.

It is crucial to abide by the manufacturer's instructions in all cases, as there are many varieties of appliance and all should be checked and set exactly as intended.

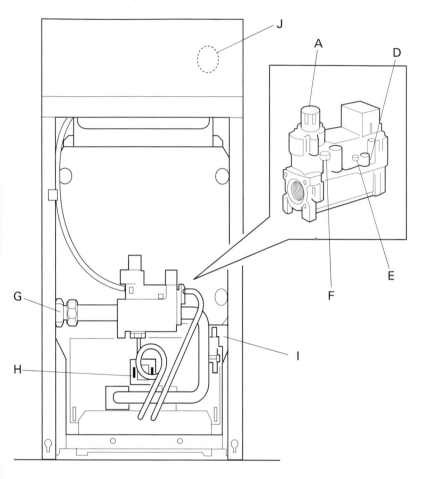

A – Gas control knob
D – Burner pressure test nipple
E – Main burner pressure adjuster
F – Inlet pressure test nipple
G – Gas service cock

H – Sightglass
I – Piezo ignition button
J – Boiler thermostat knob

Figure 6.31 Burner pressure adjustment/controls to gas boiler

3. Checking gas rates

Even when we have checked the meter regulator working pressure and the actual burner pressure, there is still one important check remaining: the gas rate. This is the checking of the actual amount (volume) of gas being burned by an appliance. The test involves checking the time it takes to pass a known volume of gas through a meter to the appliance. The result can then be compared to the appliance's data plate.

It can be seen that burner pressure alone is not a good enough check, as the injectors could be partially blocked, giving a less than satisfactory output or, even worse, the injectors could be wrongly oversized, giving a higher gas rate than the appliance is designed for. This is therefore an important check.

It is also important to consider meter types. The most common is the U6, which still measures gas in cubic feet. This is gradually being replaced by the G4 or E6 electronic meters, both of which measure gas in cubic metres. Measuring the gas rate with these is very simple, but we also need to be able to use U6 meters. To calculate the gas rate we need to carry out a calculation procedure.

Imperial U6 meter

We need to know the Calorific Value (CV) of the gas. This is usually:

- natural gas average CV is 38.76 MJ/m³ Gross or 1040 btu/ft³
- propane average CV is 93.1 MJ/m³ Gross or 2496 btu/ft³
- butane average CV is 121.8 MJ/m³ Gross or 3265 btu/ft³.

The formula for checking gas rate is:

$$\frac{\text{Seconds in one hour (3600)} \times \text{the CV of the gas (1040 btu/ft}^3)}{\text{Number of seconds for one revolution of the dial}}$$

Example

The time taken for one revolution of the test dial is 76 seconds:

$$\frac{3600 \times 1040}{76 \text{ seconds}} = 49{,}263 \text{ btu/hr}$$

Then convert btus to kW by dividing by a constant of 3412:

$$\frac{49{,}263}{3412} = 14.44 \text{ kW gross}$$

Try another boiler rated at 11.5 kW, which has been timed for one complete revolution for 62 seconds:

$$\frac{3600 \text{ multiply by } 1040}{62 \text{ seconds}}$$

$$= \frac{60{,}387}{3412} \text{ btu/hr} = 17.7 \text{ kW}$$

Clearly this amount is well over the stated gas rate of 11.5 kW and therefore indicates something seriously wrong with the injectors. To show the importance of gas rates it must be emphasised that the burner pressure alone would not have detected this problem.

Did you know?

A much easier way, of course, is simply to refer to a gas rating chart. These are available in a variety of publications, including the CORGI Manual

Metric G4/E6 meter

The calculation procedure for establishing the gas rate with these types of meters is slightly different owing to the fact that they do not have a test dial. Here we determine the amount of gas burned across a fixed test period (usually two minutes) by identifying the quantity of gas used in m^3 during that period. Let's have a look at the calculation:

$$\frac{3600 \times m^3 \times \text{CV of gas in MJ}/m^3/3.6}{\text{time in seconds (120)}}$$

This can be simplified further, as the CV of natural gas is usually fixed at:

$$38.76 \div 3.6 = 10.76$$

So this can usually be used as a constant figure.

Example

To work out the gas rate of a boiler, take the amount of gas measured at the meter over a two-minute period:

- first meter reading = $45324.010m^3$

- second meter reading = $45324.052m^3$ (which is 0.042 higher).

$$\frac{3600 \times 0.042 \times 10.76}{120} = 13.56\text{kW}$$

This can then be checked against the data plate on the appliance.

Just one quick additional point. You have calculated what is called the gross rating of the appliance. Some data plates quote these figures as a net figure, so in this case you would divide your final figure (13.56) by a constant figure of 1.1 for natural gas to give a net figure of 12.22 kW.

The constant figures for LPG fuels are:

- propane – 1.09

- butane – 1.08.

Combustion

At the end of this section you should be able to:

- describe the combustion process
- state the characteristics of combustion
- recognise types of burner
- understand complete/incomplete combustion
- recognise symptoms of CO poisoning
- identify signs of incomplete combustion.

When we light burners it is important to know about the principles of combustion and the reasons why the correct mixture of gas and air is needed. This correct mixture is called **complete combustion**, and this is essential for gas safety.

Combustion is a chemical reaction that needs *three* elements:

Fuel + Oxygen + Ignition

The reaction of this process causes heat and products of combustion (POC). Gases such as natural gas (methane) and LPG (propane and butane) are carbon-based gases, and if the combustion process is not correct then **carbon monoxide (CO)** can be produced. *This is a killer!* CO is a highly toxic gas.

So it is very important that the correct amount of oxygen is mixed with the gas to ensure complete combustion. There is only about 20 per cent of oxygen available in air, which explains why air vents need to be so big. The other 80 per cent of air is mainly nitrogen, which plays no important part in the process; it just adds to the bulk of products going out of the flue.

This ideal mixture of gas and air is sometimes referred to as the **stoichiometric mixture**.

Figure 6.32 demonstrates the correct combustion process, including the products of combustion.

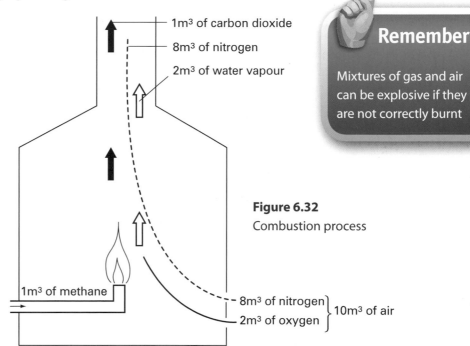

Figure 6.32
Combustion process

> **Remember**
>
> Mixtures of gas and air can be explosive if they are not correctly burnt

Properties of gases

Let's take a look at some of the key properties of the gases that we use as fuels as detailed in Table 6.7.

Characteristic	Natural gas*	Propane	Butane	Notes
Specific gravity (SG of air = 1.0)	0.6	1.5	2.0	Methane will rise but propane and butane will fall to low level
Calorific value	39 MJ/m³	93 MJ/m³	122 MJ/m³	Appliances are designed to burn a particular gas
Stoichiometric air requirements	10:1	24:1	30:1	Methane requires 10 volumes of air to 1 volume of gas LPG requires more
Supply pressure	21 mbar	37 mbar	28 mbar	Appliances must be matched to the gas used
Flammability limits	5 to 15% in air	2 to 10% in air	2 to 9% in air	Ranges within which gas/air mixtures will burn
Flame speed	0.36 m/sec	0.46 m/sec	0.45 m/sec	This is the speed at which a flame will burn along a gas mixture
Ignition temperature	704°C	530°C	408°C	Approximate temperatures
Flame temperatures	1930°C	1980°C	1996°C	Approximate temperatures

* Methane

Table 6.7 Key properties of gases

Specific gravity

When we compare the weight of natural gas (NG) to air, which has a specific gravity (SG) of 1.0, we find that NG is 0.6, just over half the weight of air. Therefore natural gas will rise. Propane has an SG of 1.5, is heavier and will fall to low level. This will have an effect on where we look for leaks.

Calorific value

This is the amount of heat given when a unit quantity of fuel is burnt. It is the amount of energy released and is expressed as Mega Joules per cubic metre – MJ/m³.

Stoichiometric air requirements

This is the amount of air (in cubic metres) required for complete combustion of one cubic metre of gas.

Flammability limits

These are the limits at which gas and air will burn. Too much or too little gas/air will not burn. There is a small range within which natural gas (5 to 15 per cent of gas in air) will burn. Remember from the purging section: if it is not burned it will also become an explosive mix.

Flame speed

To keep flames stable on a burner, the pressure of the gas being supplied and the injector size must be correct. Gas supplied too fast would cause **flame lift**, while too slow may cause **light back** into the burner tube. These are described a little later.

Figure 6.33 Warning! Explosion risk!

Some other terms used

Families of gases

The three main families of gases are:

 (i) Family 1: Manufactured gases

 (ii) Family 2: Natural gas

(iii) Family 3: Liquid petroleum gases.

Wobbe number

An indication of the heat produced from a burner for a particular gas, the **Wobbe number**, is found by dividing the CV by the square root of the SG:

$$\frac{CV}{\sqrt{SG}} = \text{Wobbe number}$$

Burners

There is a wide variety of burners used for domestic gas appliances, but most work on the principle of the 'pre-aerated' flame, as shown in Figure 6.34. This is identical to a Bunsen burner, with the primary air port open.

Burners can be further classed as:

- atmospheric – natural draught domestic burners

- forced draught – mainly used for commercial burners, but a growing number of domestic burners operate on this principle, using a zero-rated governor.

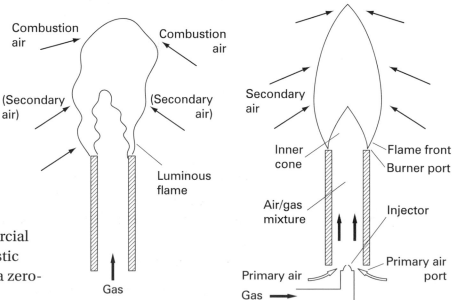

Figure 6.34 Post-aerated flame and pre-aerated flame

Item	Component
A	Injector
B	Primary air port
C	Aeration control shutter
D	Throat restrictor
F	Venturi throat
G	Burner port
H	Retention port

Table 6.8 Main components of atmospheric burner in Figure 6.35

We are only going to deal with atmospheric burners here, but you should be aware of the main components of the forced draught burner and how these help to make a steady, stable flame.

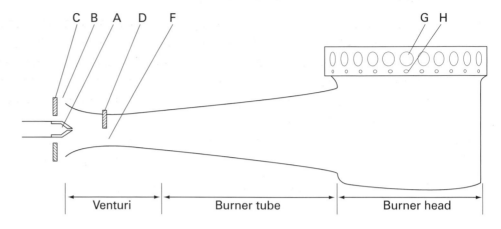

Figure 6.35 Main parts of atmospheric burner

Common problems that exist are lighting back and flame lift.

Lighting back

This is caused when the speed of the gas/air mixture is reduced (such as with too low a pressure). The flame will burn its way back down the burner tube to the injector, sometimes called 'striking back'.

Flame lift

This is the opposite of lighting back. If the speed of the mixture is too great then the flame is pushed away from the burner ports. You will notice that the drawing of the gas burner in Figure 6.35 has **retention ports**: these help prevent lift-off.

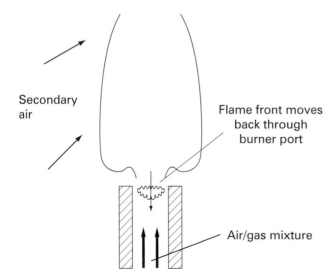

Figure 6.36 Lighting back

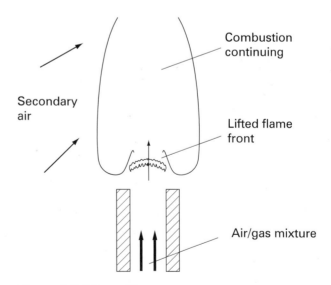

Figure 6.37 Flame lift

Incomplete combustion

If combustion is not correct then we have what is called 'incomplete combustion' and products can be released from the flame.

The most dangerous product of incomplete combustion is the toxic gas carbon monoxide. A concentration of only 0.04 per cent in air can be fatal within a few minutes. Carbon monoxide combines much more easily with the haemoglobin in the blood stream than oxygen, which is what the blood really needs. The blood normally picks up oxygen from the lungs and carries it to the tissues of the body. If the haemoglobin (red cells) become saturated with carbon monoxide then the tissues cannot take oxygen in. The body is quickly poisoned and the person dies. Even if death does not occur there may be serious brain damage due to the lack of oxygen.

Table 6.9 shows the effects of carbon monoxide on adults, with the saturation of haemoglobin shown as a percentage.

% CO	Symptom
0 to 10%	No obvious symptoms
10 to 20%	Tightness across the forehead, yawning
20 to 30%	Flushed skin, headache, breathlessness and palpitation on exertion, slight dizziness
30 to 40%	Severe headache, dizziness, nausea, weakness of the knees, irritability, impaired judgment (see Note 1), possible collapse
40 to 50%	Symptoms as above with increased respiration and pulse rates, collapse on exertion (see Note 2)
50 to 60%	Loss of consciousness, coma
60 to 70%	Coma, weakened heart and respiration
70% or more	Respiratory failure and death

Table 6.9 Effects of carbon monoxide intake on adults

Note 1: Mental ability is affected so that a person may be confused and on the verge of collapse without realising that anything is wrong.

Note 2: Any sudden exertion would cause immediate collapse, and therefore an inability to escape from the situation.

These two factors explain why CO poisoning is frequently fatal. Incomplete combustion is very dangerous, so let's have a look at some of the causes and visual signs.

Causes of incomplete combustion

- Lack of oxygen: air may be restricted to the burner or there may be a partial blockage of the flue outlet.

- Overgassing: incorrect burner pressure and/or wrong injector size giving more gas than the appliance was designed for.

- Chilling: this occurs when a flame touches a cold surface or is exposed to a cold draught; the flame pattern is disturbed and sooting may occur, causing even more problems.

- Flame impingement: when flames touch each other or touch a cold surface they 'impinge' and this may cause poor combustion, e.g. burner not positioned correctly in the appliance.

- Vitiation: reduced oxygen levels in a room will cause the air to become 'vitiated' (made impure) and will affect combustion.

Visual signs of incomplete combustion

- Yellow flames: burners with insufficient air will burn with a yellow flame (please note that some gas fires are designed to give yellow flames as a live fuel effect).

- Sooting: unburnt carbon (soot) will be seen on the appliance heat exchanger or radiants.

- Staining: this may be seen around the flue or draught diverter, and may also be due to 'spillage' of flue products due to a poor flue.

All of these problems need close investigation by a qualified, CORGI-registered operative, and the customer must be advised *not* to use the appliance as it could prove fatal!

Any appliance showing symptoms of incomplete combustion must not be used, and the correct unsafe situations procedure must be followed – we'll look at this a little later.

Safety tip

It is vital that any gas-related work is carried out to the Regulations and Standards. The role of the operative working on gas is an extremely responsible one

FAQ

If the flame is alight and not yellow, is it safe?

Not necessarily. The flame could be blue, but may be lifting off, for example. If you are not sure, it may help to read the last chapter again.

Ventilation

At the end of this section you will be able to:

- understand requirements for ventilation
- recognise the ventilation needs of different appliances
- calculate vent sizes
- identify correct installations
- state the requirements for compartment ventilation.

In the last section on combustion we saw how important it was to ensure complete combustion to prevent a danger to any user of a gas appliance. This section details the ventilation requirements for any appliance up to 70 kW using 1st, 2nd or 3rd family gases. Details are given in each appliance manufacturer's instructions, and these are to be followed closely. If the manufacturers do not state anything specific, then the guidance available in BS 5440-2-2000 should be followed. We shall look at the main points of ventilation here.

It is essential that anyone carrying out gas work is competent to do so, and this includes checking and working out the ventilation for appliances.

First, let us look at the various appliances and their differing needs for ventilation:

- **Flueless appliances** such as cookers and some water heaters require a constant supply of fresh air to prevent the air in the room from becoming vitiated. The smaller the room, the greater the problem. Remember: if the air becomes vitiated then we get incomplete combustion.

- **Open-flued appliances** need air for combustion, and the movement of products up the flue will cause air to be taken from the room. This needs to be replaced for the appliance to work safely.

- **Appliances in compartments** need air for combustion, and there is a need to cool the compartment down to stop the appliance overheating. Even room-sealed appliances, which take combustion air from outside, need to have compartment ventilation.

Flueless appliances

The Building Regulations also state that all rooms with a flueless appliance must have an opening window or similar opening direct to outside. Table 6.10 from BS 5400 shows whether permanent openings are required.

Type of appliance	Max. appliance rated input (net)	Room volume (m³)	Permanent vent size cm³	Openable window or see note b
Domestic oven, hotplate, grill or any combination thereof	None	< 5 5 to 10 > 10	100 50 (a) see below Nil	Yes
Instantaneous water heater	11 kW	< 5 5 to 10 10 to 20 > 20	Installation not allowed 100 50 Nil	Yes
Space heater in a room	45 W/m² of heated space		100 plus 55 for every kW (net) by which the appliance rated input exceeds 2.7 kW (net)	Yes
Space heater in an internal space	90 W/m² of heated space		100 plus 27.5 for every kW (net) by which the appliance rated input exceeds 5.4 kW (net)	Yes
Space heaters conforming to BS EN 449:1997 in a room	45 W/m² of heated space		50 plus 27.5 for every kW (net) by which the appliance rated input exceeds 1.8 kW (net)	Yes
Space heaters conforming to BS EN 449:1997 in an internal space	90 W/m² of heated space		50 plus 13.7 for every kW (net) by which the appliance rated input exceeds 3.6 kW (net)	Yes
Refrigerator	None		Nil	No
Boiling ring	None		Nil	No

Notes:
(a) If the room has a door direct to outside then no permanent vent is required.
(b) Alternatives include adjustable louvres, hinged panel etc. that open directly to outside.

Table 6.10 Permanent openings required

Open-flued appliances

Where an open-flued appliance is installed with a rated input of more than 7 kW, the room it is in must have an air vent with a free area of: 5 cm² for every kW input in excess of 7 kW.

It is assumed that a room can provide adequate ventilation for an open-flued appliance up to 7 kW. This is due to natural ventilation through cracks in floorboards, windows and doors etc. This is often called **adventitious ventilation**.

All calculations are now based on net kW ratings, so there may be a need to convert the figure if the kW rating is given in gross. This can be done simply by dividing the gross rating by 1.1 to give the net kW rating, e.g.:

9.0 kW Gross ÷ 1.1 = 8.1 kW net

Example for working out ventilation requirement

Work out the ventilation requirement for a natural gas boiler rated at 15 kW gross heat input. 15 kW gross/1.1 converts to 13.5 kW net heat input.

If we need to calculate the size of air vent for this boiler then we need to subtract the 7 kW (adventitious air) and then multiply by 5 cm² for each kW:

13.6 kW – 7 kW = 6.6 kW

6.6 × by 5 cm² = 33 cm²

This means that the boiler requires an air vent of 33 cm² free area.

It is worth noting that when a range-rated boiler of, say, 11 kW to 15 kW is used, the air-vent size should be calculated on the maximum setting.

Multiple appliances

You may encounter situations where more than one appliance is installed in the same room. Ventilation is calculated on the basis of the following:

1 One or more appliance totalling in excess of 7 kW (net). In a single room, internal space, through room, lounge/diner etc: 5 cm² per kW (net) of total rated heat input above 7 kW.

2 Two or more gas fires, up to a total rated heat input of 7 kW (net or gross) each (14kW). In a through room, lounge/diner etc: ventilation not normally required as adventitious ventilation will usually provide sufficient air for combustion. For a higher kW rating allow an additional 5 cm²/kW (net) above 14 kW.

3 Two or more appliances. Single room or internal space: calculate the total ventilation requirements of all the appliances based on the greatest of the following:

   (a)   total rated heat input of flueless space heating appliances

   (b)   total rated heat input of open flue space heating appliances

   (c)   maximum rated heat input of any other type of appliance.

Remember

Provide 5 cm²/kW (net) in excess of 7 kW. For multi-appliance installations only one appliance should be considered for the adventitious air allowance

Example for calculating ventilation

A gas boiler of 15 kW (net) is installed in the same room as a gas fire rated at 3 kW (net).

The ventilation requirement is as follows:

15 kW + 3 kW = 18 kW

18 kW – 7 kW (adventitious air) = 11 kW (only deduct adventitious air for one appliance)

11 kW × 5 cm² = 55 cm² of free area of air vent.

Air vents

There is a wide variety of air vents in use, and the most important features are:

- They should be non-closable.

- No flyscreen less than 5 mm² should be fitted.

- They should be corrosion-resistant and stable.

- The actual free area of the air vent is the size of slots or holes used (this applies to both sides of the ventilation arrangement, i.e. air vent and outside air brick) – this is really important!

On the job: Servicing a conventional flue gas boiler

As part of the servicing of a conventional flue gas boiler Dave measures the air vent to ensure it is big enough and complies with the Regulations. It turns out that after seven years the boiler has been working trouble-free but that the air vent is slightly too small.

1   Is it OK to leave the air vent as it is?

2   What may be the consequences of leaving the vent undersized?

Examples

For air vents the free area is measured correctly by checking the actual width and length of the slots accurately. So, as an example, if the width of the slots on an air vent (B and C) below is 75 mm and the depth (D) is 10 mm, what is the free area?

Free area = 75 mm × 10 mm × 10 slots = 7500 mm²

There are 100 mm² in a cm². So free area = 75 cm².

Here's another example: it is proposed that a 20 kW (gross boiler) will be ventilated by an air brick with a free area of holes measuring 8 mm × 8 mm. There are 48 holes. Is the air brick suitable?

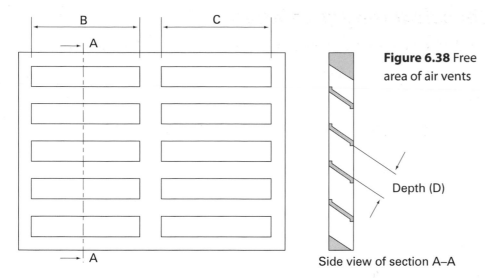

Side view of section A–A

Figure 6.38 Free area of air vents

Depth (D)

Convert boiler heat input from gross to net:

$= 20 \div 1.1 = 18\,kW$

Free area required:

$= 18\,kW - 7\,kW$ adventitious air $= 11\,kW \times 5\,cm^2 = 55\,cm^2$

Actual free area of air brick:

$= 8\,mm \times 8\,mm \times 48$ holes $= 3072\,mm^2 / 100 = 30.7\,cm^2$

The air brick is unsuitable for the boiler heat input requirement.

Air vents should not be positioned where they may easily be blocked by leaves or snow, and should not be positioned where they cause the occupants discomfort through draughts.

The inner and outer grilles must be connected by a liner to prevent anything from within the cavity interfering with the free air flow. It has been known for 'cavity foam fill' to block the flow of air completely.

Air vents should be positioned so that there is a minimum separation distance from any flue terminal, as shown in Table 6.11.

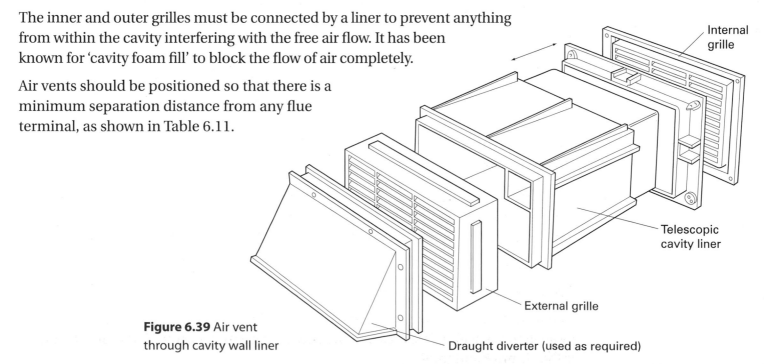

Internal grille

Telescopic cavity liner

External grille

Draught diverter (used as required)

Figure 6.39 Air vent through cavity wall liner

Air vent position	Appliance input (kW)	Room sealed		Open flue	
		Natural draught	Fanned draught	Natural draught	Fanned draught
		Separation distance (mm)			
A Above a terminal	0 to 7	300	300	300	300
	> 7 to 14	600	300	600	300
	> 14 to 32	1500	300	1500	300
	> 32	2000	300	2000	300
B Below a terminal	0 to 7	300	300	300	300
	> 7 to 14	300	300	300	300
	> 14 to 32	300	300	300	300
	> 32	600	300	600	300
C	0 to 7	300	300	300	300
	> 7 to 14	400	300	400	300
	> 14 to 32	600	300	600	300
	> 32	600	300	600	300

Table 6.11 Minimum separation distance from any flue terminal

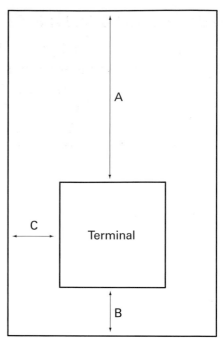

Figure 6.40 Distance of vents from terminal

Air-vent positions in rooms (open flues)

Figures 6.41 and 6.42 show air-vent positioning requirements within rooms for open-flued appliances.

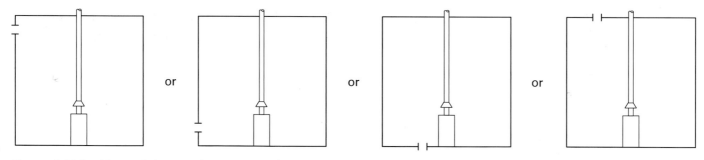

Figure 6.41 Positions of air vents direct to outside or to a ventilated floor/loft

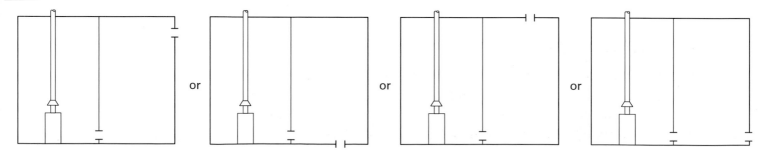

Figure 6.42 Vent positions via another room or space

Air vents in series

Where air vents cannot be taken direct from the outside, they need to connect to an air vent through another room. There are some general rules regarding the sizing of air vents that are 'in series' – not direct to outside. These are usually at least 50 per cent above the requirements of a single vent, as shown in Figures 6.43, 6.44 and 6.45.

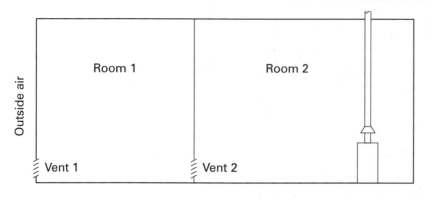

Figure 6.43 Two vents in series

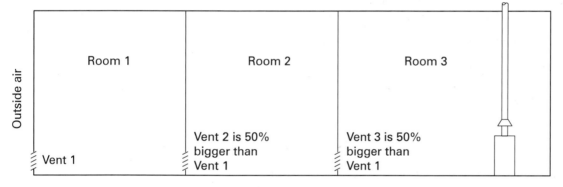

Figure 6.44 Three or more vents in series

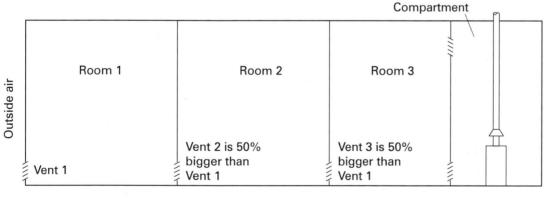

Figure 6.45 Three or more vents in series feeding a compartment

Effects of extractor fans

Care must also be taken to check if any extractor fans are fitted that may affect the performance of an open-flued appliance. The fan must be put onto its maximum setting and a check for spillage carried out – we'll cover this later. Additional air vents may be necessary to prevent any spillage from taking place.

Appliances in compartments

Room-sealed or open-flued boilers may be installed in compartments. So what is a compartment? Well, generally it's an enclosure designed to house a gas appliance. It could also be a small room such as a coal house, outside WC, cupboard, lobby to a hallway… Essentially it's a pretty small room that because of its size may be subject to significant heat build-up. This room therefore needs suitable air circulation provided by high- and low-level vents to outside air or to another room in the building.

The ventilation to a compartment is in addition to permanent ventilation for correct combustion of open-flued appliances – room-sealed appliances do not require permanent ventilation for combustion but do require compartment ventilation.

So let's look at the size of compartment vents that must be provided.

Open-flued appliances

Vent position	Appliance compartment ventilated	
	To room or internal space (see note (a))	Direct to outside air
	cm² per kW (net) of appliance Maximum rated input	cm² per kW (net) of appliance Maximum rated input
High-level	10	5
Low-level	20	10
(a) A room containing an appliance compartment for an open-flued appliance will also require ventilation		

Table 6.12 Compartment vents required for open-flued appliances

An appliance compartment with an open-flued appliance must be labelled to warn against blockage of the vents and advise against use for storage. The label should read – 'IMPORTANT – DO NOT BLOCK THIS VENT – Do not use for storage'.

So let's look at an example. A 24 kW (gross) open-flued boiler is to be sited in a compartment. What is the compartment ventilation requirement if it is to be ventilated to outside air?

Convert boiler to net input: 24 kW ÷ 1.1 = 21.6 kW

Compartment vent:

High-level: 5 cm² × 21.6 kW = **108 cm²**

Low-level: 10 cm² × 21.6 kW = **216 cm²**

Room-sealed appliances

Here you'll need to consult the manufacturer's installation instructions. With the advances in boiler design some room-sealed appliances do not need ventilation in compartments. If there is no guidance provided then ventilation must be provided in accordance with Table 6.13.

Vent position	Appliance compartment ventilated	
	To room or internal space	Direct to outside air
	cm² per kW (net) of appliance Maximum rated input	cm² per kW (net) of appliance Maximum rated input
High-level	10	5
Low-level	10	5

Table 6.13 Ventilation for room-sealed appliances

Air-vent positions in compartments (open flues)

Figures 6.46 and 6.47 are examples of ventilator positions for open-flued boilers in compartments.

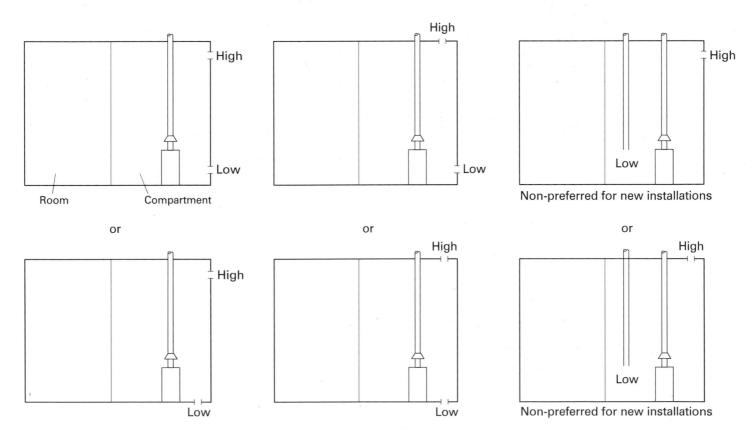

Figure 6.46 Examples of high and low vents direct to outside air or ventilated floor/loft

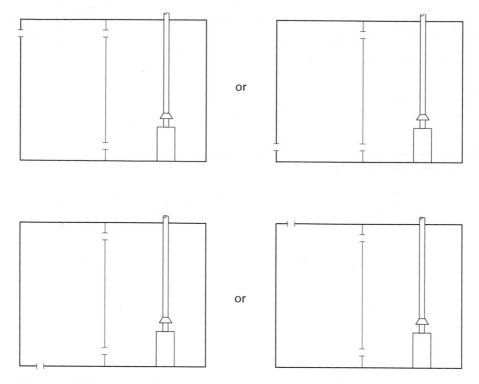

Figure 6.47 Examples of high- and low-vent positions via a room or space

Air-vent positions (room sealed)

Figures 6.48 and 6.49 are examples of ventilator positions for room-sealed boilers in compartments.

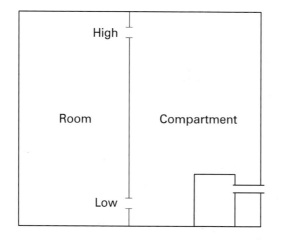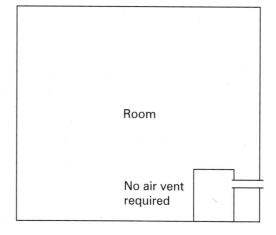

Figure 6.48 Examples of ventilator position via a room

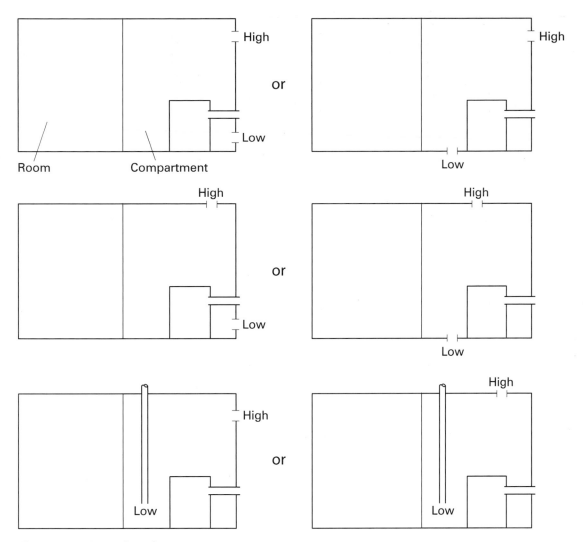

Figure 6.49 Examples of compartments

FAQ

A customer has taken off the internal vent in their lounge and blocked the vent with cardboard. Is this acceptable?

No. Ventilation is provided for safe conbustion to take place. Blocking vents in any way may cause incomplete combustion, which is a major hazard.

Open-flued systems

At the end of this section you should be able to:

- recognise appliance types
- state the main parts of a flue
- select suitable materials for given installations
- identify flue terminations
- inspect flue systems
- recognise faults in flues.

Just as with other work on gas, it is essential that persons carrying out work on the flues for gas appliances are competent to do so, and any work that is subject to the Gas Safety (Installation and Use) Regulations must comply with the requirements. Here we are not going to cover just open-flued appliances; we'll look at categories of flueless and room-sealed appliances as well.

All appliances must be connected to a suitable flue system as detailed in the appliance manufacturer's installation instructions. Appliances should not be connected unless the correct manufacturer's instructions are available to the installer.

There are three main types of appliance, which are grouped according to how they discharge their products of combustion:

- flueless
- open-flued
- room-sealed (balanced flued).

These are then further classified according to flue types, as shown in Table 6.14.

		Natural draught	Fan downstream of heat exchanger	Fan upstream of heat exchanger
A – Flueless		A1*	A2	A2
B – Open-flued	B1 – with draught diverter	B11*	B12* B14	B13*
	B2 – without draught diverter	B21	B22*	B23
C – Room-sealed	C1 – Horizontal balanced flue/inlet air ducts to outside air	C11	C12	C13
	C2 – Inlet and outlet ducts connect to common duct system for multi-appliance connections	C21	C22	C23
	C3 – Vertical balanced flue/inlet air ducts to outside air	C31	C32	C33
	C4 – Inlet and outlet appliance connection ducts connected to a U-shaped duct for multi-appliance system	C41	C42	C43
	C5 – Non-balanced flue/inlet air-ducted system	C51	C52	C53
	C6 – Appliance sold without flue/air-inlet ducts	C61	C62	C63
	C7 – Vertical flue to outside air with air-supply ducts in loft. Draught diverter in loft above air inlet	C71	C72* (Vertex)	C73* (Vertex)
	C8 – Non-balanced system with air-supply from outside and flue into a common duct system	C81	C82	C83

Common types in the UK are shown with an asterisk *

Table 6.14 Appliances classified by flue type

Typical appliances of type A and B

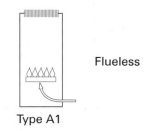

Flueless

Type A1

Open-flued types

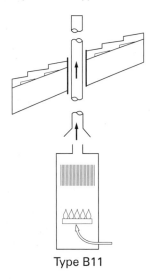

Type B11

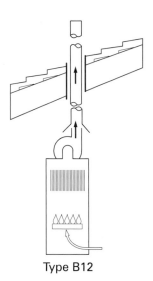

Type B12

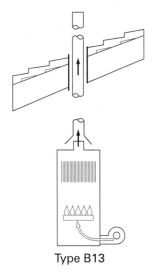

Type B13

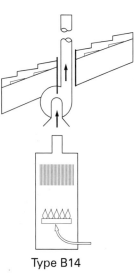

Type B14

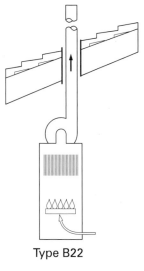

Type B22

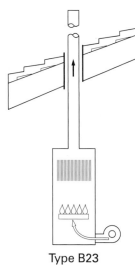

Type B23

Figure 6.50 Typical appliances of types A and B

Find out

Try giving a description to the following appliance types, and check your answer with Table 6.14

Flue type

1 B13

2 B22

Typical appliances of type C

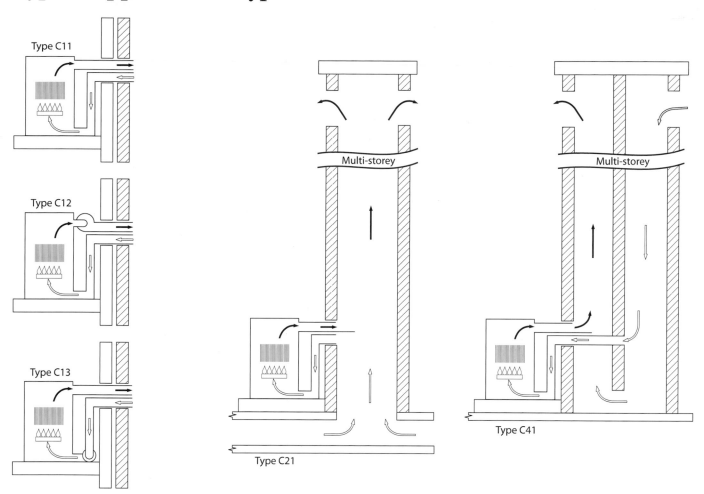

Figure 6.51 Typical appliances of type C

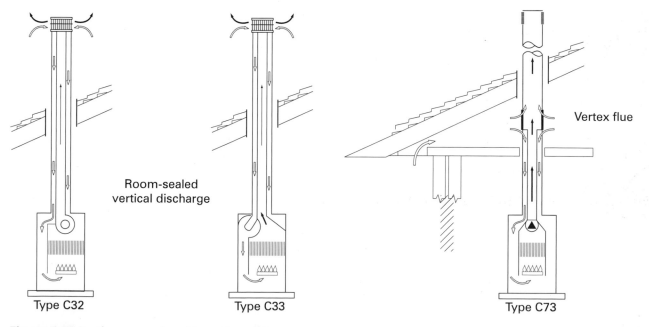

Figure 6.52 Further examples of type C appliances

We will start with open flues; room-sealed flues will be covered in the next section. First, a mention of flueless appliances: these are self-explanatory, so there is no need to spend time on them here. Provided they are installed correctly and sufficient ventilation is available they should not be a problem.

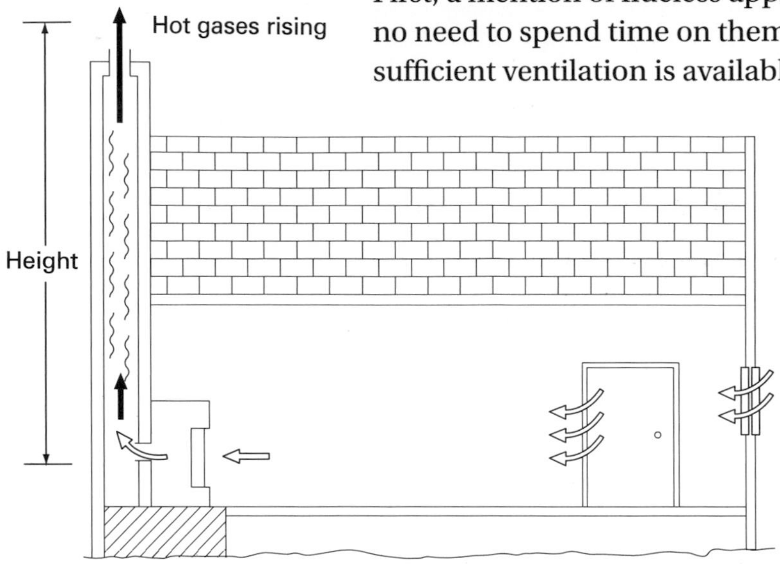

Open flues

How do they work? Well, combustion air is taken from the room and the products of combustion travel up the flue by natural draught. This 'flue pull' is caused by the difference in the densities of hot flue gases and the cold air outside.

The flue pull or draught is increased when the flue gases are hotter or if the flue height is increased. It is also true to say that cooler flue gases will slow it down, as will 90-degree bends and horizontal flue runs – so these must be avoided.

Figure 6.53 Flue draught

Flue draught is created by natural means and is quite slight, so care is needed to design/install a flue to give the necessary updraught. Fans can be fitted in flues to overcome problems and allow more flexibility; these are dealt with later.

Open flues are sometimes referred to as conventional flues and have four main parts:

1   primary flue

2   draught diverter

Both of these are normally part of the appliance, while the:

3   secondary flue and

4   terminal

are installed on the job to suit the particular position of the appliance.

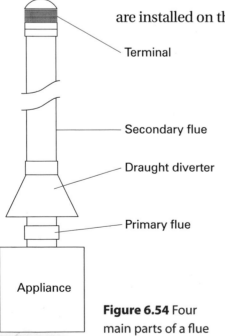

Figure 6.54 Four main parts of a flue

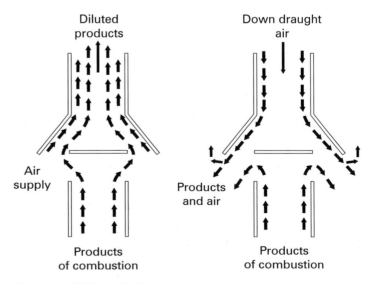

Figure 6.55 Draught diverter – section through

The primary flue is part of the appliance and creates the initial flue pull to clear the products from the combustion chamber.

The draught diverter does three things:

- diverts any downdraught from the burner
- allows dilution of flue products
- breaks any excessive pull on the flue (i.e. in windy weather).

The secondary flue passes all the products up to the terminal and should be constructed to give the best possible conditions for the flue to work properly. Some key points are:

- Avoid horizontal/shallow runs.
- Keep bends to a minimum of 45°.
- Keep flues internal where possible (warm).
- A 600 mm vertical rise from the appliance to the first bend should be provided.
- The flue size must be at least equal to the appliance outlet and as identified by the manufacturer.

The terminal is fitted on the top of the secondary flue. Its purpose is to:

- stop rain, birds and leaves etc. entering the flue
- minimise downdraught
- help the flue gases discharge from the flue.

Only approved terminals are to be used, as these have been checked for satisfactory performance and have limited openings of not less than 6 mm but not more than 16 mm (except for incinerators, which are allowed 25 mm).

There are many types available, but Figures 6.56 and 6.57 show some examples of open-flued terminals.

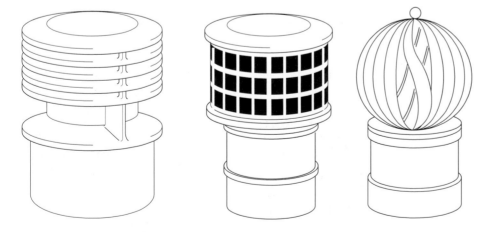

Figure 6.56 Types of flue terminals

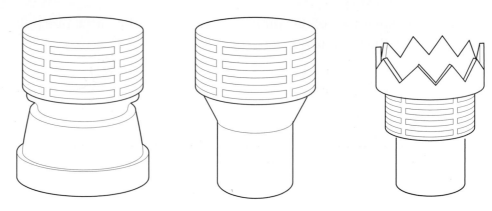

Figure 6.57 More examples of flue terminals

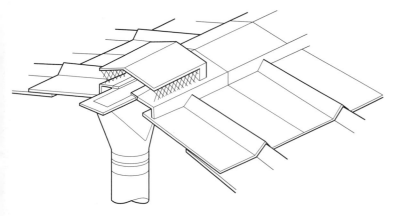

Figure 6.58 Ridge terminal

Ridge terminals (see Figure 6.58) are also allowed and are very popular on new-build properties.

The position of the terminal is critical, as we must avoid any possibility of downdraught or pressure zone around the terminal, which would cause products of combustion to spill back into the room.

Details of flue-terminal positions can be found in BS 5440 – Part 1 (2000); some of these are shown in Figures 6.59, 6.60 and 6.61. Terminals are not required for chimneys with a flue size greater than 170 mm.

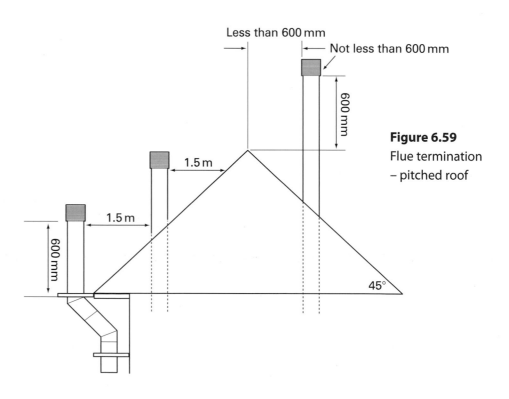

Figure 6.59
Flue termination
– pitched roof

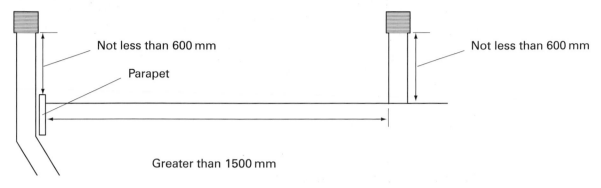

Figure 6.60 Flue termination – flat roof with parapet wall

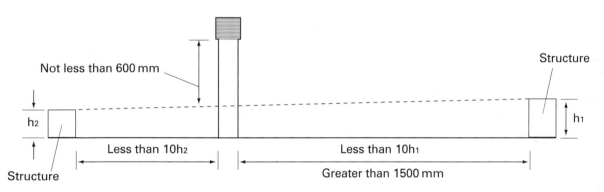

Figure 6.61 Flue termination – flat roof with structures (envelope method)

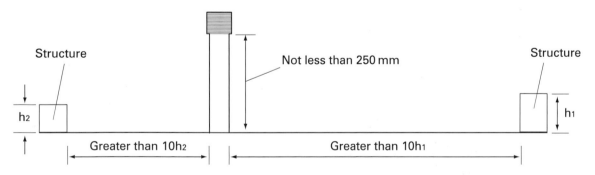

Figure 6.62 Flue termination – flat roof with flue outlet more than 10 times the height of nearby structures away

Prohibited zones

There are prohibited zones on or adjacent to buildings where open flues must not terminate, primarily to prevent downdraught and reversal of the flue-pipe operation, which would spill the products of combustion into the building. Figures 6.63 and 6.64 show where flues must not be sited.

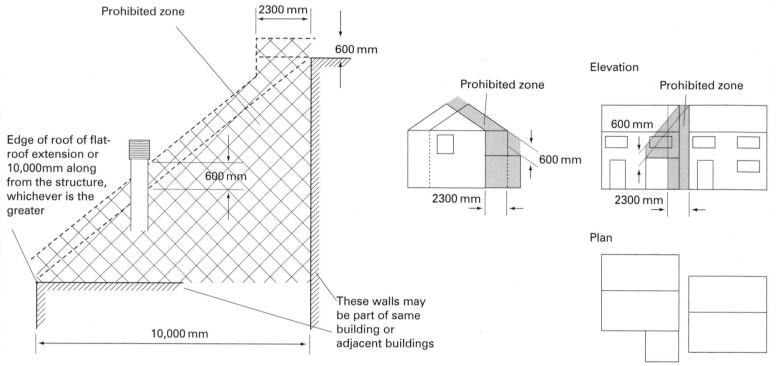

Figure 6.63 Example of prohibited zone near adjacent building

Figure 6.64 Prohibited zones for flues

If **A** is less than 600 mm then **B** to be not less than 600 mm

Figure 6.65 Locations of terminals for pitched roof with structures

Open-flued terminal locations near structures and openings on pitched roofs

Figures 6.65 and 6.66 illustrate acceptable open-flued terminal positions. Please note that measurements are taken from the bottom of the terminal.

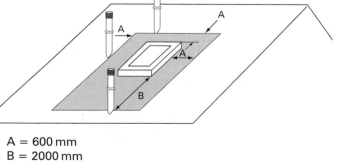

A = 600 mm
B = 2000 mm
Flue not to penetrate shaded area of roof

Figure 6.66 Terminal distances adjacent to windows or openings on pitched roof should be no less than shown

Flue construction

The most commonly used materials for open flues in existing properties are:

- twin-wall metal
- vitreous enamelled (internal only)
- flexible stainless-steel flue liner
- pre-cast concrete blocks
- clay blocks to BS 1289.

Twin-wall metal flues

These are available in a variety of lengths and diameters, with a vast range of fittings and brackets to suit every installation. At this point consult manufacturers' information booklets to familiarise yourself with the range of products available.

Twin-wall flues are provided in two types: the first is with an air gap only; this is only suitable for internal use or externally for lengths less than 3 m. For all other external situations fully insulated twin-wall pipe should be used.

You will note that the joints are designed to be fitted with the 'male' or spigot end uppermost. Where a pipe passes through a combustible material like a floor/ceiling, a sleeve must be provided to give an annular space of 25 mm minimum.

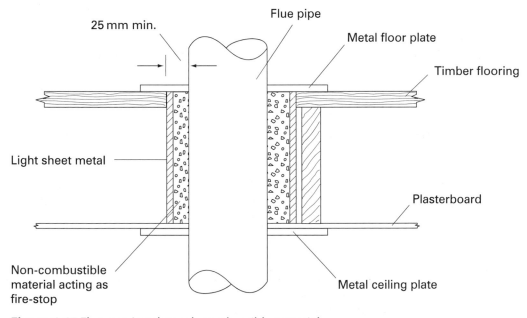

Figure 6.67 Flue passing through combustible material

Where a flue pipe passes through a tiled sloping roof, a purpose-made weathering slate is required with an upstand of 150 mm minimum.

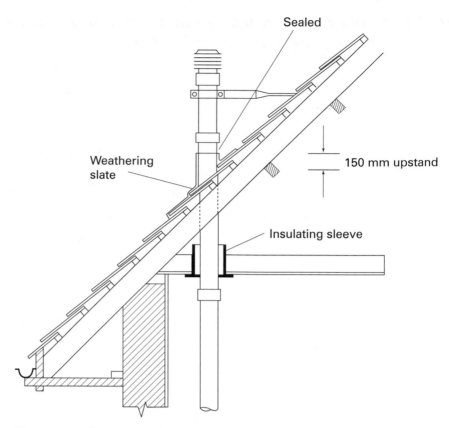

Figure 6.68 Flue passing through sloping roof

Sealed

Weathering slate

150 mm upstand

Insulating sleeve

Vitreous-enamelled steel flues

Vitreous-enamelled steel is a single-skin pipe available in many lengths and sizes, although it can be cut to any length. It is often used as the connection between an appliance and the main flue and may include a disconnecting collar to allow appliance removal. It is interesting to note that the socket on single-wall pipe is fitted uppermost, unlike the twin-wall previously described.

Flexible stainless-steel flue liners

These are used to line an existing chimney that does not have a suitable clay lining as part of the original building construction. Liners are also used when the existing chimney or flue has given unsatisfactory performance in the past. It is essential to seal the top and the base of the chimney; the liner must be of one continuous length without tight radius bends.

A sealing plate must be included at the base of the flue system to prevent debris from falling into the appliance opening and onto the appliance.

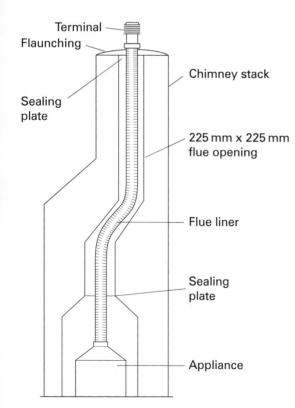

Terminal
Flaunching
Sealing plate
Chimney stack
225 mm x 225 mm flue opening
Flue liner
Sealing plate
Appliance

Figure 6.69 Flexible stainless-steel liner

Figure 6.70 shows the correct method of connection of a flue to a gas fire-back boiler – note the disconnection socket and the sealing plate.

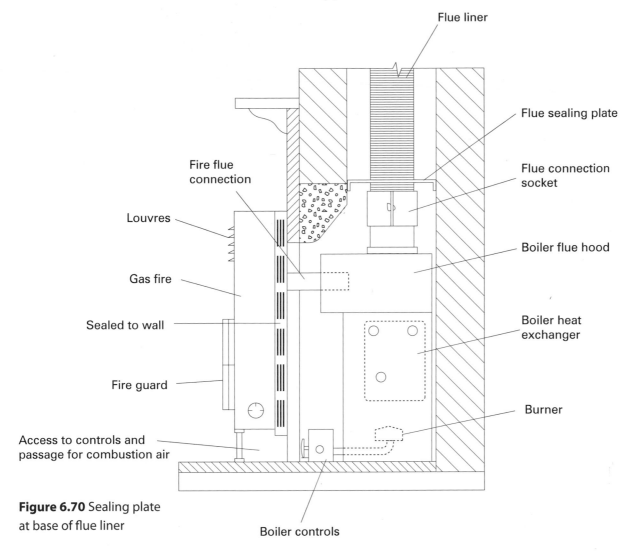

Figure 6.70 Sealing plate at base of flue liner

Pre-cast concrete blocks

These are specially designed for building into walls of domestic properties during construction. This is called the bonded type, while there is a non-bonded version which is more suited to existing properties.

It is important that the gas-flue blocks conform to BS 1289, and they must be fitted to manufacturers' instructions – with excess cement carefully removed from the block during construction and no air gaps left. Particular attention should be given to the connection from the flue blocks to the ridge terminal, which should be in twin-wall insulated pipe using the correct fittings.

Further guidance is given in BS 5440 – Part 1:2000, and examples of the flue blocks are shown here in Figure 6.71.

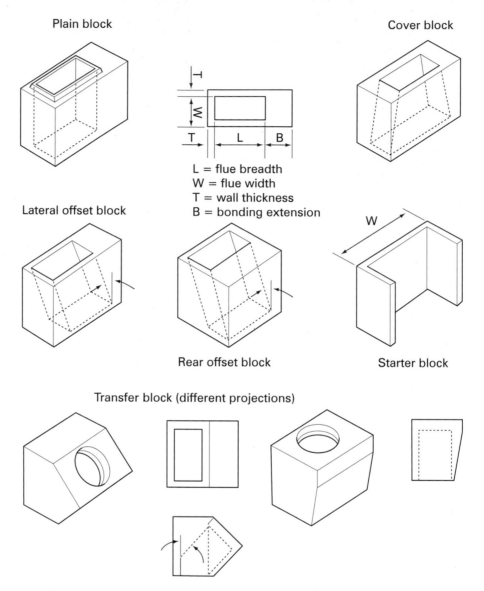

L = flue breadth
W = flue width
T = wall thickness
B = bonding extension

Figure 6.71 Flue blocks

Figure 6.71 shows a detail of a completed installation with flue blocks.

Just a word of warning about pre-cast flue blocks: due to poor installation standards in the past, they are often associated with poor flue performance and spillage of the products of combustion. As with any open flue, always ensure that a thorough visual inspection of the flue system is conducted and adequate spillage tests are carried out prior to handover.

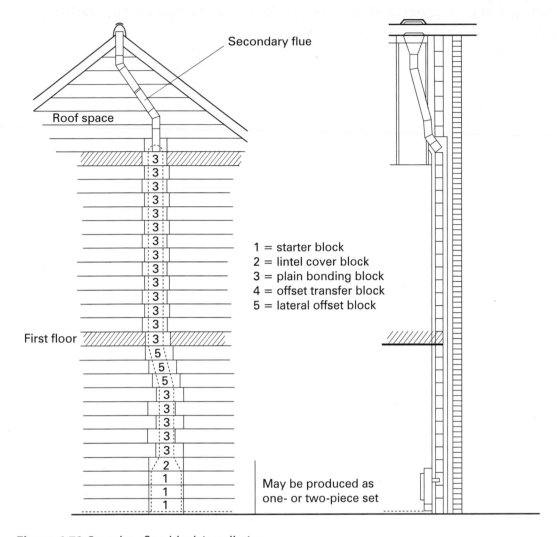

Figure 6.72 Complete flue-block installation

Diagram labels:
- Secondary flue
- Roof space
- First floor
- 1 = starter block
- 2 = lintel cover block
- 3 = plain bonding block
- 4 = offset transfer block
- 5 = lateral offset block
- May be produced as one- or two-piece set

Condensation

An open flue should be installed to keep flue gases at their maximum temperature and avoid problems of excessive condensation forming in the flue. An example of this is that single-wall flues are not allowed externally except where they project through a roof, and twin-wall flues, with only an air gap for insulation, are only allowed up to 3 m in length when used externally.

Table 6.15 shows the maximum lengths of open flue used with a gas fire in order to avoid condensation.

Flue exposure	Condensate-free length		
	225 mm² brick chimney: or pre-cast concrete block flue of 1300 mm² *	125 mm flue pipe	
		Single wall	Double wall
Internal	12 metres	20 metres	33 metres
External	10 metres	Not allowed	28 metres

* see BS 5440-1 for more details

Table 6.15 Maximum lengths of open flue used with a gas fire in order to avoid condensation

When chimneys exceed certain lengths they may need to be lined, depending on the type of appliance fitted, as shown in Table 6.16.

Appliance type	Flue length
Gas fire	> 10 m (external wall)
	> 12 m (internal wall)
Gas fire with back boiler	Any length
Gas fire with circulator	> 10 m (external wall)
	> 12 m (internal wall)
Circulator	> 6 m (external wall)
	> 1.5 m (external length and total length > 9m)
Other appliance	Flue lengths greater than in Table 6.13

Table 6.16 Required flue lining for different appliances

Shared flues

In special cases it is permissible to connect two or more appliances into the same flue, but special rules apply. BS 5440-Part 1 2000 identifies five important rules:

1　Each appliance shall have a draught diverter.

2　Each appliance shall have a flame-failure device.

3　Each appliance shall have an atmospheric sensing device.

4　Flue must be sized to ensure complete removal of the products of the whole installation.

5　Chimney must have access for inspection and maintenance.

Room-sealed flue systems

At the end of this section you should be able to:

- state the principles of room-sealed flues
- identify correct termination positions
- state the requirements for balanced compartments
- recognise alternative flue systems.

We have seen that open-flued appliances provide many difficulties, as the route and terminal position are critical to ensure that safe dispersal of the products take place. Furthermore, the provision of air for combustion gives us more problems, as customers can block sites and create unsafe situations.

A room-sealed or balanced-flue appliance provides the solution to this, as air for combustion is taken directly from outside and, as the name implies, the appliance is 'room-sealed', so there is no danger of products of combustion entering the room.

Room-sealed appliances are therefore preferable to open flues. It is necessary to fit them within the vicinity of an external wall or roof termination. The flueing options are increased greatly with these types of appliance, where a fanned-flue system is chosen. These will be covered in more detail later.

Find out

Fan-flued appliances have some advantages over natural draught. What are they?

Operating principle of room-sealed appliances

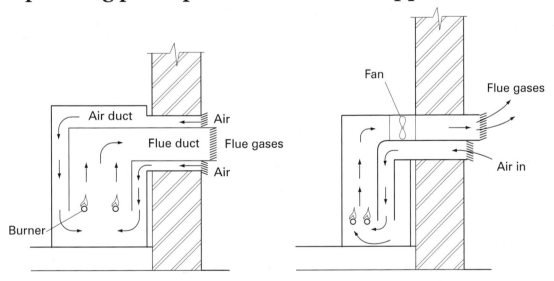

Figure 6.73 Principles of balanced-flue operation

It can be seen that the products of combustion outlet and the air intake are at the same point and are therefore at equal pressure, whatever the wind conditions. This is why it is called 'balanced' flue. The special terminal that is part of the appliance must be fitted in such a position as to:

- prevent products from re-entering the building

- allow free air movement

- prevent any nearby obstacles causing imbalance around the terminal.

BS 5440 Part 1 details acceptable positions for flue terminals on buildings, as shown in Figure 6.74 (larger appliances need greater distances).

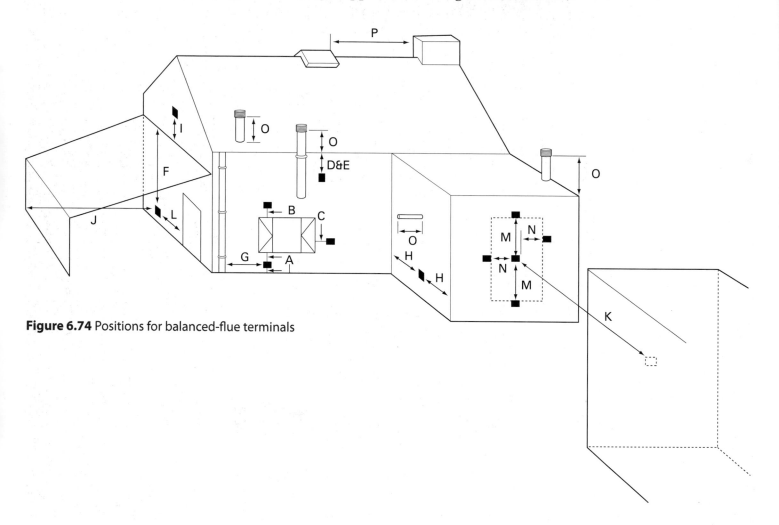

Figure 6.74 Positions for balanced-flue terminals

Dimension	Terminal position	Heat input (kW net)	Natural draught	Fanned draught
A – see note 1	Directly below an opening, air brick, opening window, door etc.	0–7 kW >7–14 kW >14–32 kW >32–70 kW	300 mm 600 mm 1500 mm 2000 mm	300 mm 300 mm 300 mm 300 mm
B – see note 1	Above an opening, air brick, opening window, door etc.	0–7 kW >7–14 kW >14–32 kW >32–70 kW	300 mm 300 mm 300 mm 600 mm	300 mm 300 mm 300 mm 300 mm
C – see note 1	Horizontally to an opening, air brick, opening window, door etc.	0–7 kW >7–14 kW >14–32 kW >32–70 kW	300 mm 400 mm 600 mm 600 mm	300 mm 300 mm 300 mm 300 mm
D	Below gutters, drain pipes or soil pipes	0–70 kW	300 mm	75 mm
E	Below eaves	0–70 kW	300 mm	200 mm
F	Below balconies or car-port roofs	0–70 kW	600 mm	200 mm
G	From a vertical drain pipe or soil pipe	0–70 kW		1500 mm – see note 4
H – see note 2	From an internal or external corner	0–70 kW	600 mm	300 mm
I	Above ground, roof or balcony	0–70 kW	300 mm	300 mm
J	From a surface facing a terminal – see note 3	0–70 kW	600 mm	600 mm
K	From a terminal facing a terminal	0–70 kW	600 mm	1200 mm
L	From an opening in the car-port into the dwelling	0–70 kW	1200 mm	1200 mm
M	Vertically from a terminal on the same wall	0–70 kW	1500 mm	1500 mm

Table 6.17 Positioning of balanced-flue terminals

continued overleaf

Dimension	Terminal position	Heat input (kW net)	Natural draught	Fanned draught
N	Horizontally from a terminal on the same wall	0–70 kW	300 mm	300 mm
O	Above intersection with the roof	0–70 kW	N/A	Manufacturer's instructions
P	Between a chimney and a ridge terminal		1500 mm (300 mm between similar ridge terminals)	

Note 1

In addition the terminal should not be closer than 150 mm (fanned) or 300 mm (natural) from an opening in the building fabric for the purpose of accommodating a built-in element such as a window frame.

Note 2

This does not apply to building protrusions less than 450 mm, e.g. a chimney or an external wall, for the following appliance types – fanned draught, natural draught up to 7 kW, or if detailed in the manufacturer's instructions.

Note 3

Fanned-flue terminal should be at least 2 m from any opening in a building that is directly opposite and should not discharge POCs across an adjoining boundary.

Note 4

This dimension may be reduced to 75 mm for appliances up to 5 kW (net) input.

Table 6.17 Positioning of balanced-flue terminals (continued)

Find out

Check in BS 5440 Part 1 to determine the minimum distance between the guard and terminal

Find out

From BS 5400 (section 10) determine four critical points for balanced compartments

It should be remembered that the outlet part of the terminal can become quite hot, and therefore a guard must be fitted if the terminal is within 2 metres of ground level or if persons have access to touch, e.g. on a balcony.

Special care must be taken when fitting room-sealed flues through walls, particularly in timber-framed buildings, for obvious reasons of protecting against fire; as always, follow the manufacturer's instructions carefully.

Basement areas, light wells and retaining walls

Terminals for room-sealed flues or fanned-draught open flues must be positioned to ensure the safe dispersal of flue gases. In general this means that no terminal shall be located more than one metre below the top level of a basement area, light well or retaining wall. The products must discharge into free open air. Further guidance is given in BS 5440.

Balanced compartments

A slight variation of the room-sealed flue is the balanced compartment. This is where an open-flued appliance is installed in a sealed, fire-resistant enclosure. Balanced compartments are generally for the larger installation and again Regulations apply to their installation.

Figure 6.75 is an example of a balanced compartment installation that meets BS requirements.

Shared flues – room-sealed

This type of flueing system is mainly for use in multi-storey buildings, but since you may work on an appliance in a domestic flat it is important that you recognise the main features.

The two types of system are the SE duct and the U duct, as shown in Figure 6.76.

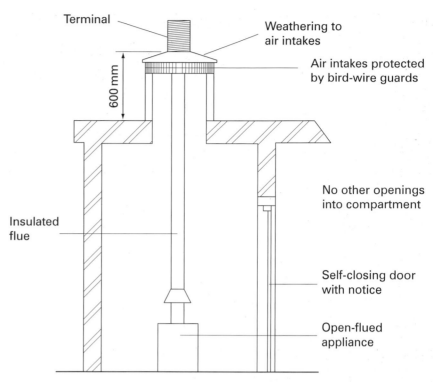

Figure 6.75 Balanced compartment

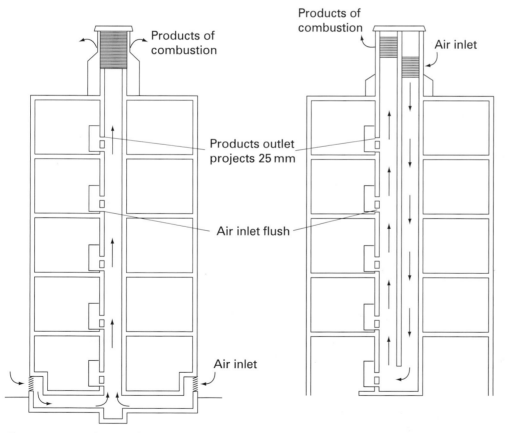

Figure 6.76 SE duct and U duct

Find out

Where shared room-sealed systems are used, a particular family of gas is not allowed. What gas is it?

Find out

Have a look at the classification of appliances in BS 5400 to determine the types of appliance used in SE and U ducts. What are they?

SE duct and U duct

Appliances used in these type of ducts are specially adapted versions of room-sealed flue appliances which are fitted into the vertical flue and air duct on each floor. Only appliances that are suitable can be used. Replacement appliances must be of the same type and suitably labelled, stating that they are fitted to a shared flue.

The responsibility for the shared flue itself is that of the landlord or the person responsible for the building. Annual checks need to be carried out on the shared flue.

Condensing appliances

Condensing boilers are normally of the fanned-flue type and are becoming more popular, owing to their increased efficiency. These appliances have a tendency to form a plume of vapour from the flue terminal which may need to be considered when siting the appliance. The disposal of condensate needs to be carefully considered too, and manufacturers' instructions must be followed. Typical condensate drainage is from plastics to avoid corrosion problems, and the position of termination may be to internal or external discharge stacks with a trap fitted.

Vertex flues (type C[7])

These deserve a special mention as they are unusual, with the air supply being taken from the roof space. The secondary flue is connected to a draught break in the attic which should be ventilated to the standard of current Building Regulations. The draught break must be at least 300 mm above the level of any insulation, and the flue above the break should be vertical for at least 600 mm before any bend is used.

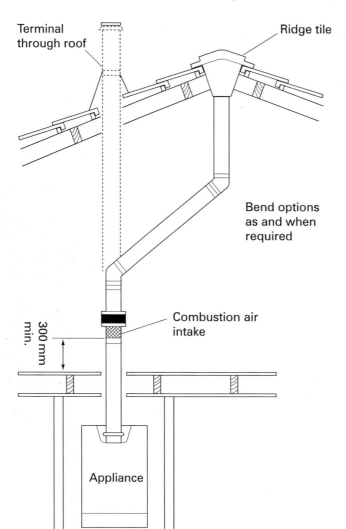

Figure 6.77 Vertex flue

Flue inspection and testing

At the end of this section you should be able to:

- carry out flue inspections
- recognise faults on flues
- perform flue-flow testing
- carry out spillage testing
- take appropriate action if defects are found.

The Regulations state that no person shall install a gas appliance to a flue unless the flue is suitable and in a proper condition for the safe operation of the appliance.

It is essential, then, for any operative to inspect and test any flue for gas appliances, not just at installation but each time the appliance is worked on, including service/maintenance.

From the preceding sections on flues and ventilation you should now have a fairly good knowledge of the installation requirements, so let us begin an inspection. It is necessary to carry out checks on the complete flue system, as follows.

Visual inspection

Common key checks are as follows:

Materials

Ensure:

- materials are suitable for the purpose as defined in BS 5440
- appliance is matched to correct flue material.

Installation

Ensure:

- flue is supported throughout its length and clips are not more than 1.8 metres apart
- draught diverter is in good condition
- flexible flue liners are sealed at top and bottom of chimney
- catchment space is correct size and sealed
- dampers or restrictions are checked for.

Corrosion

- Note that metal flue pipes can become corroded, while other flues may be prone to cracking.

Figure 6.78 Visual inspection of flues

Bends

Bends should not be more than 45° unless it is a pre-cast flue, where the minimum is 30°.

Firestop spacing

The outer skin of the flue pipe should be at least 25 mm from any combustible material; some insulated flues may be exempt – check manufacturer's literature. Pay particular attention to timber-framed buildings, all wooden floors and boxing in.

Termination

Ensure:

- terminal is of correct type
- position of terminal complies with BS 5440.

Seals on room-sealed appliances

- Casing seals must be in good condition to prevent any products from entering the room; also check seals on fan-flued joints.

Signs of spillage

Look for staining around the draught diverter and above gas fires as this could indicate that spillage may be occurring.

Once the condition of the flue has been checked it is essential to carry out both a flue-flow test and a spillage test on open-flued appliances.

Flue-flow testing

Safety tip

When a person does any work on a gas appliance, flue-flow and spillage tests must be carried out where applicable

Flue-flow testing – no pull

Flue-flow testing – pulling

This test is to confirm the soundness of the flue by using a smoke pellet. Where a gas fire is fitted to a chimney it is necessary to remove the fire first.

Procedure

- Check there is an adequate air supply.

- Close all doors and windows in the room/adjoining room.

- Check with a match or taper for some pull; the flue may need warming before lighting the smoke pellet. It is pointless lighting a pellet unless there is some draught available, as the room would fill with smoke.

- Position pellet at base of flue, and light. Note: the pellet should produce at least $5\,m^3$ of smoke in a 30-second burn time.

- Check entire length of flue to ensure that there is no leakage of smoke into the property, including the loft space.

- Check that smoke is seen to come out from only one terminal and that the termination is correct.

- Any faults must be rectified and the flue re-tested before the appliance is fitted/refitted and put into operation.

If the flue meets the above requirements then it is safe to use.

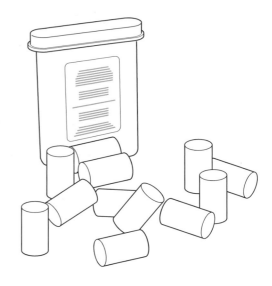

Figure 6.79 Smoke pellets

Find out

Check BS 5440 to see if the closure plate to a gas fire needs to be removed for a spillage test

Spillage testing

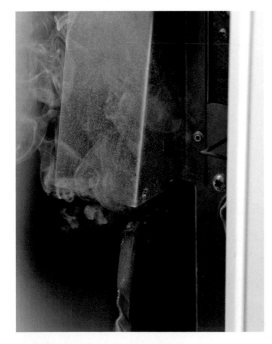

Spillage test being carried out

Figure 6.80 Smoke matches

Find out

How long must you leave the appliance running before carrying out the test? Refer to BS 5440

This test is done with a smoke match while the appliance is running.

Procedure

- Close all doors and windows in the room.

- Switch any fans in the room to maximum. Don't forget fans exist in appliances such as tumble driers!

- Turn on the appliance to full setting and leave on for five minutes.

- Light a smoke match and check for spillage at the draught diverter or canopy; manufacturer's instructions will give exact details of positioning.

- All smoke, apart from the odd wisp, must be drawn into the flue.

- If the test fails it is permissible to leave the appliance running for a further 10 minutes and then re-test; a satisfactory test is essential.

- If fans are present in adjoining rooms, then the interconnecting doors are opened, fans switched to maximum and the previously satisfactory spillage test must now be repeated.

- Any appliance found to be spilling products of combustion is 'immediately dangerous' and must be disconnected. More about this in the next section.

Room-sealed appliances

As you've probably gathered there is not much to test here – key checks are the visual checks:

- Is the flue terminal position correct?

- Is the flue assembly correctly fitted and jointed in line with the manufacturer's requirements?

- Is the casing seal made correctly?

Particular attention, however, does need to be given to fanned-flue appliances with the fan on the intake – this means that the air pressure in the combustion chamber is higher than atmospheric pressure. If the casing seals are defective then the products of combustion will be discharged into the room. This type of fanned-draught appliance does not tend to be used with new appliances, but you may well find it on some old appliances. Key checks here are:

- Carry out a visual inspection of the boiler-case seal for any potential leakage points and ensure that an airtight seal can be made.

- Light the appliance and check round the entire case seal with leak-detection fluid or a lighted taper to identify any leak sources.

- If a leak is detected and it cannot be rectified then the appliance cannot be used.

Unsafe situations

At the end of this section you should be able to:

- explain the categories of risk
- recognise unsafe situations
- identify correct notices and labels
- follow correct procedures.

When gas operatives carry out any work on a customer's premises they are expected to make sure that the installation and appliances are installed and commissioned in line with the Gas Regulations and manufacturers' instructions. If this is not possible then the installation/appliance must not be left working and may need to be disconnected from the gas supply, labelled accordingly and capped off.

Regulation 34 deals with the safe use of appliances, and the operative is expected to make a decision as to whether the installation is safe and, if not, to judge the level of risk so that the necessary action can be taken.

In addition to checking manufacturers' instructions, you may have to consult British Standards or Codes of Practice to check that the installation complies.

Remember

The gas operative will be making very important decisions, and the first priority is to safeguard life and property

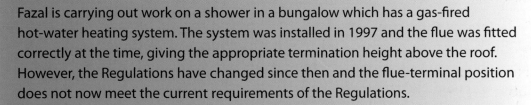

On the job: Complying with the Gas Regulations

Fazal is carrying out work on a shower in a bungalow which has a gas-fired hot-water heating system. The system was installed in 1997 and the flue was fitted correctly at the time, giving the appropriate termination height above the roof. However, the Regulations have changed since then and the flue-terminal position does not now meet the current requirements of the Regulations.

1   What should Fazal do to make sure that he complies with the requirements?

Immediately dangerous (ID)

This is an installation that is an *immediate danger* to life or property if it is operated or left connected to the gas supply. What to do:

1 Where possible, with user permission, try to rectify the fault.

2 Failing rectification of fault, explain the danger to the user.

3 Disconnect the appliance and cap off, with the owner's permission.

4 Attach a 'DO NOT USE' label.

5 Complete a warning notice.

6 Ask a responsible person to sign the paperwork.

Find out

What should you do if the user refuses to let you disconnect the gas supply?

What should you do if the user will not sign a warning notice?

At risk (AR)

This is an installation which *could* create a risk to life or property. What to do:

1 Where possible, with user permission, try to rectify the fault.

2 Failing rectification of the fault, turn off the appliance and explain that continued use is at user's own risk.

3 Attach a 'DO NOT USE' label.

4 Complete a warning notice.

5 Ask a responsible person to sign.

Not to current standards (NCS)

Other situations that can be identified as faults but are not immediately dangerous (ID) or at risk (AR) are classed as 'Not to current standards' (NCS).

The gas operative has a duty to take appropriate action regarding unsafe situations and advise the 'responsible person'; this is normally the owner or occupier of the premises, and, in the case of rented property, this is the landlord or managing agent. This means notifying the customer or landlord of deficiencies in the gas system even if they do not meet the At Risk or Immediately Dangerous classifications. An example could be a substandard installation brought about by a change to the Gas Safety Regulations.

RIDDOR

There is just one other important set of regulations you need to be aware of : RIDDOR – the Reporting of Injuries, Diseases and Dangerous Occurrences Regulations. There is a requirement for operatives to report certain types of dangerous situations to the HSE. This allows them to investigate reports, gauge the scale/nature of the problems and decide on future enforcement action.

Get your tutor to show you an example of the RIDDOR Form F2508G and look at the detail that needs to be completed on the form. We will move on now to look at the classification of various unsafe situations. Keep an eye on the RIDDOR box, as this indicates whether a situation should be reported to the HSE.

Tables 6.18 to 6.27 show some common examples to help you take the correct course of action. Full details can be found in the CORGI 'Unsafe Situations Procedure' booklet.

Unburnt gas escapes

Situation	Action	RIDDOR	Additional actions
From a primary meter, bulk vessel or cylinder	ID		Inform gas emergency call centre
Downstream of a primary meter: 1. Requiring property evacuation 2. Outside tolerance of tightness test 3. Within test but smell of gas	ID ID ID	✔ ✔	Make safe Repair if authorised by user Inform gas emergency call centre if smell persists

Table 6.18 Action to take in the event of unburnt gas escape

Safety tip

It is worth noting here that should two or more NCS faults be found on a natural-draught open-flue or flue-less appliance then the installation is treated as At Risk

---

Meters and regulators

Situation	Action	RIDDOR	Additional actions
Meter showing signs of corrosion, damage or contact with electrical equipment	AR		Inform gas emergency call centre for primary meters or 'responsible person' for secondary meters
No pressure regulation at primary meter	ID		Inform gas emergency call centre Note LPG meters may not have a regulator
No primary meter where required	AR		Refer the responsible person to gas supplier
Installation pipework not sealed within meter box entering property	AR		Seal if requested by responsible person

Table 6.19 Action to take on meters and regulators

Reports of fumes

Situation	Action	RIDDOR	Additional actions
CO detector sounds or fumes reported due to spillage/ leakage of products on flued appliances or evidence of poor combustion on flueless appliances	ID		Check flame pictures and test for spillage on all appliances in property

Table 6.20 Action to take in the event of fumes

Installation pipework

Situation	Action	RIDDOR	Additional actions
Pipework showing signs of corrosion or damage likely to affect safety	AR		
Pipework with an open end, connected to a gas supply	AR	✔	Seal with appropriate fitting
Pipework undersized and preventing manufacturer's gas input rating	AR		
Pipework of inappropriate material, e.g. plastic pipe/hose	AR	✔	

Table 6.21 Action to take in the event of problems with pipework

Air supply

Situation	Action	RIDDOR	Additional actions
Appliances requiring purpose-provided permanent combustion air supply – None Provided	AR		Appliances that are working satisfactorily but have an air supply more than 10% undersized or incorrectly configured (including compartments) should be regarded as NCS

Table 6.22 Action to take in the event of problems with air supply

Open flues

Situation	Action	RIDDOR	Additional actions
Flues leaking products of combustion into buildings	ID	✔	
Spillage occurring or signs of spillage	ID	✔	Particular attention is needed with compartments
Appliance connected to an unlined chimney which is in poor condition	AR		
Two or more appliances connected to one flue and: 1. Appliance with no FSD fitted 2. Appliances in separate rooms ventilated from different sides of building 3. The flue is not designed for the purpose	AR AR AR		**Items 1 and 2** Disconnect and attach warning labels to all except the highest-priority appliance Seal flue connections of each disconnected appliance Retest for spillage **Item 3** Regard whole installation as At Risk
Any natural-draught wall-faced termination	AR		Not acceptable for any natural-draught open-flue installation
Appliance flued into loft space	AR		
Incomplete or damaged flue, poor fixings or sealing	AR		
Manual damper in place and not secured in open position	AR		

Table 6.23 Action to take in the event of problems with open flues

Room-sealed flues

Situation	Action	RIDDOR	Additional actions
Flue terminating in an internal space, e.g. a conservatory	ID	✔	

Table 6.24 Action to take in the event of problems with room-sealed flues

Water heaters

Situation	Action	RIDDOR	Additional actions
Open-flued or flueless water heaters without a built-in ASD installed after 31 October 1998	AR	✔	Appliances installed before this date that are safe and working satisfactorily should be classed as NCS

Table 6.25 Action to take in the event of problems with water heaters

Gas fires/fireplaces/back-boiler units

Situation	Action	RIDDOR	Additional actions
Builder's opening not sealed correctly	AR		
No closure plate where required or incorrectly sealed	AR		
Gas fire fitted to letterbox opening or inadequate catchment space	AR		
Gas fire fitted on carpet, burner less than 225 mm above carpet	AR		
Builder's opening not sealed around pipes/flue liner	AR		

Table 6.26 Action to take in the event of problems with gas fires/fireplaces/back-boiler units

Appliances (general)

Situation	Action	RIDDOR	Additional actions
Multiple situations on an installation which are NCS	AR		
Appliances that should be flued but are not	ID	✔	
Appliances that are unsafe due to lack of servicing	ID	✔	
Flued appliances that are spilling or leaking products into an internal space or room	ID	✔	Includes faulty or missing case seals/sight-glass window
Flueless or open-flued appliance in a bathroom or shower room	AR	✔	Appliances installed before November 1984 that are safe and working satisfactory should be classed as NCS
Flueless or open-flued appliance over 14 kW or under 14 kW without an ASD installed after 1 January 1996 in bedroom or bedsit room	AR	✔	Appliances installed before 1 January 1996 that are safe and working satisfactorily should be classed as NCS
Appliances installed in a room that has been converted into a bedroom after 31 October 1998 where appliances do not comply with current requirements relating to bedrooms in rented accommodation	AR	✔	Appliances installed before 31 October 1998 that are safe and working satisfactorily should be classed as NCS
Evidence of damage to adjacent combustible materials	AR		
Appliance not suitable for the gas being supplied	ID	✔	
Safety devices such as FSD rendered inoperative or failing to danger	ID	✔	
Flexible connection to a flued gas appliance	AR		
Appliance not secure/stable	AR		Not really intended for a cooker without a stability bracket unless the appliance is unsafe

Table 6.27 Action to take in the event of problems with appliances (general)

Bi-metal strip

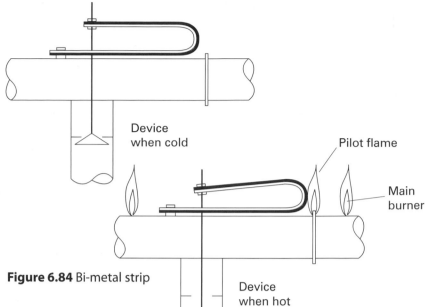

Figure 6.84 Bi-metal strip

Because the metals are fixed together, any heat applied causes the bending effect which then opens the gas valve. In the past, bi-metal strips were used on back-boilers and water heaters but are becoming less common.

Flame conduction and rectification

These electronic flame-protection methods are now being used much more on domestic appliances. The chemical reaction in a flame produces 'ions', which are electrically charged particles. These ions, together with an electrode, rectify the a.c. and produce a small d.c. output, which in turn operates a relay which then activates the gas valve.

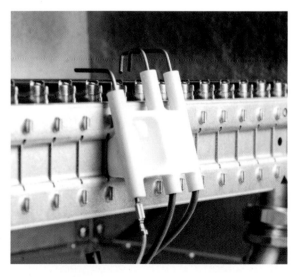

Flame rectification

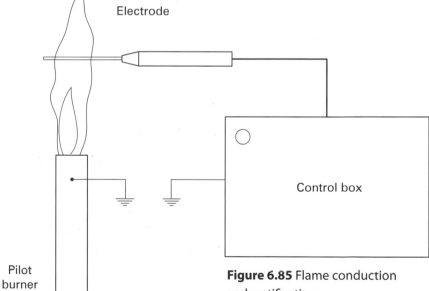

Figure 6.85 Flame conduction and rectification

The shut-down of the gas valve is instantaneous in the event of flame failure. Faults in the system can be identified using a suitable test instrument (usually a multimeter). Readings should be in line with manufacturer requirements.

Vapour-pressure device

These are used on room heaters, water heaters and cooker ovens. When liquids are heated enough they will turn into a vapour, which has a much greater volume. This increase in volume results in a higher pressure that can be used to operate a valve, as shown in Figure 6.86.

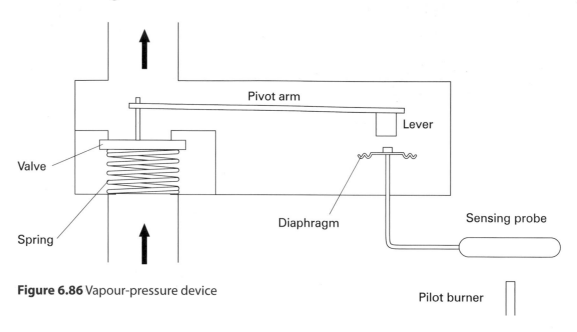

Figure 6.86 Vapour-pressure device

Thermo-electric devices

These work on the principle of heat applied to a 'thermocouple' producing a small amount of electrical energy. The thermocouple is simply a loop of two dissimilar metals joined at one end, which is heated by a pilot flame. The other ends are connected to a magnet in a spring-loaded gas valve, which holds the valve open as long as the pilot produces heat. If the thermocouple goes cold (as in the failure of a pilot), the magnet is de-energised and the valve closes.

It is therefore essential that the tip of the thermocouple is properly positioned in the pilot assembly to work correctly. The thermocouple can be checked to see if it is working correctly by measuring the electrical current generated when it is heated between its tip and gas valve connection point using a multimeter. The reading should generally be between 10 mA and 30 mA.

The operation can be seen clearly in Figures 6.87, 6.88 and 6.89.

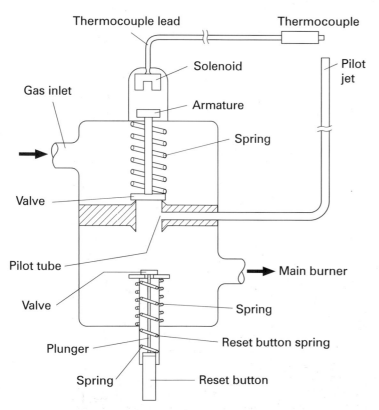

Figure 6.87 Thermo-electric valve – closed position

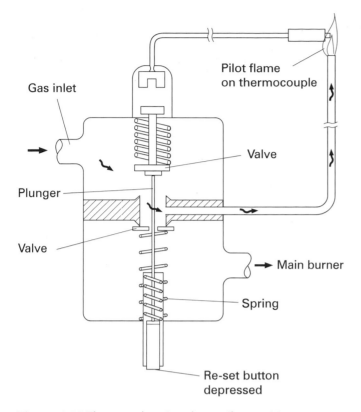

Figure 6.88 Thermo-electric valve – pilot position

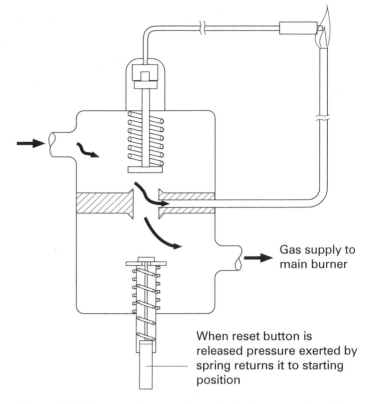

Figure 6.89 Thermo-electric valve – 'main burner on' position

Other safety cut-off devices

Atmospheric sensing device (ASD)

These are also known as **vitiation-sensing devices** or **oxygen-depletion devices (ODD)**. They are fitted to appliances that can be a danger in terms of products of combustion affecting the air in a room. Typically these are flueless space or water heaters or open-flued appliances such as gas fires/back-boiler units.

The principle of operation is that if the combustion air becomes contaminated, the oxygen level falls and the specially designed pilot lifts away, cooling the thermocouple and eventually closing the thermo-electric valve.

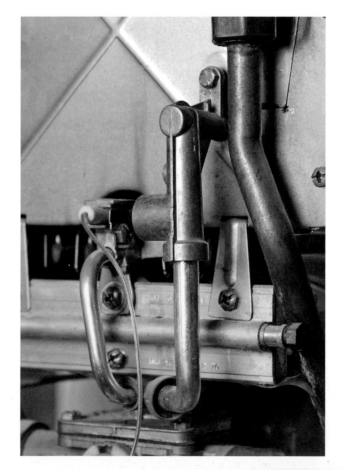

Atmospheric sensing device

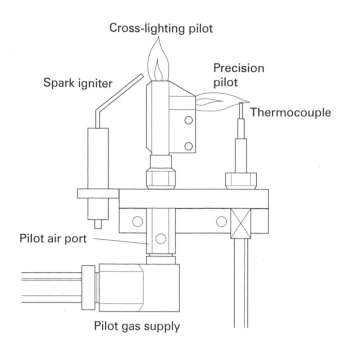

Figure 6.90 Satisfactory air supply, pilot working normally

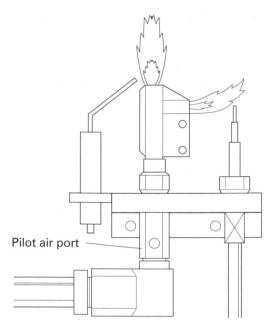

Figure 6.91 Lack of oxygen to pilot thermocouple will cool and gas valve will close

Temperature control – thermostats

Thermostats are designed to give control and maintain a steady temperature of, for example, the water in a boiler or the air in an oven. Most thermostats give a clear on/off action, while some types can give a gradual change of gas rate by restricting the flow according to the temperature setting.

Thermostats may be of the fixed or adjustable type and may be activated either by different expansion of metals or by expansion of liquids/vapour. Alternatively they can be electrically operated.

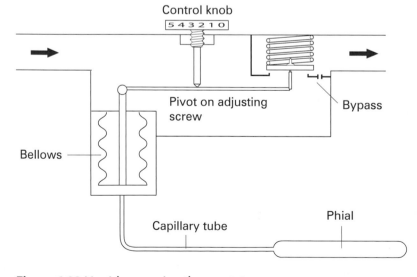

Figure 6.92 Liquid-expansion thermostat

Liquid-expansion thermostat (mechanical thermostat)

Found mainly on gas cookers. The temperature is sensed by the remote phial, which contains liquid. It can be seen from Figure 6.92 that, on heating, the bellows expand and close the gas valve to the burner. Take note of the bypass too, as this maintains a small amount of gas to keep the oven burner lit even if the main valve is fully closed. When the oven cools, the thermostat phial detects this; the bellows contract, therefore opening the gas valve again.

Electrical thermostat

This differs from a mechanical thermostat as it does not have any gas passing through it. It may be built into a boiler on a similar expansion principle as the oven thermostat, but the bellows expanding will in this case activate an electrical contact to switch power on/off to a solenoid valve.

Other electrical thermostats that have on/off contacts can be remote, usually for room or domestic water-temperature control. Many boilers have a high-limit thermostat, which again is simply an electrical switch to deactivate the boiler should the temperature become unusually high.

Solenoid valve

Although not a thermostat, this valve works in conjunction with electrical thermostats/switches to control the flow of gas to a burner – it's essentially an on/off valve.

Solenoid valves will open when energised, as the current flowing into the coil will cause the soft iron core to be magnetised, lifting it open against the spring. Loss of current allows the valve to spring to the closed position. Solenoid valves are used in multifunction controls but can also be used as valves in a gas line to a burner.

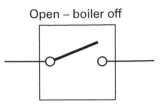

Open – boiler off

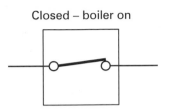

Closed – boiler on

Figure 6.93 Electrical thermostat – switch off and switch on

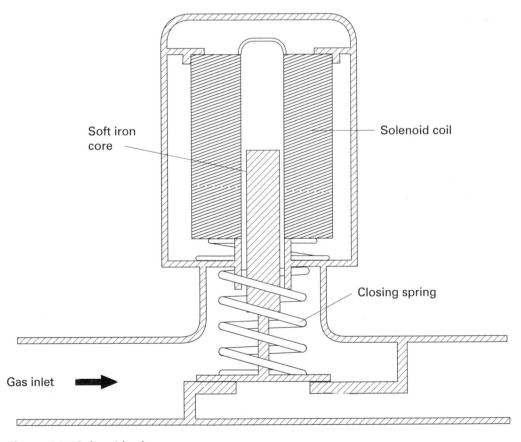

Figure 6.94 Solenoid valve

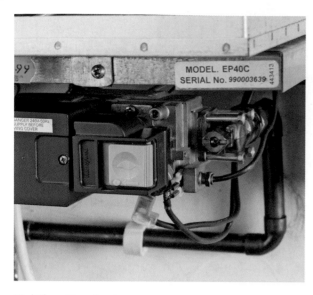

Multifunctional valve

Multifunctional control valves

A 'multifunctional control' is a combination of several control devices all contained in one unit and fitted to an appliance, the most common of which is a boiler.

The multifunctional control often includes:

- main control cock
- constant pressure regulator (adjustable)
- solenoid valve
- flame-failure device (thermocouple)
- thermostat
- igniter.

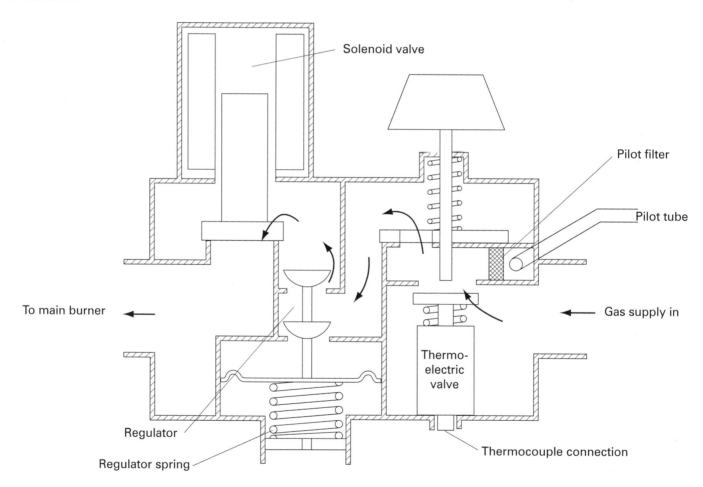

Figure 6.95 Section through a multifunctional control valve

The operation is as follows:

- To light pilot, turn the knob to pilot position and depress.
- This keeps the main gas supply closed but opens the gas supply to the pilot.
- Light pilot and hold knob down for 20 seconds to energise thermocouple.
- The thermocouple energises the thermoelectric valve and holds it open.
- Release knob and turn to 'ON' position.
- The mains gas supply can now flow through to the regulator.
- Finally the solenoid is energised by a boiler thermostat/time switch.
- Gas now flows to the main burner and is ignited by the pilot.

Ignition devices

There are three main types of ignition device:

- permanent pilots
- spark igniters
- filament igniters.

Permanent pilots

These are simple and not dependent on other power sources. However, in the interests of energy conservation they are gradually being phased out in favour of spark ignition.

Spark igniters

There are two main types:

- **Piezo-electric**: used on gas fires and for lighting pilots on boilers/water heaters. A simple 'push button' activates crystals to produce about 6000 volts at the igniter.
- Mains transformers: a step-up transformer connected to an electrode provides the spark, while modern electronic versions can provide pulse sparks at up to 8 sparks per second of 15,000 volts. The danger from contact with the igniter, however, is minimal.

Filament igniters

Filaments are gradually being phased out but are still in use. They comprise a filament (small coil of thin wire) which glows red hot when energised from batteries or a transformer. The filament lights the pilot, which is often activated by the same control knob as the filament.

Zero-rated governor

Zero-rated governor

These are becoming increasingly popular on forced-draught boiler burner units. The principle of operation is that suction caused by the fan activates the governor to allow gas to pass to the burner. It follows that if the fan does not function correctly then the governor will not open and no gas will be allowed through. Therefore the appliance is safe.

Decommissioning gas systems

At the end of this section you should be able to:

* state the correct procedure for removing appliances/pipework
* state the correct procedure for removing meters.

Appliance removal

When appliances are removed it is essential that the installation pipe is permanently capped off and the complete installation tested for tightness.

Installation pipes that are no longer to be used should be removed to prevent unwanted dead legs, which contain live gas, becoming a future danger from accidental damage etc. Care must be taken to ensure that any electrical bonding is maintained, and the advice of a qualified electrician should be sought.

Meter removal

A meter cannot be removed without the authority of the gas supplier, the meter owner or the owner of the premises.

When removal of a primary meter has been authorised, the emergency control and the installation pipe must be capped. Similarly, when a secondary meter is removed, the meter control and installation must also be capped.

The live gas pipe that has been capped must be clearly labelled to show that it is a LIVE GAS PIPE. Where the meter or a section of pipe is removed, a permanent equipotential bond of suitable size must be fitted.

Remember

Even removing an appliance is classed as 'work' on a gas installation and therefore the operative becomes responsible for the complete installation

Liquefied petroleum gas (LPG)

At the end of this section you should be able to:

- identify suitable materials for use on LPG
- state the requirements for storage of LPG
- determine installation criteria for high- and low-pressure pipework
- indicate suitable pipe sizes
- describe the test procedures
- state the purging and commissioning procedure.

Many aspects of LPG and the requirements for installation are exactly the same as for natural gas. However, there are some important differences that need to be understood. This section gives a general overview of LPG installations for permanent dwellings. There are other legal requirements concerning LPG in caravans and boats etc., but these are not mentioned here.

BS 5482 Part 1 deals with the specifications for permanent dwellings.

Materials

Generally pipework and fittings used for natural gas are the same as for LPG, but operatives should be aware of the main differences:

- LPG will 'search out' weaknesses in joints – more so than natural gas.
- Parallel threaded joints are more prone to leaks on LPG.
- Natural rubber and some plastics are affected by LPG, which is aggressive.
- Threaded valves on cylinders have left-hand threads.
- Appliance taps and valves need to be designed for LPG.
- Oil-based lubricants on non-suited valves will be dissolved by LPG.

Storage

Storage of LPG is either by means of a bulk tank or by cylinder supply, using a number of cylinders depending on the gas rate required. Bulk tank installations are carried out by specially trained installers who run a service pipe to the boundary, terminating with an emergency control. Cylinder installations can be carried out by suitably qualified and registered installers.

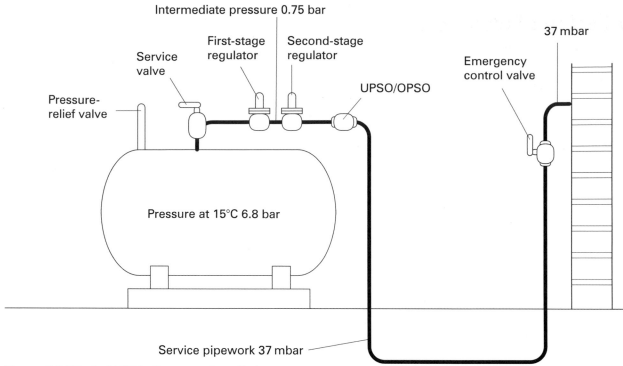

Figure 6.96 Typical LPG bulk storage installation

UPSO/OPSO valve (LPG)

UPSO/OPSO valves

There is a requirement in the Gas Safety Regulations that where gas is supplied from a bulk vessel or four or more cylinders, a device must be fitted to prevent the installation from being subjected to extreme pressures. The UPSO/OPSO valve carries out this function:

- UPSO – Under pressure shut-off

- OPSO – Over pressure shut-off.

The UPSO/OPSO valve is normally constructed as a single unit and can be mounted on the vessel or adjacent to where the supply enters the external wall of a building.

The UPSO can be reset manually but the emergency control must be in the OFF position in order to do so.

The OPSO can only be reset by a competent registered person.

The valve may also contain a regulator that operates at 37 mbar (plus or minus 5 mbar); this then eliminates the need for a second-stage regulator.

The UPSO/OPSO valve also contains a **limited-capacity relief valve**. This is a built-in valve that allows excess pressure to discharge to atmosphere sufficiently to trigger the valve to a closed position.

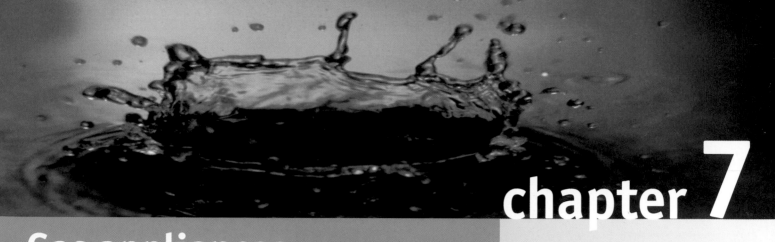

chapter 7

Gas appliances

OVERVIEW

This section deals with gas appliances and links very closely with Chapter 6 Gas safety and supply, which is preparatory work towards the Gas ACS Scheme Assessment CCN1. Domestic gas appliance installation is an important area of work for the plumbing industry. Some firms specialise in service and maintenance, while others do the full range. Whatever happens, you will need to be competent on the appliances that you work on. As with core gas safety, appliances are designed to be used either on natural gas (NG) or liquefied petroleum gas (LPG) (propane and butane). In this chapter we will deal only with natural gas; the key LPG differences have been dealt with in Chapter 6.

What you will learn about in this unit:

- **Domestic gas central-heating boilers**
- **Domestic gas fires and wall heaters**
- **Domestic gas cookers**
- **Water heaters**

Domestic gas central-heating boilers

At the end of this section you should be able to:

- recognise various types of boiler and controls
- state the installation criteria
- carry out pre-commissioning checks
- state the commissioning procedure
- explain the procedure to carry out servicing.

There are basically five types of domestic gas-fired boiler:

- floor-standing (free-standing)
- wall-mounted
- back-boiler (behind a fire)
- combined heating and hot-water units
- combination (combi) boilers.

These types of boilers can then be further categorised as:

- open-flued
- room-sealed – natural-draught or fan-flued.

Further descriptions will then give more information about the boiler, for example 'condensing' or 'low water content'. An example is 'wall-mounted fan-flued boiler' – this really describes the type and category.

Domestic boilers have an output of between 9 kW and 70 kW, and are designed to meet the maximum heating load in winter and provide hot water at the same time. Many boilers have 'range-rated' burners, which means they can be adjusted to suit the size of the system. This is done by altering the burner pressure in accordance with the manufacturer's instructions or data plate.

Floor-standing boilers

Typical floor-standing boilers were fitted in kitchens with a cast-iron heat exchanger and an open flue. The trend now, however, is towards wall-mounted fan-flued boilers with low water content and integral pumps and control systems.

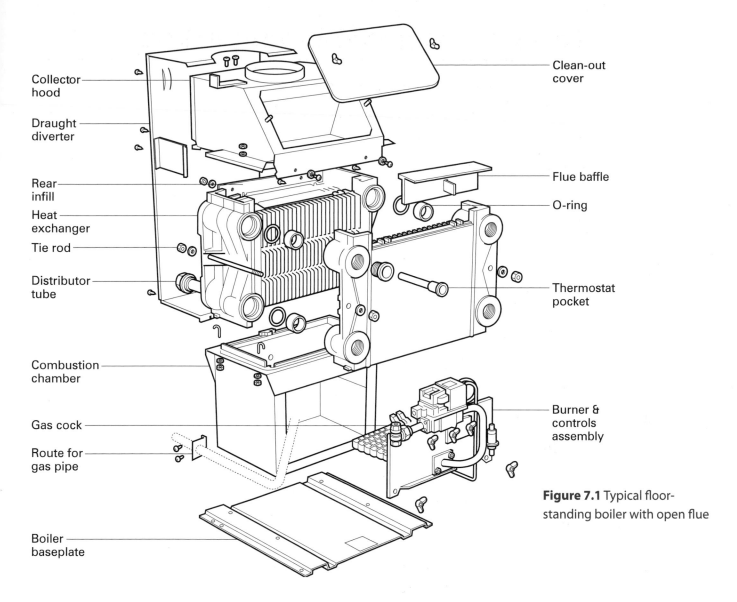

Collector hood

Draught diverter

Rear infill

Heat exchanger

Tie rod

Distributor tube

Combustion chamber

Gas cock

Route for gas pipe

Boiler baseplate

Clean-out cover

Flue baffle

O-ring

Thermostat pocket

Burner & controls assembly

Figure 7.1 Typical floor-standing boiler with open flue

Wall-mounted boilers

These are often fitted with light, compact heat exchangers that may contain as little as one litre of water. They are known as 'low water-content' boilers and require a positive flow of water at all times. If the water circulation slows down too much the water can overheat and boil. This produces a noise like the 'singing' of a kettle, which is known in the trade as **kettling**. A bypass is often fitted to these types of boiler to ensure that the flow of water is always above the minimum and therefore prevents overheating. The bypass is usually a 15 mm pipe with a regulating valve or automatic bypass valve fitted normally between the main flow and return, as shown in Figure 7.3.

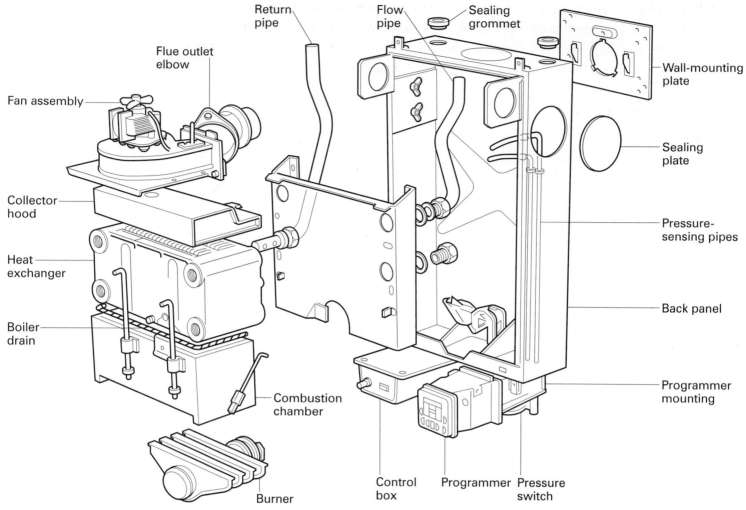

Return pipe

Flue outlet elbow

Fan assembly

Collector hood

Heat exchanger

Boiler drain

Burner

Combustion chamber

Flow pipe

Sealing grommet

Wall-mounting plate

Sealing plate

Pressure-sensing pipes

Back panel

Programmer mounting

Control box

Programmer

Pressure switch

Figure 7.2 Typical wall-mounted boiler with fanned flue

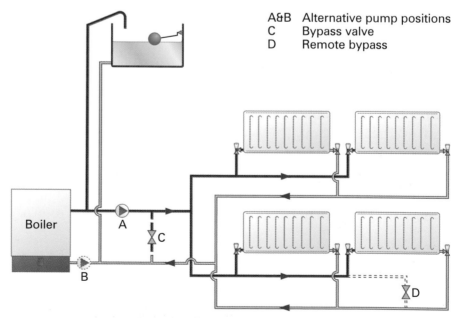

A&B   Alternative pump positions
C   Bypass valve
D   Remote bypass

Boiler

A

B

C

D

Figure 7.3 Installation with bypass fitted

A further problem can be found when a boiler shuts down but the residual heat (in the cast-iron heat exchanger) continues to heat the water and boil. In this case a **pump overrun thermostat** is fitted to allow the pump to run after the boiler has shut down to remove the residual heat from the heat exchanger.

Back-boilers

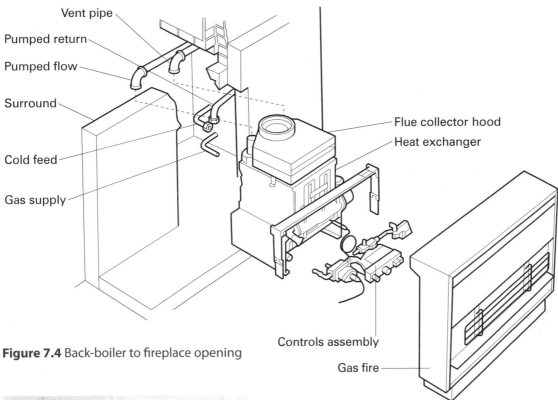

Vent pipe

Pumped return

Pumped flow

Surround

Cold feed

Gas supply

Flue collector hood

Heat exchanger

Controls assembly

Gas fire

Figure 7.4 Back-boiler to fireplace opening

Back-boiler with combustion cover removed

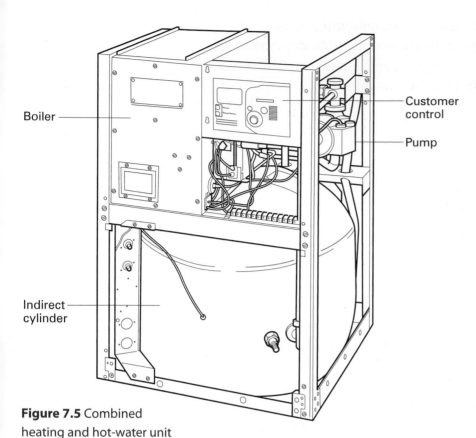

Boiler

Customer control

Pump

Indirect cylinder

Figure 7.5 Combined heating and hot-water unit

These are fitted into the fireplace opening behind a fireplace surround to which it is normal to have a gas fire fitted on the face. Although they are concealed and save wall- or floor-space, they have a limited heat output of approx 15 kW.

Combined heating and hot-water units

These have a boiler and hot-water storage cylinder all contained within one unit; some may also contain the feed-and-expansion cistern or a sealed expansion vessel. All controls and connections are built into the unit, as shown in Figure 7.5.

Combination (combi) boilers

These appliances supply central heating via a radiator system and domestic hot-water on an instantaneous water-heater basis.

The central heating normally uses a sealed system, but the important feature is that any hot-water demand takes priority over the central-heating demand. A big advantage of these appliances is that there is no space needed for a hot-water cylinder. All hot water is heated instantly; this does, however, have limitations on the hot-water delivery rate to the taps, and of course another disadvantage is that the heating system does not get any heat while hot water is being drawn off.

The combi is really only a central-heating boiler that becomes an instantaneous water heater on demand of any hot water. Figures 7.6 and 7.7 help show the two different circuits that are controlled by a diverter valve activated by opening a hot tap.

There can be sizeable differences in the controls used in combination boilers to separate and provide heat to the two circuits – here we can only provide an overview. It can be beneficial to attend manufacturer training courses to cover specific combination boilers, as their service and maintenance can often be quite complex.

When there is no demand for central heating, the boiler only fires when the domestic hot water is drawn off.

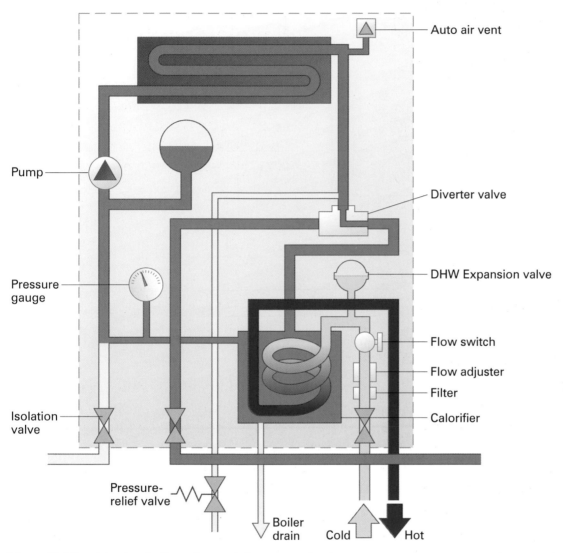

Figure 7.6 Combination boiler – domestic hot-water circuit

If there is a demand for central heating, the boiler supplies the heating system. When hot water is drawn off, the full output from the boiler is directed via the diverter valve to the calorifier to give maximum supply to hot-water.

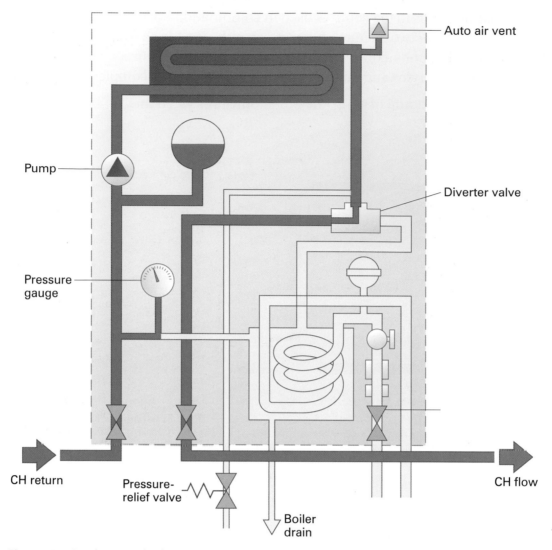

Figure 7.7 Combination boiler – central-heating circuit

Condensing boilers

Condensing boilers are approximately 94 per cent efficient in their condensing mode. They achieve this higher efficiency by using the heat from the normally wasted flue gas temperature, which is in the region of 175 to 250°C. Extra tubes/heat exchangers take this extra heat from the flue and use it to heat the water entering the boiler. Since the flue gases are now as low as 75°C, the majority of water vapour in the flue products becomes liquid giving up its latent heat, hence the term 'condensing'. The installation notes later in this section give ways of dealing with the condensate, which is a weak acid that must be discharged safely to waste.

Due to the acid that is produced, the materials from which the boiler is manufactured will usually be different from their traditional counterparts – typically stainless steel is used. The flue system must always fall back to the boiler to permit the condensate to fall back and be discharged through the condensate pipe.

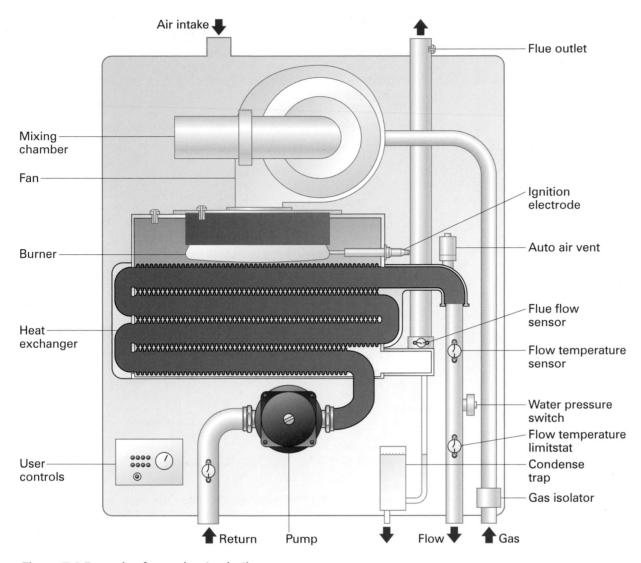

Air intake
Flue outlet
Mixing chamber
Fan
Ignition electrode
Burner
Auto air vent
Flue flow sensor
Heat exchanger
Flow temperature sensor
Water pressure switch
Flow temperature limitstat
User controls
Condense trap
Gas isolator
Return
Pump
Flow
Gas

Figure 7.9 Example of a condensing boiler

Burners and controls

All boilers use a combination of mechanical and electrical control devices. The most modern boilers now use electronic ignition, and many light using an intermittent pilot, although most existing ones use a permanent pilot.

Permanent pilot

This works as follows:

- Gas is supplied through a service cock to the inlet of the multifunctional valve.

- A piezo igniter is used to light the pilot.

- The pilot heats a thermocouple which connects to the multifunctional valve (gas valve).

- A thermostat calling for heat then gives power to the gas valve, which opens.

- The main burner receives gas, which is lit by the pilot.

- If the pilot fails then the thermocouple cools, the gas valve is de-energised, closes and fails safe.

- The boiler thermostat controls the temperature of the water by opening and closing the gas valve according to the temperature reached in the boiler.

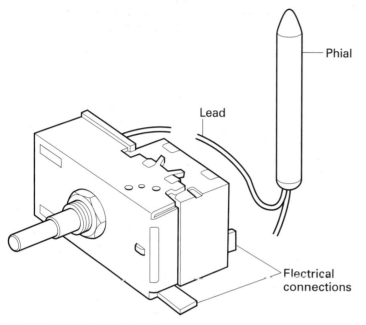

Figure 7.10 Typical boiler thermostat

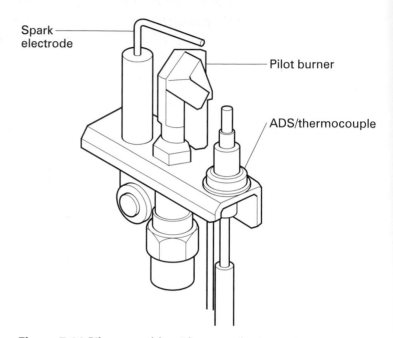

Figure 7.11 Pilot assembly with atmospheric sensing device (ASD)

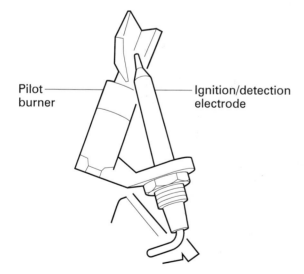

Figure 7.12 Intermittent pilot assembly

Intermittent pilot

- The boiler thermostat calls for heat.

- The control board/box activates spark ignition.

- Spark occurs at the pilot assembly.

- The pilot solenoid valve opens to let gas through to pilot.

- The pilot ignites and the ignition/detection electrode detector senses flame.

- The spark stops and the main solenoid opens.

- Gas flows to the main burner and is lit by the pilot.

Air-pressure switch

Fan-flued boilers use an air-pressure switch as a safety cut-off device should the fan fail at any time. Because the fan is used to assist with the removal of products of combustion and the supply of combustion air, it is critical to the safe operation that the fan is working satisfactorily at all times. The main feature of the air-pressure switch is that it prevents the boiler lighting sequence in the event of the fan not operating and causing sufficient air pressure.

Air-pressure switch

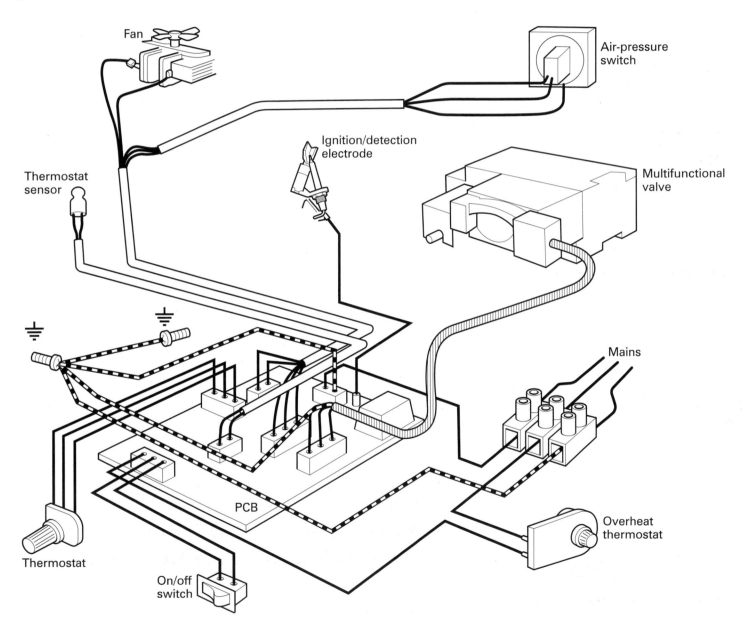

Figure 7.13 Boiler controls circuit

Figure 7.13 shows how all the components link to the printed circuit board within a typical boiler.

Overheat thermostat

Figure 7.13 also shows an overheat thermostat; this is fitted as standard on most low water-content boilers. The overheat stat is wired 'in series' with the normal temperature control stat to provide added appliance protection in the event of a control-stat failure.

In addition, with combination boilers and many low water-content boilers there are other thermostats installed to sense temperature at various points in the appliance. There can also be water-pressure switches installed, which control the operation of components such as the pump. Where multiple controls such as this are included in boilers they are usually controlled by a number of **printed circuit boards (PCBs)**.

When you attend college your tutor will provide you with a practical demonstration of the range of boilers, including an overview of key boiler controls.

Now that we have looked at types of boilers and the main controls, let us look at important installation restrictions; we will then learn commissioning procedures and finish up with servicing.

Installation criteria

The British Standard that deals with the installation of gas boilers not exceeding 70 kW is BS 6798. You should refer to a copy of this, as it also covers inspection and commissioning for all boilers using 1st, 2nd or 3rd (lpg) family gases.

Here are some of the main points, although for full details consult the BS.

General

- All new boilers must carry the CE mark or conform to appropriate BS requirements.
- The installer should check the data plate to confirm it is suitable for the gas used.
- Terminals and condensate pipes from condensing boilers must be installed as per manufacturers' instructions.

Materials

- All materials and components shall conform to BS or European Standards.

Installation and location

- A boiler using 3rd family gas (LPG) shall not be installed in a basement or cellar.
- Room-sealed boilers can be installed in any room but it is recommended *not* to fit them in bath/shower rooms/bed/bedsits or cloakrooms.
- Boilers fitted in garages or loft spaces should have frost protection.
- A boiler must have an adequate air supply for combustion and ventilation as per BS 5440-2.
- The space around the boiler shall be at least the minimum stated by the manufacturer's instructions or a minimum of 75 mm from any combustible material.

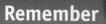

Remember

It is essential that an overheat thermostat is fitted with any boiler that is used for sealed water-heating systems

Boiler compartments

- Any boiler compartment must be of a fixed, rigid structure.
- The compartment must have air vents for cooling in addition to air required for combustion (where required for open flues).
- Where open-flued boilers are used the air vents or door must not communicate via a bathroom/shower room.
- Open-flued installations that have air vents communicating with a bedroom are to be 14 kW maximum.
- Easy access is required for inspection and servicing.
- A notice shall be fixed in a prominent position warning against use as a storage cupboard.

Understairs cupboard installations

- For premises of no more than two storeys the cupboard must comply with the same requirements as compartments (see above).
- For premises over two storeys the internal surfaces of the cupboard shall be non-combustible or suitably lined to give a minimum of half an hour's fire resistance.
- Air vents must go directly to outside.

Bathroom/shower-room installation

- Only room-sealed appliances are allowed; however:
  - The electrical controls must be out of reach of bath/shower users.
  - Avoid bathroom installation unless there is no alternative location.

Bedrooms and bedsitting rooms

- A maximum of 14 kW gross input is allowed.
- Open-flued appliances must have an ASD.
- Avoid installations in bedrooms if possible.

Roof-space installations

- Flooring to give adequate access is required.
- Fixed lighting is required.
- A loft ladder or purpose-designed access must be provided.
- A guard to prevent contact with stored articles should be fitted.

General points

- Boilers need to be fitted with either an open vent or a sealed system to ensure safe operation at all times.

- Sealed systems need a good-quality safety valve.

- Gas supply to boiler must be sized correctly to give minimum working pressure.

- A gas isolating valve must be provided adjacent to the boiler.

- On back-boiler installations in fireplaces all pipework must be protected against corrosion from falling soot etc.

- All electrical supplies must conform to manufacturers' instructions.

- Electrical isolation is to be by a double-pole fused switch/spur or a fused three-pin plug to an unswitched socket outlet.

Condensing boilers installation

These are becoming increasingly popular due to government initiatives on energy efficiency, and are likely to be the only type of appliance permissible in the near future. Condensate pipes from condensing boilers must be installed as per manufacturers' instructions. There are special requirements, as shown in these drawings from BS 6798 (Figures 7.14 to 7.18).

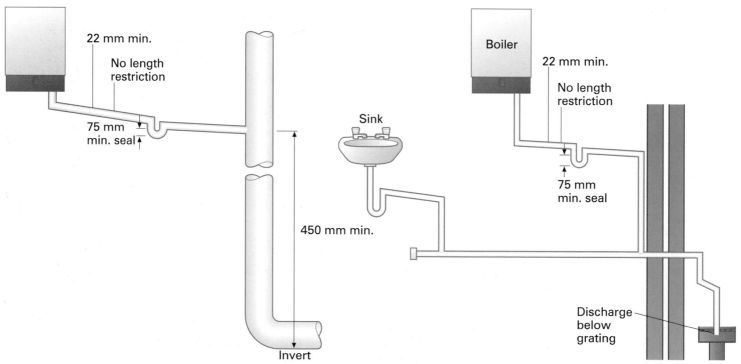

Figure 7.14 Condensate pipe to a soil/vent stack

Figure 7.15 Condensate pipe to a gully via a sink discharge pipe

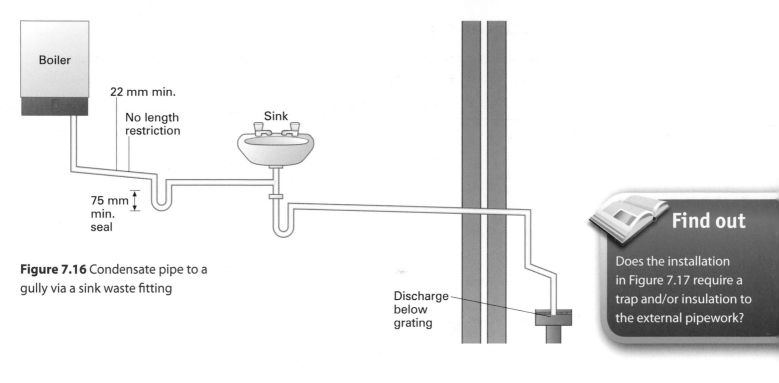

Figure 7.16 Condensate pipe to a gully via a sink waste fitting

Find out

Does the installation in Figure 7.17 require a trap and/or insulation to the external pipework?

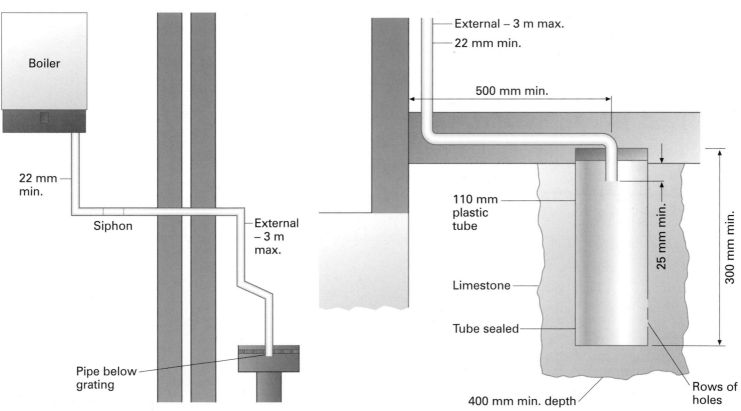

Figure 7.17 Condensate pipe to a gully via a condensate siphon

Figure 7.18 Condensate pipe to a purpose-made condensate absorption point

Preferably all condensate drainage pipes should run internally as far as possible, to prevent freezing. All connecting drainage pipework should have a fall of at least 2.5 degrees to the horizontal or approximately 50 mm per metre run of pipe run.

Commissioning

Before any boiler is commissioned it must be inspected to ensure that all work has been carried out to the manufacturer's instructions, the British Standards and the relevant sections of the Gas Safety Regulations. Some of the main checks are:

- combustion air is adequate
- the flue is correctly constructed
- general condition and suitability of the boiler
- the installation work is satisfactory and complies
- gas fittings and other works for the supply of gas are adequate – the system is gas-tight
- electrical fittings and the supply of electricity are satisfactory.

Before lighting the boiler the heating system should also be checked, tested and flushed out. Now we come to the actual lighting:

- Light boiler as per manufacturer's instructions.
- Check and adjust burner pressure as stated on data plate.
- Check gas rate.
- Check operation of flue (open flues require a spillage test).
- Check boiler controls and safety devices are functioning correctly.
- Demonstrate correct operating procedure to customer.
- Provide user with manufacturer's instructions.
- Advise customer of need for regular servicing.

When servicing/maintaining a boiler it is essential to follow the exact procedure as laid out in the manufacturer's instructions. The following procedure gives a general overview, although special appliances like condensing boilers will require additional checks.

Servicing and maintenance

1 Control cover down

2 Front casing removed

3 Combustion covers removed

4 Cleaning heat exchanger

5 Setting burner pressure

6 Flue-gas analysis

Preliminary checks

- Check installation conforms to Regulations/Standards.
- Light appliance and check correct operation/signs of spillage.
- Check flue and terminal type/position.
- Ensure ventilation is adequate (open flues).
- Ensure compartment ventilation is suitable (where appropriate).
- Check ASD if fitted.

Servicing

- Isolate gas/electric supplies to appliance.
- Remove outer casing/panels to access burner.
- Remove burner and control assembly.
- Remove burner injector and blow clean.
- Take out pilot/thermocouple/ignition assembly and clean.
- Clean main burner and lint arrestor.
- Take flue cover off and clean flue ways.
- Clean flue ways through heat exchanger.
- Remove all debris from combustion chamber.
- Check condition of all components and replace if necessary.
- Reassemble the components, checking any gas connections.
- Ensure all appliance seals are OK.

Recommissioning

- Re-light boiler as per manufacturer's instructions.
- Check burner pressure as stated on data plate.
- Check gas rate.
- Check operation of flue (open flues require a spillage test).
- Check boiler controls and safety devices are functioning correctly.
- Report back to user with manufacturer's instructions.
- Advise customer of the ongoing need for regular maintenance.

In the event that a parts failure occurs on the appliance, the manufacturer's instructions will provide a detailed fault-diagnosis and testing procedure for key system components – so refer to the instructions. A complete range of parts is usually available as replacements in the event of a failure. Again, a key document to refer to is the installation instructions.

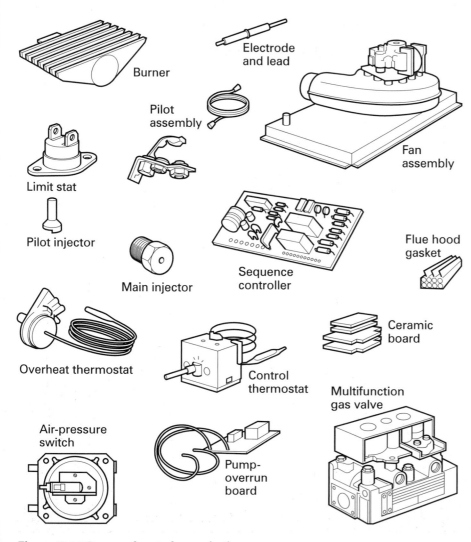

Figure 7.19 Range of parts for gas boilers

Domestic gas fires and wall heaters

At the end of this section you should be able to:

- recognise various types of gas fires, stoves and wall heaters
- check appliances are suitable for given locations
- state the main installation criteria
- carry out pre-commissioning checks
- state the commissioning procedure
- explain the procedure to carry out servicing
- check operation of related controls.

Gas fires and wall heaters are classed as individual space heaters that can provide heat to rooms according to the needs of individuals. The types available are:

- Gas fires – these may be radiant or radiant/convector, natural or fanned flue.
- Convectors – normally wall-mounted and room-sealed but may be open-flue type. Flueless convectors are less common now as they caused problems with condensation.
- Cabinet heaters (mobile) – these are flueless radiant heaters using LPG, not really part of a plumber's work.

Gas fires

These are best classed as the three types detailed in the British Standard.

1. Radiant or radiant/convector gas fire to BS 5871 Part 1

- Flue size: minimum of 125 mm across axis of flue normally required.
- Location: normally in front of closure plate, which is fitted to fireplace opening.
- Ventilation: purpose-provided ventilation not normally required up to 7 kW input.

For this type of appliance, the radiating surface can be either radiant(s) or imitation fuel giving a live-fuel effect.

Radiant convector fire

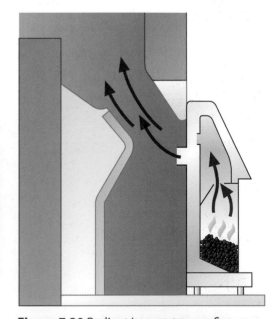

Figure 7.20 Radiant/convector gas fire

2. Inset live-fuel effect (ILFE) gas fire to BS 5871 Part 2

- Flue size: minimum of 125 mm across axis of flue normally required.

- Location: either fully or partially inset into builder's opening or fireplace recess. (For a recess, the chairbrick at the back of the opening might have to be removed, depending on the appliance design.)

- Ventilation: purpose-provided ventilation not normally required up to 7 kW input.

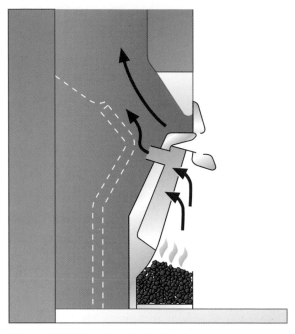

Figure 7.21 Inset live-fuel effect

LFE fire

3. Decorative fuel-effect (DFE) gas appliance to BS 5871 Part 3

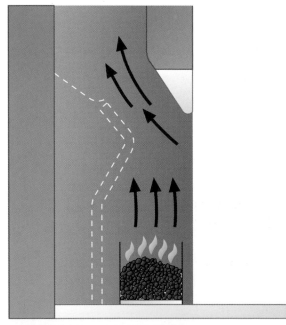

Figure 7.22 Decorative fuel-effect fire

Decorative fuel-effect fire

Heating stoves

These are available in two basic types: free-standing with a top flue outlet and those fitted with a rear flue spigot to a closure plate similar to a gas fire. Some examples of installations are shown in Figures 7.31–3.

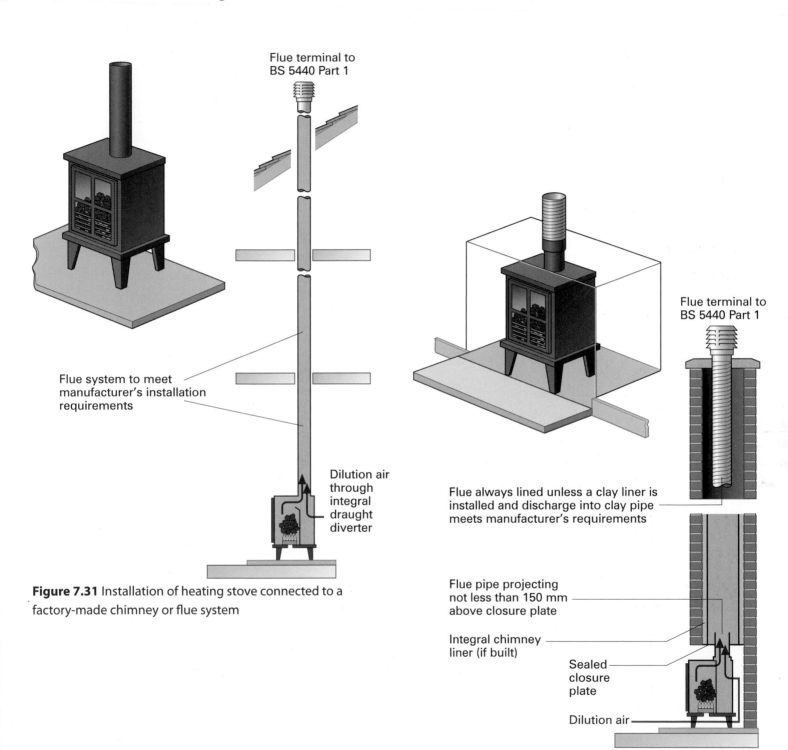

Flue terminal to
BS 5440 Part 1

Flue system to meet
manufacturer's installation
requirements

Dilution air
through
integral
draught
diverter

Figure 7.31 Installation of heating stove connected to a factory-made chimney or flue system

Flue terminal to
BS 5440 Part 1

Flue always lined unless a clay liner is
installed and discharge into clay pipe
meets manufacturer's requirements

Flue pipe projecting
not less than 150 mm
above closure plate

Integral chimney
liner (if built)

Sealed
closure
plate

Dilution air

Figure 7.32 Installation of a free-standing heating stove fitted in a builder's opening

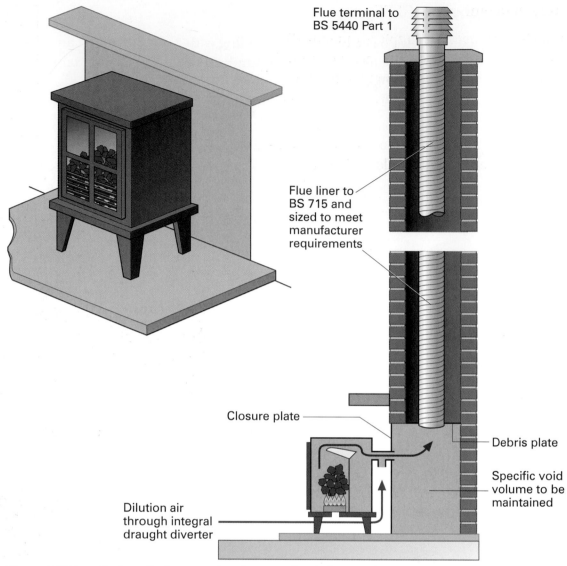

Flue terminal to
BS 5440 Part 1

Flue liner to
BS 715 and
sized to meet
manufacturer
requirements

Closure plate

Debris plate

Specific void
volume to be
maintained

Dilution air
through integral
draught diverter

Figure 7.33 Installation of a free-standing heater using a closure plate

Live fuel-effect fires

There are several types of gas fire that use a flame effect, and these are becoming
increasingly popular. The radiants are replaced by realistic coals or logs; these give
a live real-fire effect. The coals/logs must be placed in position exactly as per the
manufacturer's instructions in order to give complete combustion. There are two
main types: live fuel-effect and decorative fuel-effect.

Inset live fuel-effect (ILFE)

Usually inset into the fireplace opening (ILFE), the flame effect is open and efficiency is only about 40 to 50 per cent. However, there is a convection chamber that helps to recover some of the heat.

Note the path of the products of combustion, the convected air and the flow of room air.

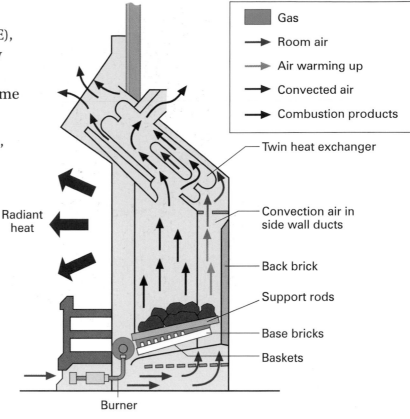

Figure 7.34 Inset live fuel-effect fire (ILFE)

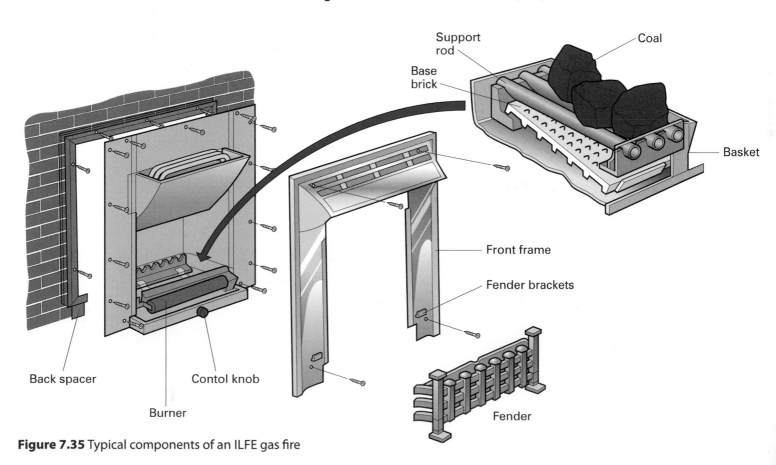

Figure 7.35 Typical components of an ILFE gas fire

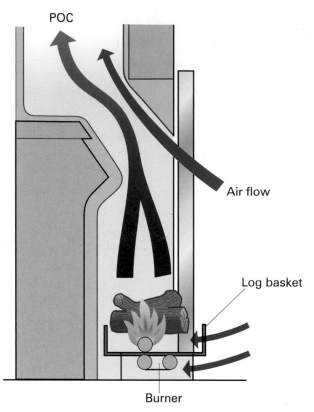

POC

Air flow

Log basket

Burner

Figure 7.36 Decorative fuel-effect fire (DFE)

Decorative fuel-effect (DFE)

This appliance produces only radiant heat and has a very low efficiency. It is extremely popular, though it is generally used as a 'decorative' focal point in a room that already has a form of heating.

The flue should have a cross-sectional dimension of not less than 175 mm unless certified in the manufacturer's instructions. Typical fireplaces into which appliances can be installed are shown in the following case types, but special installation considerations apply, as shown on the following pages.

1 Builder's opening

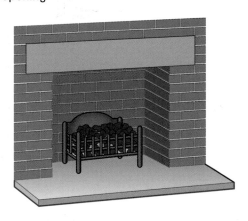

2 Fireplace recess

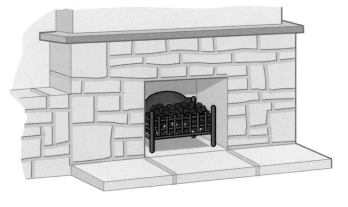

3 Raised builder's opening

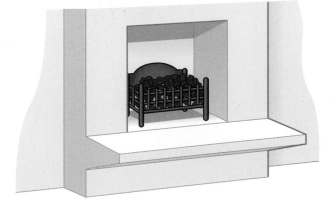

4 Builder's opening with independent canopy

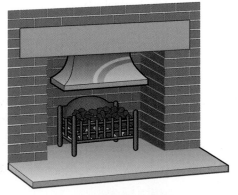

Figure 7.37 Typical fireplaces

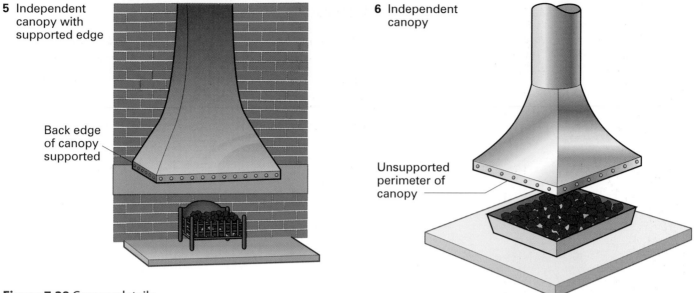

5 Independent canopy with supported edge

Back edge of canopy supported

6 Independent canopy

Unsupported perimeter of canopy

Figure 7.38 Canopy details

It is also becoming increasingly popular for DFE fires to be fitted under a canopy, as shown in example 4 in Figure 7.37 and examples 5 and 6 in Figure 7.38. Particular care should be taken with the constraints of these, as shown in Figure 7.39.

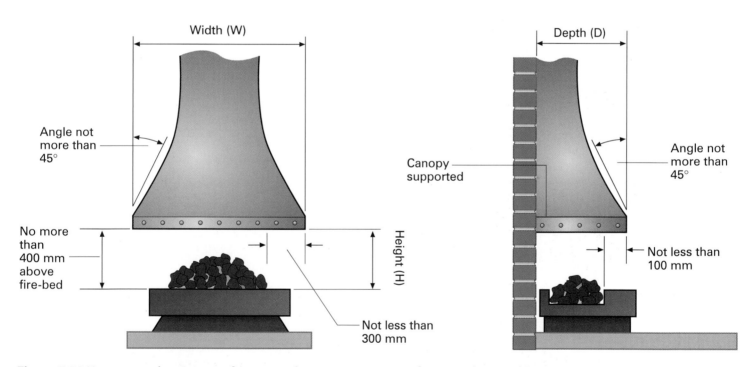

Width (W)

Angle not more than 45°

No more than 400 mm above fire-bed

Height (H)

Not less than 300 mm

Depth (D)

Canopy supported

Angle not more than 45°

Not less than 100 mm

Figure 7.39 Unsupported perimeter of canopy: relevant measurements for examples 4 and 5

Refer to BS 5871 Part 3 and take note of how to size the canopy for example 6. If you don't understand this one, you'll need to speak to your tutor.

There is one more type of appliance that we need to consider before we look at installation guidelines: this is the condensing gas fire.

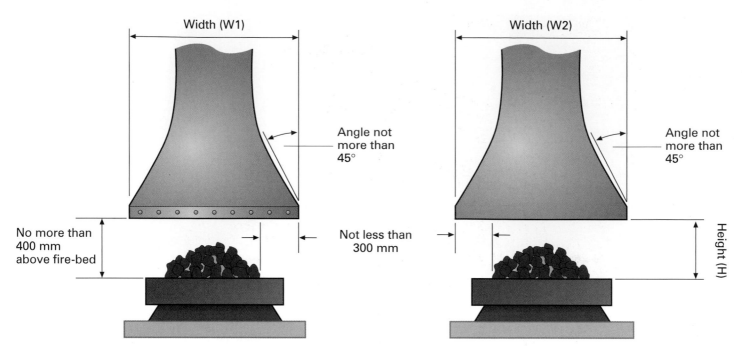

Figure 7.40 Unsupported perimeter of canopy: relevant measurements for example 6

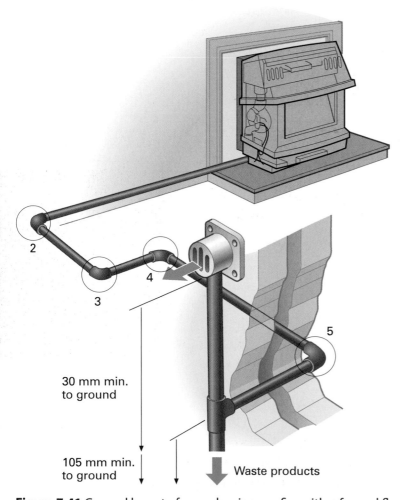

Figure 7.41 General layout of a condensing gas fire with a fanned flue

Condensing gas fire

This appliance relies on a fan flue to entrain combustion air from the room, through the heat exchanger and out through a small diameter (28 mm) CPVC plastic flue pipe. As with all fan-flue appliances, there must be a safety device, such as a pressure switch, fitted to shut down the burner in the event of fan failure.

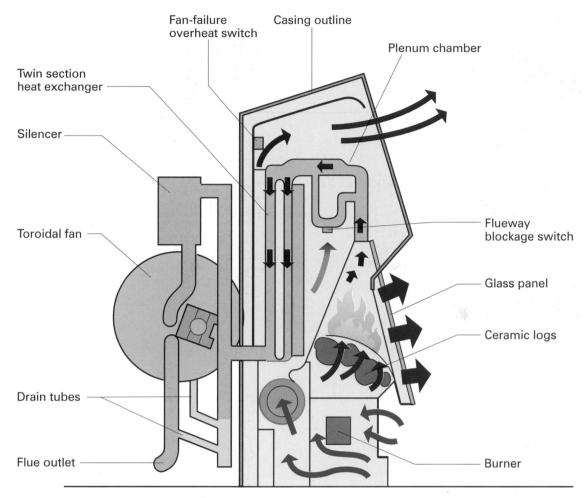

Fan-failure overheat switch
Casing outline
Plenum chamber
Twin section heat exchanger
Silencer
Toroidal fan
Drain tubes
Flue outlet
Flueway blockage switch
Glass panel
Ceramic logs
Burner

Figure 7.42 Section through a condensing fire, showing air, combustion and condensate flows

General installation criteria

When fitting gas fires and wall heaters it is essential to follow the manufacturer's instructions for the appliance. Reference to the British Standards can be made as follows:

- BS 5781 Part 1: Gas fires and convector heaters

- BS 5781 Part 2: Inset live fuel-effect fires

- BS 5781 Part 3: Decorative fuel-effect gas fires.

In addition, it is necessary to satisfy the requirements of the Gas Safety (Installation and Use) Regulations. It is important to note that:

- Open-flued or flueless appliances must NEVER be installed in bathrooms or shower rooms.

- Appliances over 12.7 kW net (this includes combined gas fire/back-boiler units) must not be fitted in a **room used for sleeping**. However, appliances have been allowed since January 1996 but only if they were fitted with an atmospheric sensing device (ASD).

Find out

What should be done with a gas fire that was fitted without an atmospheric sensing device (ASD) prior to 1996?

Some of the other crucial requirements are that the installer must ensure that the flue to which the appliance is fitted is suitably constructed and is in a proper, safe condition for the operation of the gas fire. The installer must also ensure that there is a gas tap for shutting off the appliance.

On the job: Installing a gas fire to a builder's opening with a closure plate

Kristian is a fourth-year apprentice helping his supervisor install a gas fire to a builder's opening with a closure plate. The builder's opening is slightly too small but the supervisor tells Kristian to fix the closure plate anyway. Kristian is aware of the minimum catchment space due to his studies at college, and he queries the decision on the grounds of safety. 'Get on with it, you will never get a job done if you query everything as you do,' says the supervisor.

The customer appears to have overheard some of the conversation, and during a tea break, while the supervisor is out at the van, she asks Kristian if the job is going to plan.

1   What should Kristian say?

2   Is the installation satisfactory?

As with other sections on gas appliances, it can be seen that a great responsibility is put on the installer to make sure the installation is correct and can operate safely. The following checklist may help when installing a new gas fire:

- Visually inspect flue and determine suitability.
- Carry out a flue-flow test.
- Check ventilation is suitable (are there other appliances in the room?).
- Test existing installation for tightness.
- Fit fire and pipework.
- Test completed installation for tightness.
- Purge and relight/check other appliances.
- Light fire and check/adjust burner working pressure.
- Check for spillage.
- Instruct customer as to correct use and need for maintenance.
- Hand over manufacturer's literature.

Remember

Be extra careful in situations where beds are put into sitting-rooms for the elderly or those suffering illness

5 Fire removed

6 Once fire and radiant replaced, set burner pressure

7 Spillage test

Servicing should include the following typical operations:

- Check ventilation requirements.
- Visual check of flue including termination.
- Light fire and check operation.
- Check for gas soundness.
- Disconnect and remove the fire.
- Remove closure plate.
- Check for soot and rubble behind the closure plate and clean out.
- Carry out a flue-flow test.
- Reseal the closure plate, ensuring that the relief opening is clear.
- Clean dust and lint from the base of the fire where the cool air enters from the room.
- Clean the burner and the injectors.
- Check the heat exchanger for corrosion damage and clean the flue ways.
- Reconnect the fire and check for gas soundness.
- Ensure fire is level and secure.
- Test the ignition, clean or renew any batteries or parts as necessary.
- Check any electrical components are properly insulated.
- Check burner pressure and adjust regulator if required.
- Check gas rate.
- Check the operation of flame-supervision device.
- Check user controls operate satisfactorily.
- Check for spillage.
- Report back to customer.

Domestic gas cookers

At the end of this section you should be able to:

- recognise various types of gas cookers
- check operation of user and safety controls
- state the installation criteria
- carry out pre-commissioning checks
- state the commissioning procedure
- explain the procedure to carry out servicing.

There are a number of categories of cooking appliances available today:

- free-standing
- slide-in
- built-in.

These are then further categorised as:

- low-level grill/oven
- eye-level grill/oven
- hob, with/without fold-down lid.

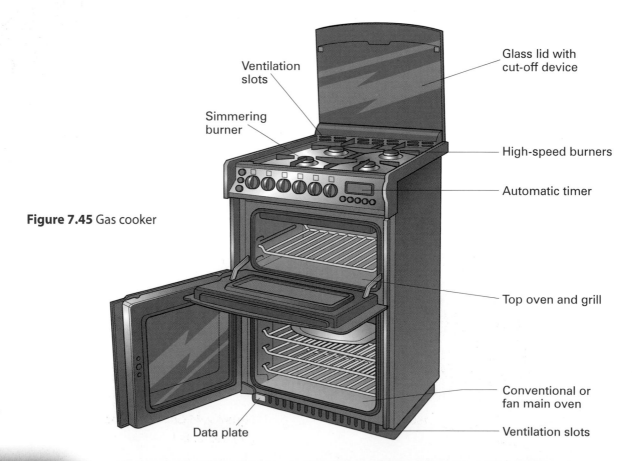

Figure 7.45 Gas cooker

Ventilation slots

Simmering burner

Glass lid with cut-off device

High-speed burners

Automatic timer

Top oven and grill

Conventional or fan main oven

Ventilation slots

Data plate

The gas cooker is now a high-tech appliance with automatic ignition and cooking time control, together with forced-convection ovens, lights and easy-clean catalytic linings.

Cookers must now carry a CE mark. Cookers manufactured before 1996 may not have this but will still comply by conforming to the Regulations, and some even carry a kite mark.

The installer must also check that the type of gas and the working pressure for which the appliance is suitable is as stated on the data plate.

In terms of cooker operation, there tend to be two main methods of controlling the operation of the oven section:

- gas mechanical controls – you've already seen the basis of these in the gas safety section on controls and devices
- electrical controls – many of these tend to be similar to the controls used in gas boilers.

Cooker installation

Now, for the installation we need to check the location is suitable and ensure that we comply with the manufacturer's instructions. Some special requirements from BS 6172 (the main standard covering gas cookers) are:

- Cookers must *not* be installed in a bathroom or shower room.
- Unless it is a single burner it must *not* be installed in a bedsitting room of less than 20 m³.
- Cookers for use on LPG must *not* be installed in basements/below ground level.

Find out

Using appropriate reference sources, research the controls used in an oven with mechanical controls and an oven with electrical controls; make drawings of the layout and positioning of the controls and describe each oven's method of operation. If you are attending college, have the details checked by your tutor before moving on

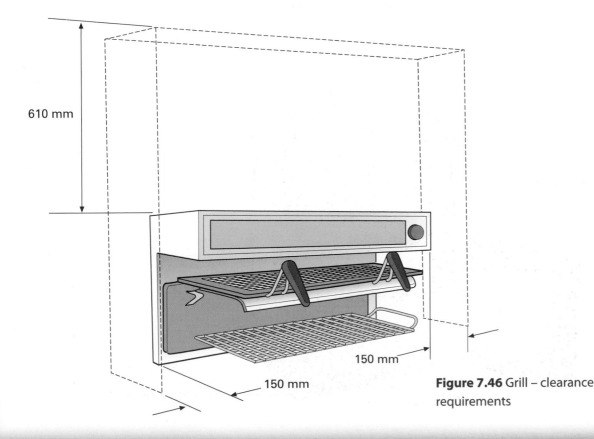

610 mm

150 mm

150 mm

Figure 7.46 Grill – clearance requirements

There are other considerations when siting a gas cooker:

- Avoid draughts by not fixing too close to doors/windows.
- Be aware of curtains and other combustible materials close by.
- Follow manufacturer's instructions regarding clearances.
- A stable base is needed for a floor-standing cooker.
- Ventilation requirements must comply with BS 5440 (reference here is to other appliances and/or use of extract fans).

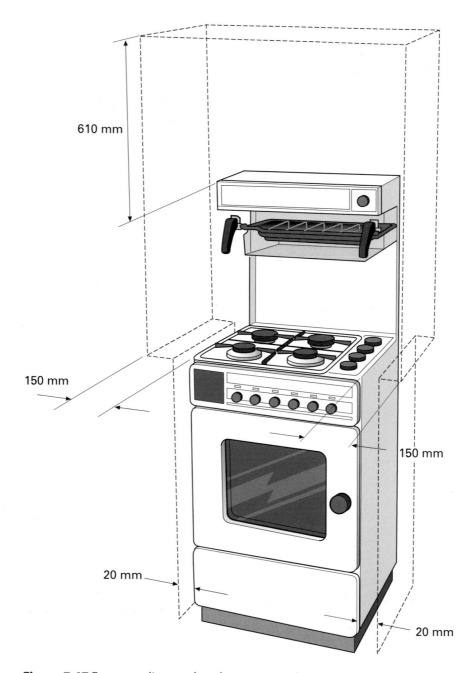

Figure 7.47 Free-standing cooker clearance requirements

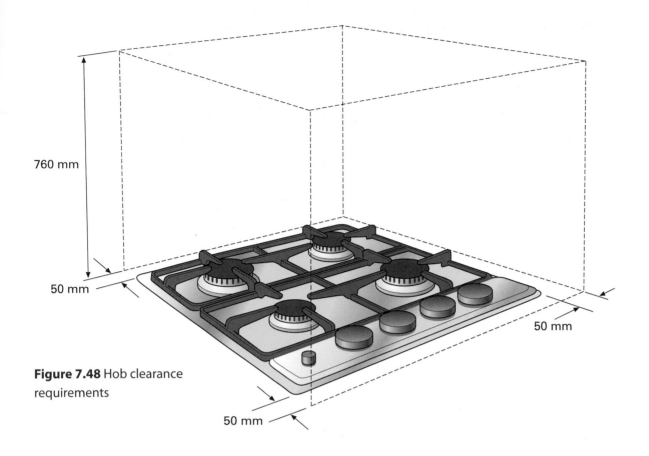

760 mm

50 mm

50 mm

50 mm

Figure 7.48 Hob clearance requirements

Gas connection

The gas installation pipe should be of an appropriate size to maintain a suitable working pressure at the appliance and to give the required heat input at the appliance. Fixed appliances will be connected by rigid pipework and an isolating valve, while slide-in appliances will normally be connected by a flexible appliance connector for use with a self-sealing plug-in device.

This flexible connector is fitted as follows:

- Do not subject to undue force.
- Keep away from excessive heat.
- Bayonet fitting should be secure to wall fitting.
- Hose should hang downwards.
- Flexible hose should not suffer from abrasion or damage.

Bayonet/flex connection

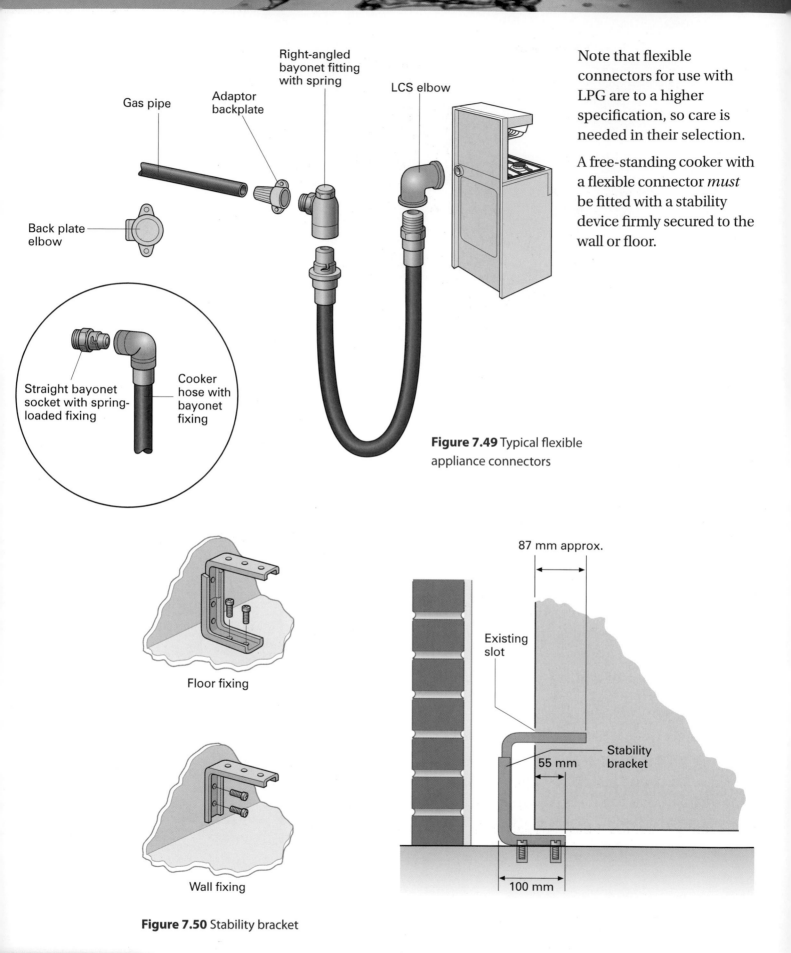

Gas pipe

Adaptor backplate

Right-angled bayonet fitting with spring

LCS elbow

Back plate elbow

Straight bayonet socket with spring-loaded fixing

Cooker hose with bayonet fixing

Note that flexible connectors for use with LPG are to a higher specification, so care is needed in their selection.

A free-standing cooker with a flexible connector *must* be fitted with a stability device firmly secured to the wall or floor.

Figure 7.49 Typical flexible appliance connectors

Floor fixing

Wall fixing

Figure 7.50 Stability bracket

87 mm approx.

Existing slot

Stability bracket

55 mm

100 mm

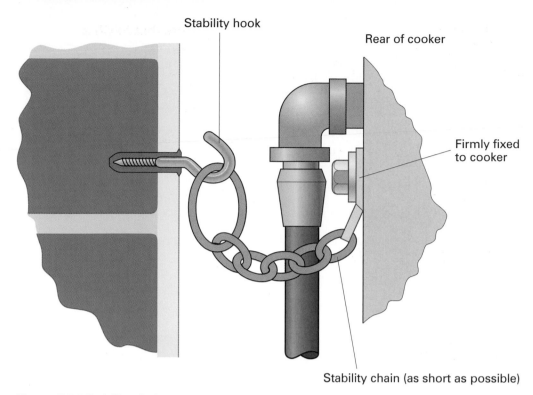

Stability hook

Rear of cooker

Firmly fixed
to cooker

Stability chain (as short as possible)

Figure 7.51 Stability chain

Electrical connection

The electrical connection should be made as per the manufacturer's instructions, with particular regard to fuse rating, continuity, polarity and earth connection. The connection point should be no more than 1.5 m from the appliance, and the flexible cable should not be exposed to hot surfaces and products of combustion.

Commissioning

Before beginning to commission the appliance some preliminary checks should be made:

- The appliance is suitable for the gas supplied.
- Room/ventilation is suitable for the appliance.
- Installation meets manufacturer's instructions.
- Gas supply/connection meets requirements and is gas-tight. The gas-tightness test must be conducted to the appliance with lid shut-off devices in the OPEN position to ensure that gas is supplied to all parts of the cooker.
- Proximity to combustible material has been checked and is acceptable.
- Stability bracket has been fitted (where required).

Now that we have made these checks and the installation complies, we can commission the cooker:

- Check electricity is on/batteries are fitted.
- Purge appliance and check gas-control taps work smoothly.
- Check igniters to all hob burners and light each burner.
- Check flames for aeration/height.
- Check burner pressure at test point and check gas rate.
- Light oven and check operation of flame-supervision device.
- Check for correct operation of oven thermostat on bypass rate.
- Instruct user and advise on regular servicing.

Servicing/maintenance

1 The cooker

2 Burners and covers removed

3 Checking grill burner/combustion

4 Checking oven burner for incomplete combustion

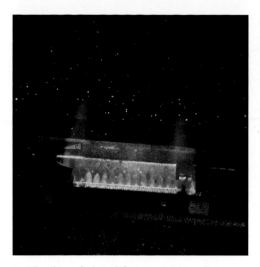

5 Checking fsd and flame picture after service

6 Checking door seal

7 Checking igniter and gas taps

When servicing/maintaining a cooker it is essential to follow the exact procedure, as laid out in the manufacturer's instructions. The following procedure gives a general overview.

- Preliminary checks:
  - Light appliance and check for correct operation.
  - Ventilation is adequate, room is suitable.
  - Appliance is suitable for gas supplied.
  - Installation meets manufacturer's instructions.
  - Gas supply/connection meets requirements and is gas-tight.
  - Proximity to combustible material is acceptable.
  - Stability bracket has been fitted (where required).

- Servicing:
  - Isolate gas/electricity supplies to appliance.
  - Check condition of flexible connector.
  - Remove burners and clean air ports.
  - Check and grease gas taps.
  - Check and clean flue ways to oven/grill.
  - Check condition of all components and replace if necessary.
  - Reassemble the components, checking any gas connections.
  - Ensure oven door seals are OK.

- Recommissioning:
  - Re-light cooker.
  - Check burner pressure as stated on data plate.
  - Check igniter function is OK.
  - Check flame supervision on oven.
  - Check oven thermostat.
  - Report back to user with manufacturer's instructions.
  - Advise customer of the ongoing need for regular cleaning/maintenance.

Water heaters

At the end of this section you should be able to:

- recognise various types of water heater
- explain the construction and operation
- state the installation criteria
- carry out pre-commissioning checks
- state the commissioning procedure
- explain the procedure to carry out servicing.

An instantaneous water heater provides instant hot water on opening a tap; there is no storage of hot water. The water is heated to the required temperature as it passes through the heat exchanger. The temperature is determined by the speed that it passes through so that on full water flow the temperature will be lower.

There are two main types of instantaneous water heater:

- single point – as the name suggests, it supplies only one point. It is normally of the flueless type
- multi-point – these are large heaters that can supply several outlets and require a flue, which is normally a room-sealed type.

Multi-point water heaters have been largely replaced now by the combination boiler, although you may well see plenty still installed in properties.

Construction and operation

Instantaneous water heaters generally consist of the following parts:

- combustion chamber and heat exchanger
- water section with automatic valve with slow ignition device and temperature selector
- gas section, with governor or throttle, main gas, pilot and burner
- ignition device and flame-supervision device.

Large multi-points are also produced with fanned-flue options. The controls associated with these are a combination of those we are about to cover, and safety controls installed in boilers, such as an air-pressure switch.

Single-point water heater – heat exchanger

Heat from the burner flames is transferred to the heat exchanger, which has a large surface area of fins to extract as much heat as possible. The slower the water passes through, the hotter the water becomes. The water heater has an open end, usually in the form of a swivel spout, so the appliance always operates safely.

The water heater also incorporates a flame-failure device, usually a thermocouple which is activated by the pilot. Should the pilot flame fail then the thermo-electric valve will close.

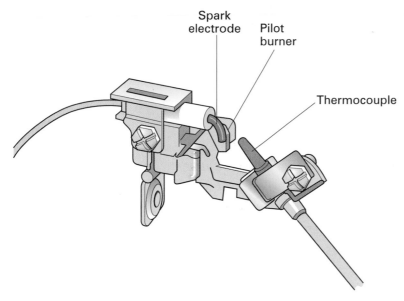

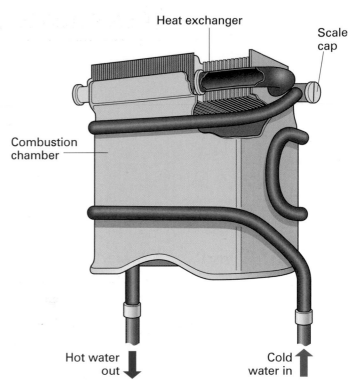

Figure 7.52 Thermocouple/flame-failure device

Figure 7.53 Heat exchanger of water heater

On most new appliances a thermal switch and ASD are built in to prevent overheating and to prevent the appliance from continuing to work if the oxygen levels fail to support complete combustion.

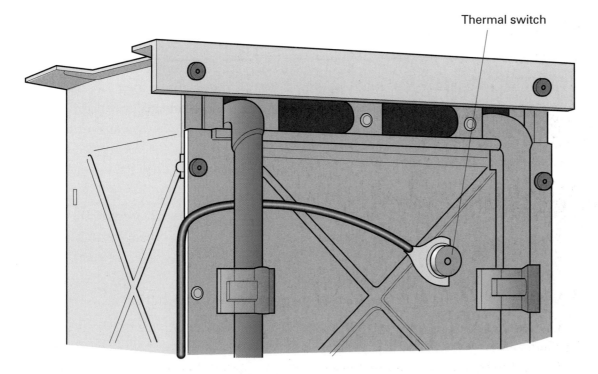

Figure 7.54 ASD/thermal switch

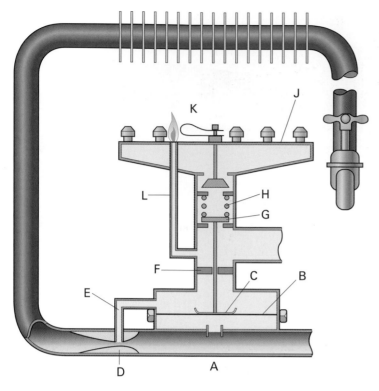

Figure 7.55 Operation – closed position

The most interesting part of an instantaneous water heater is the automatic valve. This turns the gas on when a tap is opened and turns the gas off when the tap is closed. Most work on the principle of 'differential pressure', using a venturi. The principle of operation can be seen in Figures 7.55 and 7.56.

Key to Figures 7.55 and 7.56

A – automatic valve/diaphragm housing

B – rubber diaphragm

C – bearing plate

D – venturi

E – low-pressure side

F – gland for push rod

G – gas valve

H – gas-valve spring

J – burner

K – flame supervision device (bi-metal strip)

L – pilot feed

The temperature of the water can be controlled by a thermostat in larger models, but the simplest form of control is the temperature selector, as shown in Figure 7.57. This reduces the amount of water flowing through the venturi and therefore reduces the amount the gas valve opens.

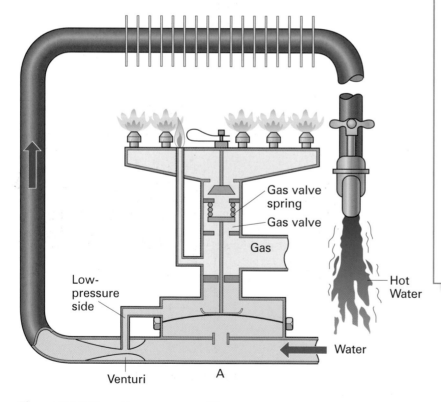

Figure 7.56 Operation – open position

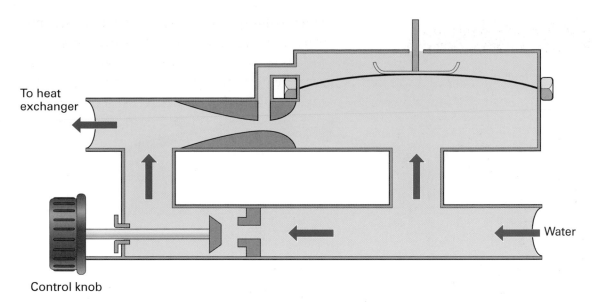

Figure 7.57 Temperature selector

Instantaneous water heaters require a substantial gas rate to heat the quantity of water to the desired temperature, and this can cause noisy ignition if the gas is not let in gradually when turned on. The 'slow ignition device' allows the main burner to light safely and quietly, and also has the advantage of closing the burner down quickly. There are several designs, some of which have adjustment. One type is the 'loose ball', usually situated in the low-pressure side of the diaphragm. Screwing the device in or out regulates the amount of water that can bypass the ball and so adjusts the rate at which the diaphragm lifts.

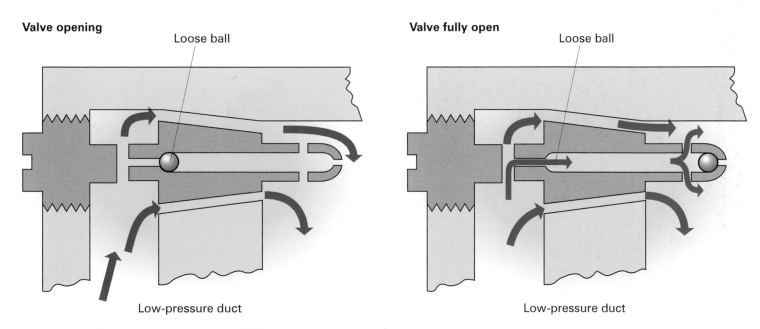

Figure 7.58 Slow ignition device, loose ball type

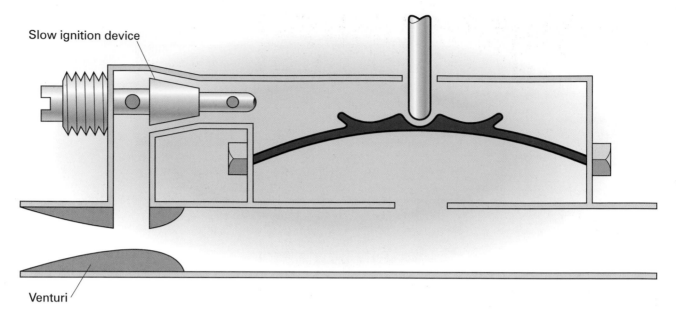

Slow ignition device

Venturi

Figure 7.59 Slow ignition device shown in automatic valve on low-pressure side of diaphragm

Now let's put together all the component parts and look at them in-situ in the full appliance.

Small instantaneous water heater

Small instantaneous water heater

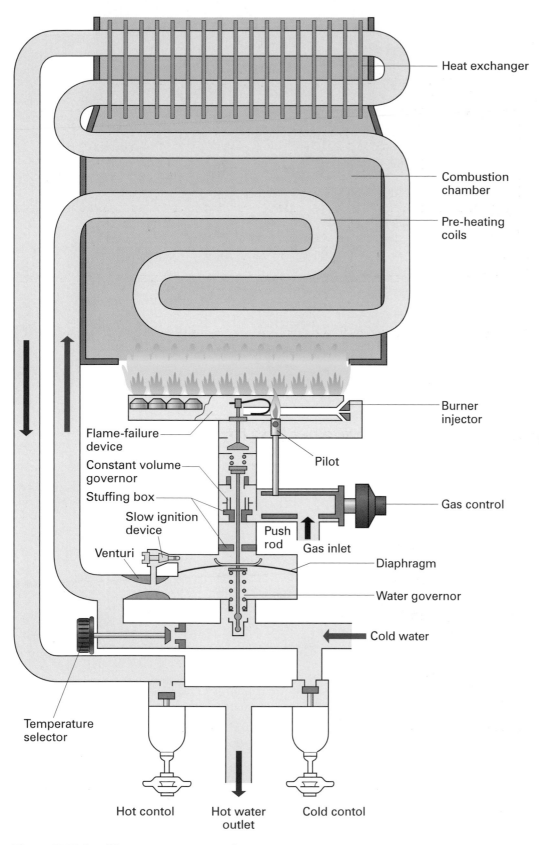

- Heat exchanger
- Combustion chamber
- Pre-heating coils
- Burner injector
- Pilot
- Gas control
- Diaphragm
- Water governor
- Cold water
- Flame-failure device
- Constant volume governor
- Stuffing box
- Slow ignition device
- Venturi
- Push rod
- Gas inlet
- Temperature selector
- Hot contol
- Hot water outlet
- Cold contol

Figure 7.60 Small instantaneous water heater

Multi-point heater

Multi-point instantaneous water heater (with thermostatic control)

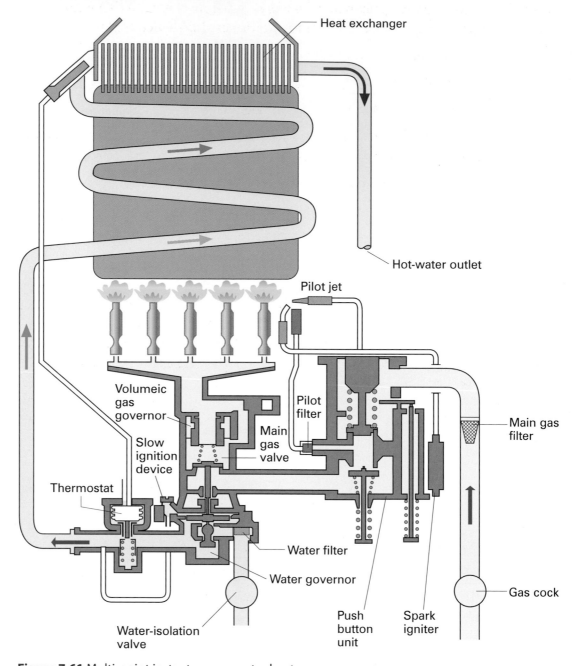

Figure 7.61 Multi-point instantaneous water heater

Installation criteria

The full requirements for installing gas water heaters are covered in BS 5546. These are the key installation points that you must take into account:

- A new water-heating appliance must carry a CE mark. Used water heaters must have a CE mark or conform to a British Standard.

- A flueless water heater must only supply an appliance in the same room.

- All instantaneous water heaters shall either be room-sealed or have a safety shut-down device to stop the appliance from working before there is a dangerous build-up of products in the room – atmospheric sensing device (ASD).

- Only room-sealed appliances are allowed in bathrooms/shower rooms.

- In rooms used for sleeping:

  - any appliance over 14 kW must be room-sealed

  - appliances less than 14 kW can be room-sealed or open-flued with safety cut-off, or flueless (12 kW gross max.) with a safety cut-off

  - rooms less than 20 m³ must have a room-sealed appliance.

- Flueless appliances must not be fitted in compartments/cupboards.

On the job: Servicing a water heater

While working with a fitter servicing a water heater you notice that the atmospheric sensing device (ASD) has broken but someone has 'fixed' it so that the heater will work and give adequate hot water without the device functioning at all. The fitter you are working with tells the customer that a new ASD is needed, but the customer refuses to accept this as the heater appears to be working OK. 'I am not paying for parts just for the sake of it,' says the customer. 'Please yourself,' says the fitter. 'I will leave it as it is then.'

1   Is this satisfactory?

2   What should be done?

Inspection and commissioning

Before it can be commissioned the water-heating installation shall be inspected to ensure that the work has been carried out as stated in the manufacturer's instructions and the Gas Safety Regulations.

Key requirements are:

- Ventilation air and combustion air are adequate.

- Flue is correctly constructed.

- General condition of the appliance and the installation are satisfactory.

- Gas fittings/installation are adequate – system is gas-tight.

- Electrical fittings/work are in accordance with BS 7671.

Commissioning

- Ensure all components and appliance are complete.
- Light pilot and check size and position of flame.
- Fit 'U' gauge to burner test point.
- Turn on heater/tap to light main burner.
- Check all gas connections/controls are sound.
- Check/adjust burner pressure.
- Compare gas rate with data plate.
- Adjust slow ignition if required.
- Carry out a spillage test (open flues only).
- Check flame supervision device.
- Check ASD if fitted.
- Determine temperature rise and flow rate of water using flow-measuring gauge.
- Demonstrate appliance to customer.
- Leave relevant documents.

Servicing

1 Water heater

2 Cover removed

3 Removing heat exchanger

4 Burner removed

5 Checking pilot assembly

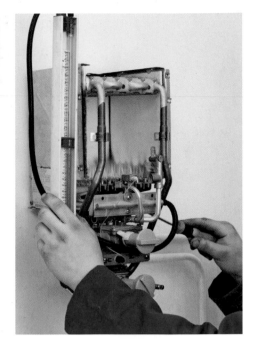

6 Setting burner pressure

- Isolate gas/electricity/water supplies to appliance.
- Remove outer casing/panels to access burner.
- Remove burner and control assembly.
- Remove burner injector and blow clean.
- Take out pilot/thermocouple/ignition assembly and clean.
- Clean main burner and lint arrestor.
- Take flue cover off and clean flue ways.
- Clean heat exchanger.
- Remove all debris from combustion chamber.
- Check condition of all components and replace if necessary.
- Reassemble the components, checking any gas connections.
- Ensure all appliance seals are OK.
- Do not forget to do a tightness test before and after the service.

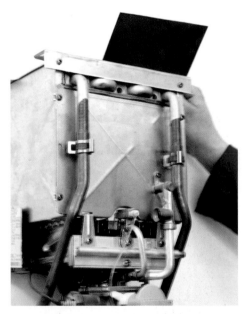

7 Checking ASD (atmospheric sensing device)

Recommissioning

- Re-light boiler as per manufacturer's instructions.
- Check burner pressure as stated on data plate.
- Check gas rate.
- Check operation of flue (open flues require a spillage test).
- Check air-pressure switch (fan flued).

- Check boiler controls and safety devices are functioning correctly.
- Check ASD where fitted.
- Report back to user with manufacturer's instructions.
- Advise customer of the need for regular maintenance.

Knowledge check

1   What does the term 'range-rated' boiler mean?

2   On a combination boiler, which has priority, hot water or heating?

3   What type of ignition is used on modern boilers?

4   State the purpose of an air-pressure switch.

5   Can an open-flue boiler be installed in a shower room?

7   What is the advantage of a duplex burner?

8   Why are room-sealed fires considered safer?

9   What is the purpose of an atmospheric sensing device fitted to a fire?

10   State three types of flue that may be used for a heating stove.

11   Why is a safety device/fan-pressure switch needed on a fan-flued fire?

12   What rooms should open-flued fires never be fitted in?

13   When a gas fire (12.7 kW Max) is fitted in a room for sleeping, what essential device must be fitted?

14   How long should a gas fire be working on full before a spillage test is carried out?

15   State the minimum ventilation needed for a DFE fire.

16   What is a lint arrester?

17   What is the common name of a 'self-sealing plug-in device'?

18   What controls need checking when commissioning/recommissioning a cooker?

19   State the two main types of water heater.

20   Can a flueless heater supply an appliance in an adjacent room?

21   What type of appliance must be fitted if the room is less than 20 m³?

22   What is an ASD?

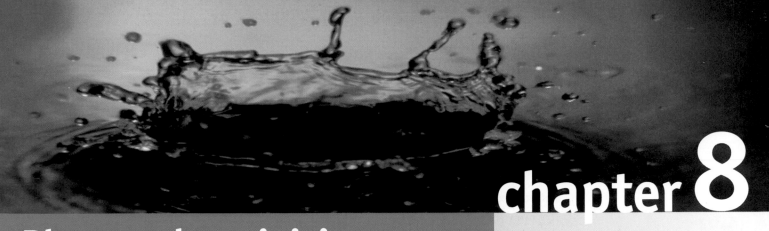

chapter 8

Plan work activities

OVERVIEW

This section is about ensuring that job information, such as plans and specifications, required for the work to be carried out, are correct and available. You need to correctly interpret and use information, including verbal instructions, to ensure that the installation complies with technical, safety and legal requirements, and that others involved in the installation process are informed of developments during the work activities, as and when necessary. Making sure that work runs to programme is key at this level and, where this is not possible, any revisions should be agreed with the correct person, for example the customer/client or supervisor.

This section underpins the following technical certificate units at Level 3:

- cold-water systems
- hot-water systems
- above ground discharge pipework systems
- central-heating systems, including boilers and controls.

What you will learn in this unit:

- **System layouts**
- **Construction details**
- **Plumbing information sources**
- **Basic design principles for cold-water and hot-water systems**
- **Design principles for above ground discharge systems**

System layouts

At the end of this section you should be able to:

- state the various types of systems and components that are appropriate for domestic applications

- make decisions affecting the choice of systems and components for domestic applications.

For the work you'll do, it's crucial that you are thoroughly briefed in all the system types and layouts. In this section you will look at the installation as a whole, because generally speaking on a new domestic dwelling job that's what you would be taking on.

Once you've gone over the various systems, you then need to be able to make informed decisions about what factors will influence the selection of a system and why you will make your choice.

Domestic installation requirements

In practice, particularly on a new installation, a plumber is likely to install all the systems, cold water, hot water, above ground drainage etc., at the same time. At Level 3, a plumber must have an in-depth working knowledge of how the system should be configured so that it will work to its design performance.

The design principles that a plumber must consider are outlined in this section. To help illustrate this you should imagine that the design considerations being discussed are for a typical new-build three-bedroom detached house with a first-floor bathroom and downstairs cloakroom. The kitchen will include a hot- and cold-fill washing machine and a cold-fill dishwasher. Gas appliances will include a gas boiler, gas fire and gas hob.

Design principles for cold-water systems

Plumbers need to know what type of cold-water system (from where it leaves the main to where it supplies an appliance or component) they are dealing with.

Direct systems

Figure 8.1 shows a typical direct system. Have a go at labelling the various components, then answer the questions.

1 What capacity should the cold-water storage cistern (CWSC) be?

2 What is the minimum and maximum depth of the outside supply pipe?

3 What valves and fittings are missing from the diagram?

The system offers the following advantages:

- it is cheaper to install because:
  - less pipework is required
  - storage cistern will be smaller (110l)
- drinking water is available from all draw-off points
- less risk of frost damage as there is less pipework
- less structural support required for smaller units.

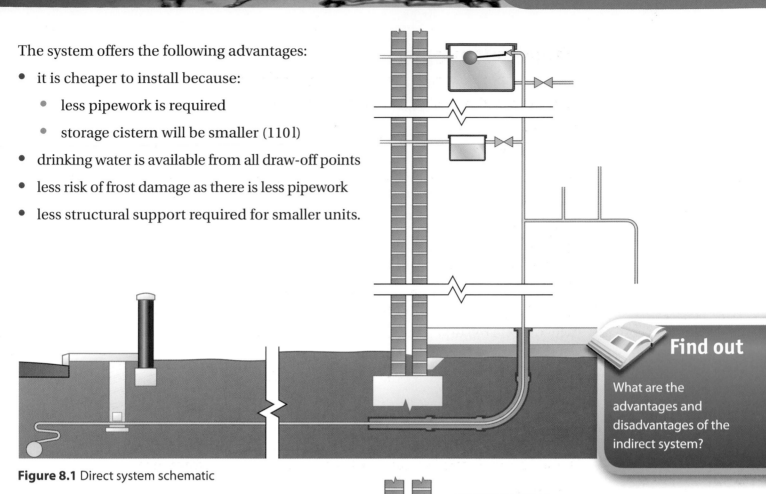

Figure 8.1 Direct system schematic

And the disadvantages are:

- higher pressure may make the system noisy
- no reserve of water if the mains or service supply is shut off
- more wear and tear on taps and valves owing to high pressure
- higher demand on the main at peak times.

The service duct should be either a flexible or a ridged sleeve formed from UPVC waste pipe, sealed at the entry to the dwelling and at the end below ground.

Indirect systems

Label the various components in Figure 8.2.

1 What capacity should the CWSC be this time?

2 What arrangements should be in place for protecting the water in the cistern?

Find out

What are the advantages and disadvantages of the indirect system?

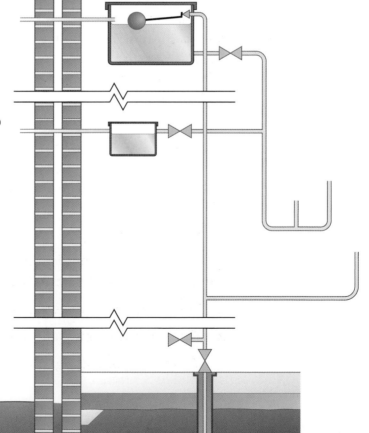

Figure 8.2 Indirect system schematic

Design principles for hot-water systems

A number of factors influence the selection and design of a hot-water system:

- amount of hot water required
- temperature of stored water at outlets
- cost of installation and maintenance
- fuel energy requirements and running costs
- economy of water and energy use
- safety of the user.

When choosing a system, the options range from a simple single-point arrangement for one outlet to the more complex centralised boiler systems supplying hot water to a number of outlets. BS 6700 sets out a number of ways of supplying hot water, which can be summarised as:

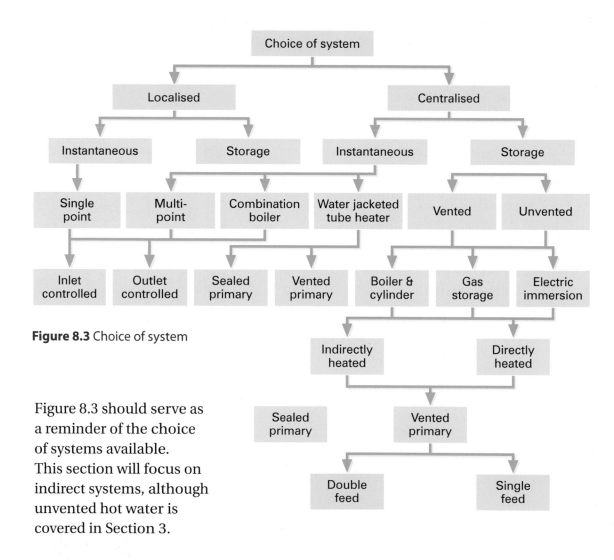

Figure 8.3 Choice of system

Figure 8.3 should serve as a reminder of the choice of systems available. This section will focus on indirect systems, although unvented hot water is covered in Section 3.

Double-feed indirect systems include:

- all hot-water supply pipework to/from appliances and components
- control and isolation valves
- appliances and components, including the boiler.

Remember, in the indirect double-feed type of system:

- the open vent and cold-feed pipes may be connected to the primary flow pipes or may be fed separately into the boiler
- where the vent pipe is not connected to the highest point in a primary circuit, an air-release valve should be fitted.

Hot- and cold-water systems – taps and valves

There is some overlap between hot- and cold-water systems, particularly in terms of taps and valves. The main purpose of taps and valves is to isolate supplies, control the flow of water, reduce the flow rate through pipework, permit drainage from systems, and provide an outlet to an appliance, i.e. sink, bath etc.

The Water Regulations require that all fittings must be:

- suitable for their purpose
- made of corrosion-resistant materials
- of sufficient strength to resist normal and surge pressure
- capable of working at appropriate temperatures
- easily accessible to renew seals and washers.

Figures 8.4 to 8.7 show a range of taps and valves.

1 Name the tap or valve.

2 Label the various parts.

3 Briefly describe where they will be located in the system and how they work.

Remember

Taps and fittings for both hot and cold water should conform to BS 1010, Parts 1 and 2, BS 1552 and BS 5433

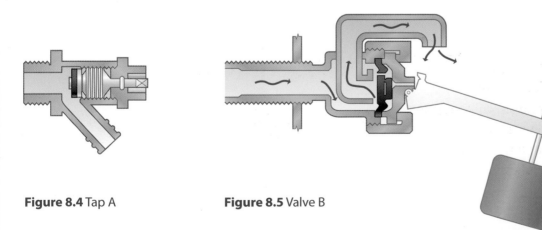

Figure 8.4 Tap A   **Figure 8.5** Valve B

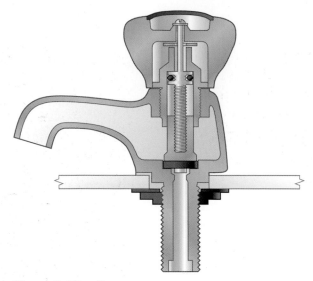

Figure 8.6 Tap C

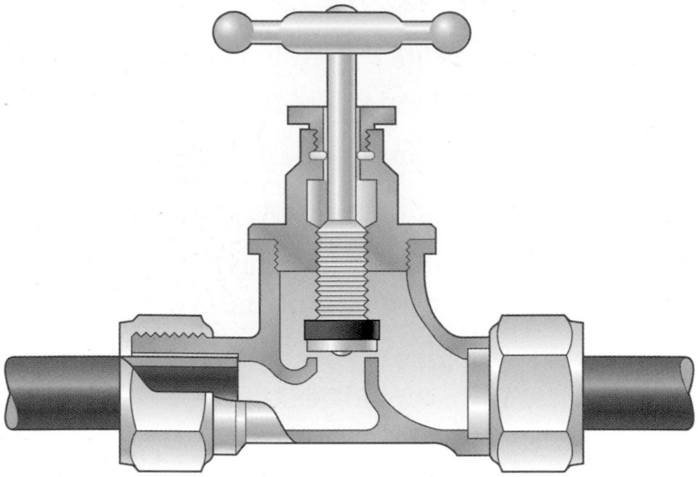

Figure 8.7 Tap D

Remember

The systems are available in either a unit or a packaged format. A unit system is where all the components are assembled by the manufacturer during the production process, and a package is where all the components, other than the temperature-activated controls, are fitted by the installer. In all cases, safety devices are factory-set

Unvented (domestic) hot-water systems (uhws)

The installation requirements of uhws in excess of 15 litres are governed by Part G of the Building Regulations. Other installation requirements are also covered in the Water Regulations and BS 6700.

Part G requires that any system must be installed by an approved installer, and that at no stage should the water reach 100°C. This is achieved by using the following safety devices:

- thermostat set at 60°C
- high-temperature thermal cut-out set to cut out at around 85°C
- thermal-relief valve designed to open at around 90°C.

The units can be supplied direct, e.g. heated by immersion heater(s) or indirect from a hot-water boiler.

Figure 8.8 shows the components required for a typical package for either an indirect or a direct system. A unit will share the same components but will be factory-fitted.

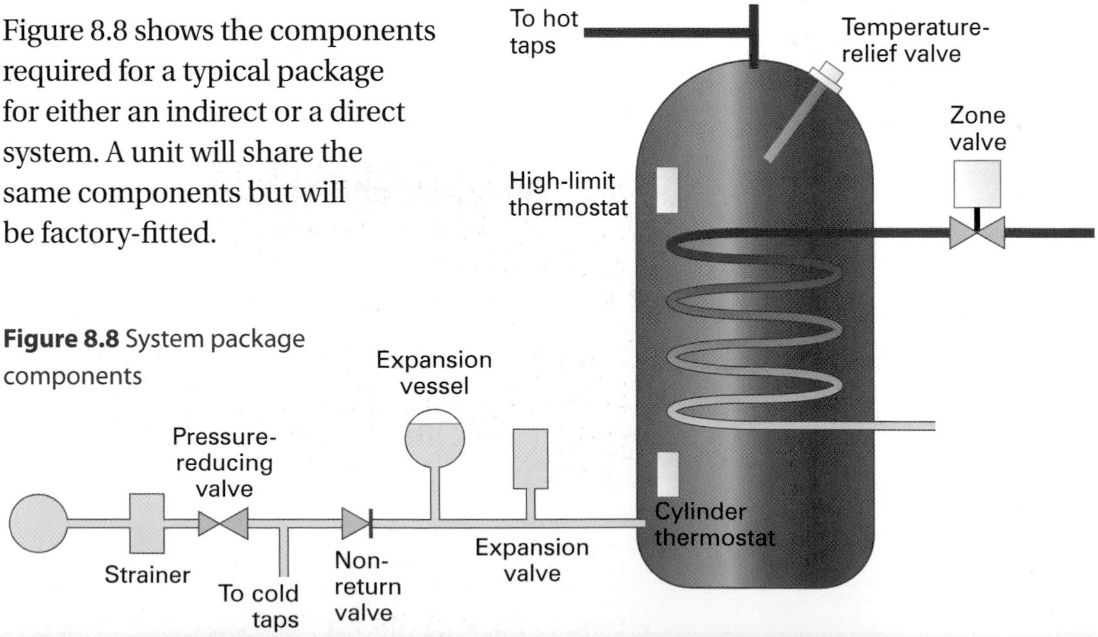

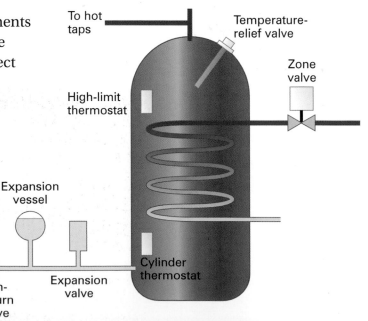

Figure 8.8 System package components

Basic principles of operation

A storage cistern is not required, so the take-up of expanded water as it is heated up from cold is done by an expansion vessel. A typical expansion vessel consists of a rubber bag inside a steel container. As the water heats up it expands into the rubber bag, and the extra capacity is taken up in the steel container, which has air on the other side of the bag. When the water cools, the air pressure in the steel container forces the water out of the bag and back into the system.

Central heating, including boilers

Figure 8.9 shows a typical system layout for a fully pumped central-heating system using a mid-position valve.

Sealed systems are often associated with combination boilers, which is a common application on domestic installations. But there are a number of other applications, including both vented and unvented secondary hot-water systems. Figure 8.10 shows a typical sealed central-heating system with vented hot water.

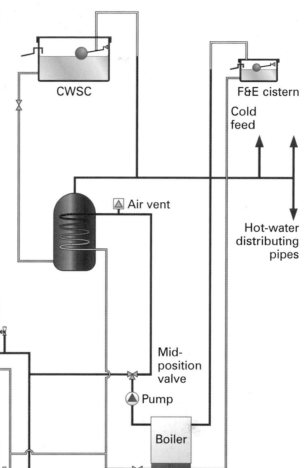

Figure 8.9 Fully pumped system schematic

Remember

The water supply to the unvented system is taken directly from the mains supply. Generally, systems will operate at around 3 bar. So it's important that the water pressure is deemed adequate prior to installation

Find out

State two other types of fully pumped control systems that could be installed

Similarly, the hot-water system could be an indirect unvented system. The sealed system, like the unvented one, requires the use of an expansion vessel to take up the increased volume when the water is heated through temperatures between 10 and 80°C. The system also requires the safety devices shown in Figure 8.10.

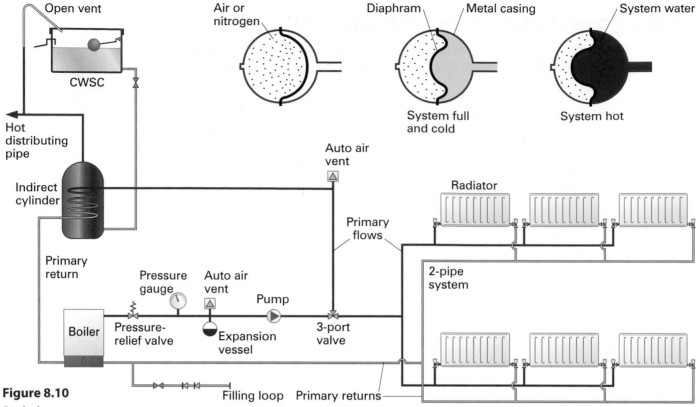

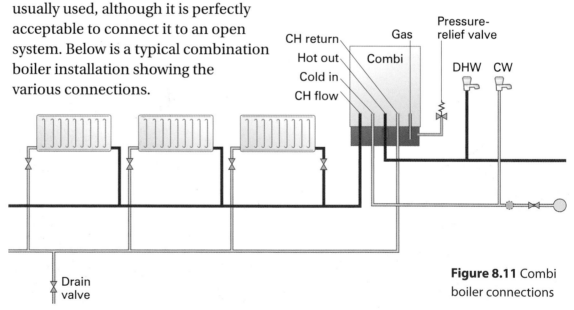

Figure 8.10 Sealed system – vented hot water

Combination boilers

This type of appliance supplies full or partial central heating through a radiator system and domestic hot water on an instantaneous basis. A sealed system is usually used, although it is perfectly acceptable to connect it to an open system. Below is a typical combination boiler installation showing the various connections.

Figure 8.11 Combi boiler connections

With this type of installation, the system should be filled using a flexible filling loop, which must be disconnected when filling is complete. Expansion of water when heated is accommodated in an expansion vessel within the boiler casing.

Some boilers incorporate an automatic bypass, but where that is not the case, provision must be made within the system as per manufacturer's instructions.

Above ground discharge systems (AGDS)

In planning to install AGDS for a domestic dwelling, you need to take another look at the primary ventilated stack system, probably the most popular system.

The primary ventilated stack system

Figure 8.12 shows a layout for a primary ventilated stack system, suitable for a typical domestic installation.

1 Label the pipe sizes for the installation pipework from each appliance.

2 Work out the minimum dimension at x, and the minimum permissible size of the discharge stack.

Building Regulations Part H1 sets out the rules about the design of the above system. As well as the pipe sizes, there are limitations to the maximum lengths of the branch connections and their gradients.

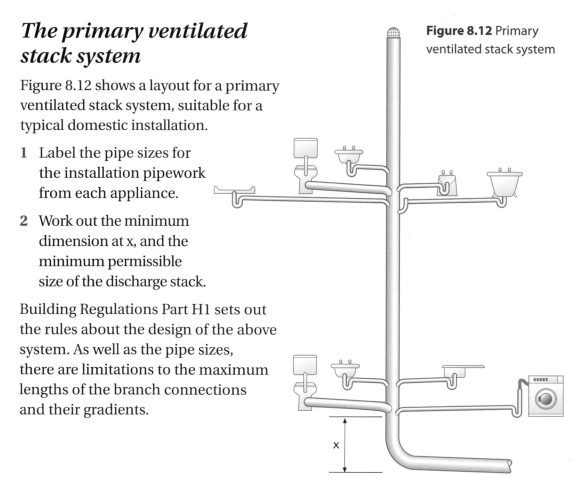

Figure 8.12 Primary ventilated stack system

Appliance	Pipe size (mm)	Max. length (m)	Slope (mm fall per metre)
Basin	32	1.7	18–20
Bath	40	3.0	19–90
*Shower	50	4.0	18–90
WC	100	6.0	18 min
*Normally 40 mm			

Table 8.1 Requirements for pipe sizes and branch connections for primary ventilated stack system

Choosing a system, including components and appliances

In some cases, the decision about the choice of system in a dwelling may be taken by someone else. Typically, this would include:

- where the customer (or their representative) has specified the system for a one-off dwelling

- on a large multi-dwelling housing development, where the systems have been specified by an architect or services design engineer.

There may be other circumstances, but as a plumber operating at Level 3, you may be required to advise a customer on the best systems to suit their needs. This will be influenced by:

- customer requirements

- budgets

- technical requirements and limitations.

Customer requirements are best determined through discussion. A good plumber will have catalogues for the customer to look at, and will listen carefully to what the customer wants in terms of specifications, quality and the location of appliances and components.

Consideration should also be given to any special needs – for example, people with disabilities, the elderly, households with toddlers etc.

Budgets can be difficult, as in some cases the customer is not sure what the cost of the materials or the installation is likely to be. Your employing organisation will provide cost options for the materials, in terms of quality of appliances and components, and at the planning stage should be able to provide a general idea of labour costs.

Once the customer is comfortable with an outlined cost, a more detailed estimate can be prepared by the organisation.

Technical requirements and limitations are probably the key planning consideration. Some customer aspirations may be technically impossible, or technical difficulties may make the cost prohibitive.

There are other technical considerations:

- Nature of the water supply – is it soft or hard? Will it mean fitting a water softener?

- Water pressure – choice of indirect cold-water system if pressure is poor, limitations for unvented systems.

- Hot-water and heating systems – initial installation costs against long-term running costs; occupancy, how long people spend in the house, zoning etc; types of controls; hot-water demand; choice of boiler, limitations of space in terms of location options; fuel availability; construction of the property, energy conservation; choice of unvented and sealed systems; aesthetic choices (how it looks), radiator design; application, e.g. micro-bore, small-bore; legislation (Part L1); and emerging technology such as solar heating.

- Above ground discharge systems – types of appliances and location within the dwelling. Would one primary ventilated stack suffice or would two stack systems be required? Connection to drainage systems, combined drainage system (existing dwellings), separate systems.

- Influence of legislation on the technical considerations outlined above – Water Regulations, Building Regulations etc.

These are all key factors that will need to be decided on prior to the work taking place. You are likely to be involved in some of these decisions. A fair proportion of the skills needed to handle these decisions comes with experience and using the experience of the people that you work with.

Protecting work in progress

On the job: Protecting materials and equipment

Martin is working on a new-build and is installing all the pipework and sanitary ware. He carries dust sheets and an old woollen blanket in his van; some of them are marked with his initials. On this job, as in others, he arranges to install appliances when other trades have finished in the area where he intends to work. After the first fixing he seals open-ended and incomplete pipework, and once the toilets are installed he tapes down their seats. He never has materials in storage longer than he has to.

1 What two things might Martin use the dust sheets and blankets for?

2 Why does he identify them with his initials?

3 Why does he prefer to install appliances after other trades have finished working in the area?

4 Why does he tape down the toilet seat and seal open-ended pipework?

5 Why does he like to minimise the storage time of materials?

6 If Martin has to leave appliances unattended while other tradespeople are working, what should he do to protect them?

7 What are the final things Martin should do before signing off the job?

Construction details

At the end of this section you should be able to:

- state the main areas of current legislation related to construction

- understand the main construction features of newly constructed buildings into which the systems or components are to be installed

- understand the differences between new-build and older traditional forms of construction

- state the provisions within buildings for the accommodation of system components

- state the requirements for entry into the building of the services to the systems

- understand how to inspect buildings to confirm that provision for the systems or components is suitable.

Planning work activities for plumbing installations will require you to have a working knowledge of how a typical domestic dwelling is 'put together'. This will help you to determine whether you need to make provision for any specific fixing materials or specialist tools. It will also make you aware of the nature of the construction into which you intend to carry out the installation work, so that you can inspect buildings to confirm that provision for the system or components is suitable. You need to be aware of the legislation that governs construction work.

Construction legislation

Just like the plumbing industry, the construction industry has to meet the requirements of Health and Safety legislation, including:

- Health and Safety at Work Act 1974

- Electricity at Work Regulations 1989

- Construction (Design and Management) Regulations 1994

- Control of Substances Hazardous to Health Regulations 2002 (COSHH)

- Confined Spaces Regulations 1997.

There are a number of other Regulations, but for the purposes of this topic area we will focus on those that affect the actual design and construction of buildings, namely the Building Regulations.

Building Regulations

Building Regulation approval is required if someone wants to:

- build a house

- carry out alterations to a property, extend it, replace or add new window frames or carry out internal alterations

- build a commercial building, shop, office, or carry out alterations, extend it, replace or add new window frames or carry out internal alterations

- convert (or change the use of) a building or part of a building

- alter, replace or install a controlled service or fitting (i.e. sanitation, drainage, new or replacement heating, a new or replacement roof, new wall surfaces [external or internal] or energy-using equipment)

- create or alter a dwelling or flat in an existing building.

The Regulations exist to provide minimum standards for building work in order to safeguard the health and safety of people in or around buildings. They also include minimum standards for easy access and facilities for disabled people, for the conservation of energy, and water use and disposal.

Prior to 1965, the construction of buildings was controlled by local building byelaws; these were superseded by national Building Regulations, which form the basis of the ones used today.

The current format (simple Regulations plus Approved Documents giving technical guidance) of the Building Regulations was created in 1985. The Regulations are divided up into 'Parts', each 'Part' dealing with a specific technical, construction or design topic.

Several major amendments were made in April 2002, affecting Parts H (Drainage and waste disposal), J (Combustion appliances and fuel storage systems) and L (Conservation of fuel and power).

On 1 July 2003 a new Part E (Resistance to the passage of sound) was introduced. This extends the range of (new and extended) buildings that must offer users relief from excessive noise, both from noise sources within a building and from adjoining buildings. All new (and extended) educational buildings must also design out excessive noise, to give acceptable acoustics in teaching areas within schools.

The current Building Regulations were issued by Parliament in December 2000 and they have recently been amended. The Building Regulations extended the range of the building work that is now controlled, with effect from 1 April 2002.

The Building Regulations do not offer general advice on how to design 'green' buildings. However, green drainage design and water disposal are encouraged and illustrated in Part H (2002 edition). The Building Regulations do not prevent the building of 'eco homes' and 'sustainable design solutions' so long as these designs meet or exceed the minimum requirements of the current Building Regulations.

Administering the Building Regulations

Local authorities administer and enforce the Regulations, but private sector **approved inspectors** can also administer them. However, the private sector approved inspectors must pass back to the local authority any job where the Building Regulations are not being complied with for enforcement action.

Did you know?

Building Regulations are legal technical requirements laid down by Parliament for controlling the construction of buildings in England and Wales. Scotland and Northern Ireland have their own sets of Building Regulations

At present, most of these private-sector professionals have insurance cover only for the inspection and supervision of commercial buildings and new-build houses. The Government plans to extend the insurance cover to encompass a wider range of building works.

Obtaining the Building Regulations

The Building Regulations (known as a **Statutory Instrument**) and the associated Guidance Booklets (called **Approved Documents**) can be bought direct from The Stationery Office or you can order them online at http://www.tso.org/.

To browse without buying, try your local public reference library or look at the various official 'Approved Documents' online on the national Government website at http://www.odpm.gov.uk/ under 'Building Regulations'.

The Approved Documents are published separately in booklet form, and are called 'Parts'. Each 'Part' deals with specific subjects:

- **Part A – Structure** (2004 edition; Part A has been consulted for revision)

- **Part B – Fire Safety** (2000 edition consolidated with 2000 and 2002 amendments)

- **Part C – Site Preparation and Resistance to Moisture** (1992 edition amended 2000; Part C is subject to consultation for amendment and revision)

- **Part D – Toxic Substances** (Cavity Insulation) (1992 edition amended 2000)

- **Part E – Resistance to the Passage of Sound** (2000 edition came into effect on 1 July 2003, together with the Mandatory Sound Resistance testing of new houses constructed after 1 July 2004)

- **Part E – Resistance to the Passage of Sound** (1992 edition) (now valid only for approved applications where the building commenced before 1 July 2003)

- **Part F – Ventilation** (1995 edition amended 2000)

- **Part G – Hygiene** (Sanitary Conveniences etc.) (1999 edition amended 2000)

- **Part H – Drainage and Waste Disposal** (2002 edition)

- **Part J – Combustion Appliances and Fuel Storage Systems** (2002 edition)

- **Part K – Protection from Falling, Collision and Impact** (1998 edition amended 2000)

- **Part L1 – Conservation of Fuel and Power in Dwellings** (2002 edition)

- **Part L2 – Conservation of Fuel and Power in Buildings other than Dwellings** (2002 edition)

- **Part M – Access to and Use of Buildings** (Facilities for Disabled People) (2004 edition)

- **Part N – Glazing – Safety in Relation to Impact, Opening and Cleaning** (1998 edition amended 2000)

- **Part P – Electric Safety** (a new Approved Document which came into force on 1 January 2005).

Obtaining Building Regulations approval

1. By depositing a 'Full Plans' application to the local authority

Where someone wishes to obtain a prior approval of their detailed scheme before they're going to start the work, they should allow up to five weeks for the Council to check in detail their Full Plans application. The Council will try to give its decision as fast as possible, but statutory consultation with other bodies can delay decisions. A Full Plans application will receive a formal approval certificate – the work will be inspected on site and, if satisfactory, a Completion Certificate will be issued.

2. By submitting a Building Notice to the local authority

An applicant may use a simple Building Notice for minor work. They may still be required to submit to the council some drawings and further details of the work.

Note that a Building Notice will not get a formal approval certificate – the work will be inspected on site and, if satisfactory, a Completion Certificate will be issued. It's not possible to use a Building Notice for a workplace, hotel, boarding house or any 'designated premises' under the Fire Precautions Act 1971. A Building Notice can't be used where a building (or piling work) will be built over or within three metres of any sewer shown on the public network map.

As a plumber, it is important that you have a basic understanding of how a domestic building is constructed. This should include a working knowledge of:

- foundations
- walls and floors
- roof construction.

Find out

BS 8000 Workmanship on building sites is a useful reference document. Try to obtain a copy of the Code of Practice to look at

Foundations

Foundation construction is covered under Building Regulations Approved Document A – Structure. A foundation is the base on which a building rests, and its job is to transfer safely the load of the building to a suitable subsoil.

It is useful to have an idea of the factors that influence the type of foundation you may come across on site. The main factors governing the type, size and depth of a foundation are:

- the nature and bearing capacity of the ground supporting it
- the structure and size of building it supports (total load).

Nature and bearing capacity of foundations

This needs to be established well before any work commences, and can be determined using:

- trial holes, followed by further investigation
- bore holes and core analysis
- local knowledge.

Clay is the most difficult subsoil to deal with, because down to a depth of about one metre it is subject to seasonal movement. That is, it swells in winter due to excess water, and shrinks in summer as the moisture dries out, thus causing substantial movement of the soil.

Soils that absorb and retain water are also subject to frost heave in periods of cold weather. This is where the subsoil swells due to the expansion of the freezing water.

Structure and size of building (total load)

The total loads of a building are taken per metre run and calculated for the worst-case scenario. Designers use the following information:

- roof material dead load on the wall, plus imposed load from snow
- floor material dead load on the wall, plus an imposed load allowance for people and furniture
- wall load on the foundations
- total load on the foundations of the above three points, and any allowances for wind in exposed conditions.

Having determined the nature and bearing capacity of the subsoil, the width of the foundation can be worked out:

- using guidance provided in tables published in Document A of the Building Regulations, or
- calculating the total load per metre run of the foundation and relating this to the safe bearing capacity of the subsoil, e.g.

$$\frac{\text{Total load of the building per metre}}{\text{min. foundation width}} = \text{Safe bearing capacity of subsoil}$$

Types of foundation

Foundations are usually made of mass, or reinforced concrete, and are classified as either:

- shallow foundation, or
- deep foundation.

For the purpose of domestic applications, we'll concentrate on shallow foundations.

Shallow foundations for the purpose of domestic dwellings includes strips and rafts. In the main, foundations are constructed using poured concrete, which in some cases, depending on the load of the building and the bearing capacity of the soil, may also be reinforced using steel bars in strip foundations or steel mesh in rafts.

Strip foundations

Strip foundations consist of a continuous strip of concrete, positioned centrally under a load-bearing wall. In domestic house construction this would be the external cavity wall carrying the weight of the floors and roof. It may also apply to some internal load-bearing walls that may be used to carry the weight of a long span of flooring.

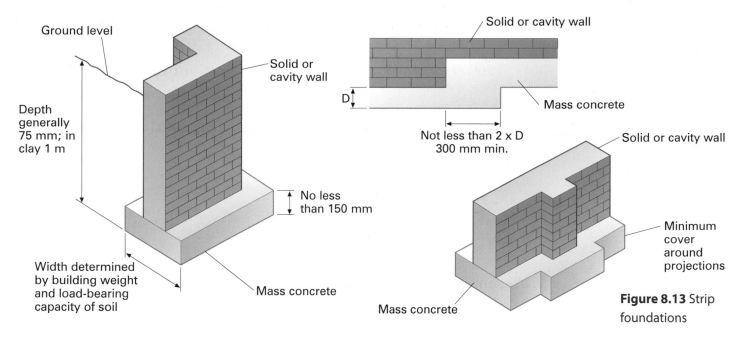

Figure 8.13 Strip foundations

Strip foundations for use in soils with poor bearing capacity are intended to take the depth of the foundation to a level where the subsoil has a better bearing capacity. It usually applies where the subsoil is of firm shrinkable clay, whose moisture content can be affected by deep-rooted vegetation for some depth below the surface. This can be achieved by increasing the depth of concrete, as in Figure 8.14, or using a short-bored pile foundation, as in Figure 8.15.

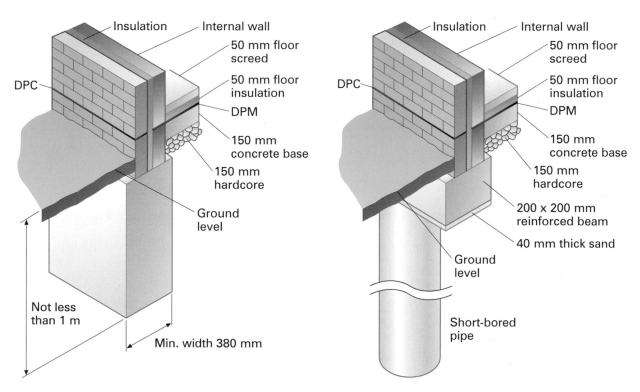

Figure 8.14 Increased foundation depth

Figure 8.15 Short-bored pile foundation

Short-bored pile foundations are basically concrete columns which are pre-cast and then driven into the ground, or can be cast into pre-drilled holes, as shown in Figure 8.15.

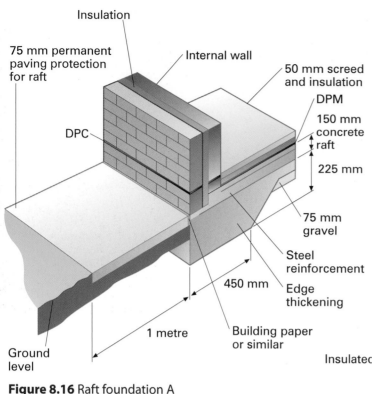

Figure 8.16 Raft foundation A

Raft foundations

These consist of a raft of reinforced concrete underneath the whole of the building. They are usually used when building on compressible ground such as very soft clay or made-up ground, where strip, deep or piled foundations would not provide a stable base without excessive excavation.

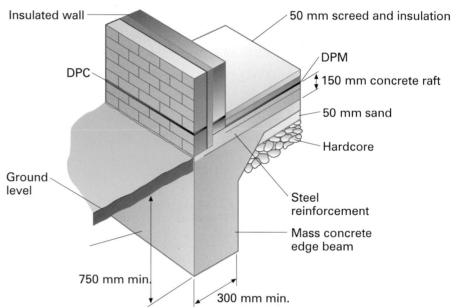

Figure 8.17 Raft foundation B

Wall and floor construction

Wall construction for domestic dwellings starts at the foundation level, which, with the exception of raft foundations, is below ground.

Walls below ground

Figure 8.18 shows a typical detail, although there are a number of variations to this theme, which are beyond the scope of your Level 3 studies. The outer skin will be constructed of common bricks or concrete blocks, the inner skin of concrete aggregate blocks. Wall ties are also used to bond the two skins together. The cavity will be filled below ground with concrete, as shown in Figure 8.18. Common bricks are manufactured without a decorative face, are cheaper to produce, and are used where the face of the brick won't be seen.

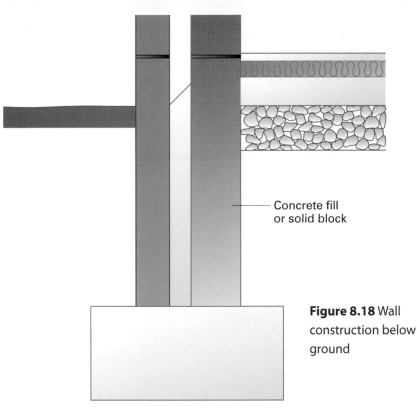

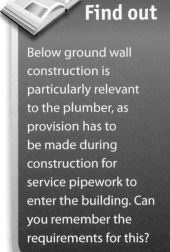

Find out

Below ground wall construction is particularly relevant to the plumber, as provision has to be made during construction for service pipework to enter the building. Can you remember the requirements for this?

Concrete fill or solid block

Figure 8.18 Wall construction below ground

Suspended timbers for flooring

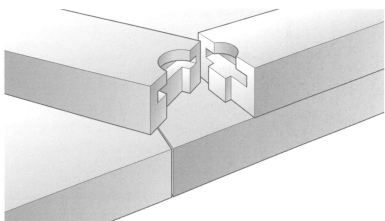

Figure 8.19 Use of trench blocks

Trench blocks are increasingly being used these days as an alternative to concrete fill between cavity walls. A trench block is a solid block which overcomes the need to use a cavity construction, concrete fill and wall ties. Some blocks on the market are **tongue and grooved** on the perpendicular end of the block, so it isn't necessary to apply mortar at that point, saving time and materials.

Ground floor construction

The next bit of the dwelling to be constructed is the ground floor. A ground floor could be:

- solid concrete
- suspended timber
- suspended pre-cast concrete (**beam and block**).

Figure 8.20 shows a typical solid concrete floor construction for a domestic dwelling.

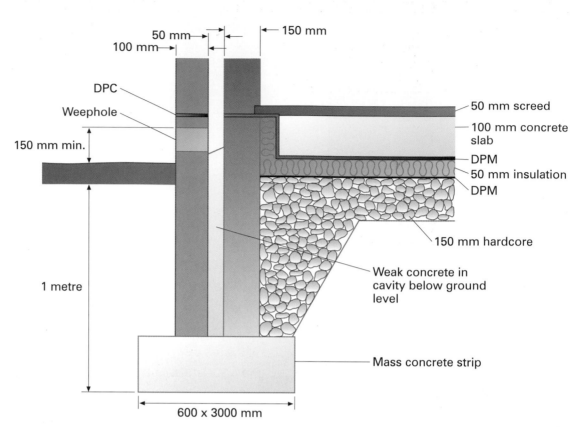

Figure 8.20 Solid concrete floor construction

Damp-proof membrane

When the site is being prepared for the foundations, the ground inside the foundation perimeter is excavated and then compacted using a mechanical vibrator plate to receive the **hardcore blinding**. The hardcore blinding is also compacted in a similar manner.

Sand blinding, usually 50 mm thick, is then spread over the hard core, which is compacted and then covered with a **damp-proof membrane (DPM)**. The DPM prevents any moisture from penetrating the concrete floor, and is also taken into the damp-proof course (DPC) in the wall to ensure complete moisture tightness. The purpose of the sand blinding is to prevent the hard core from puncturing the DPM.

The concrete slab is then poured, usually to a thickness of 100 mm, and once set is finished off with a smooth **screed** of sand and cement 50 mm thick.

Current Building Regulations set high standards for the thermal insulation of buildings, in order to reduce the heat loss and thus conserve energy. Figure 8.21 shows two methods of insulating solid concrete floors. The insulation should also be continued where the floor butts up to the wall. Expanded polystyrene is a typical insulating material.

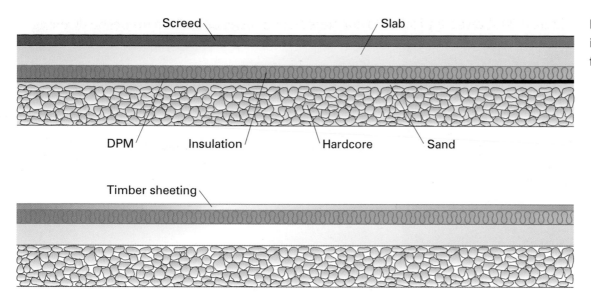

Figure 8.21 Types of insulated solid concrete floor

Suspended timber floor

Preparation is similar to that of the solid ground floor, in that hard core is compacted over the ground. This is then covered with **oversite concrete**, usually 100 mm thick. The purpose of the oversite is to provide a sound and level surface on which to build the **honeycomb walls**, used to provide interim support for long spans of timber floor joists.

A DPM isn't necessary, as moisture penetration will be prevented by using a damp-proof course between the honeycomb brickwork and the timber wall plate, as shown in Figure 8.22.

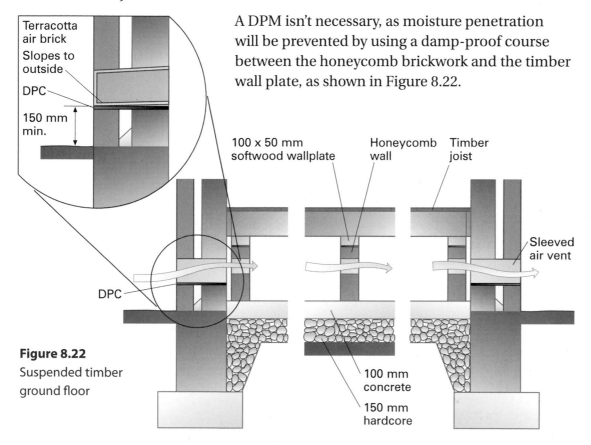

Figure 8.22 Suspended timber ground floor

Wallplates

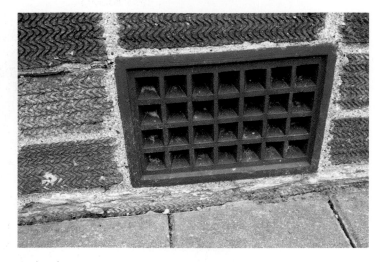

Air brick

A timber **wallplate** is used to spread the load of the floor equally across the length of the honeycomb wall. Floor joists are then laid in the opposite direction to the honeycomb wall and wallplates, and covered either with tongue-and-grooved floorboards or chipboard sheets.

Figure 8.23 Thermal insulation of floor

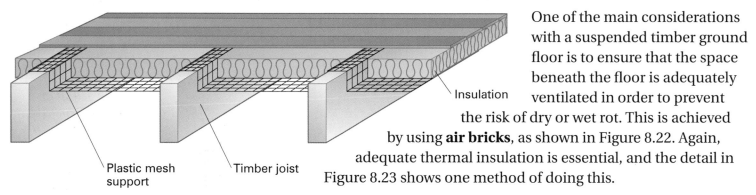

Insulation

Plastic mesh support

Timber joist

One of the main considerations with a suspended timber ground floor is to ensure that the space beneath the floor is adequately ventilated in order to prevent the risk of dry or wet rot. This is achieved by using **air bricks**, as shown in Figure 8.22. Again, adequate thermal insulation is essential, and the detail in Figure 8.23 shows one method of doing this.

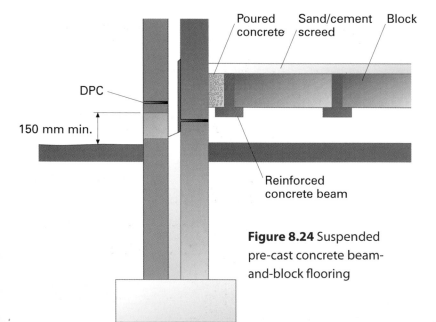

Poured concrete   Sand/cement screed   Block

DPC

150 mm min.

Reinforced concrete beam

Figure 8.24 Suspended pre-cast concrete beam-and-block flooring

Suspended pre-cast concrete floor

This method of ground floor construction uses a number of pre-cast reinforced concrete beams that are basically an inverted T shape in order to provide a lip to receive concrete blocks. The blocks are placed lengthways onto the lips of the beam, and this forms the basis of the floor surface. The floor is finished with a sand and cement screed, or alternatively tongue-and-grooved or chipboard flooring, as shown in Figure 8.25 – which also shows the thermal insulation requirements. A DPM is also included in the floor.

There are two possible options for construction, as shown in Figure 8.25.

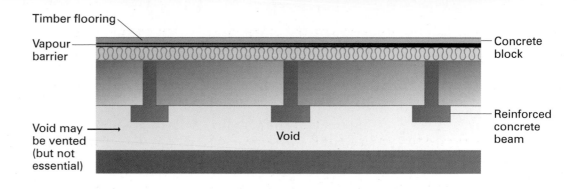

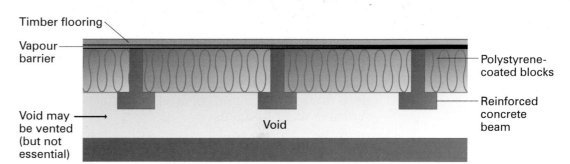

Figure 8.25 Floor construction options

External walls

External walls are the 'envelope' of the building, and are described as part of the superstructure. To function properly the external wall needs to be:

- stable
- resistant to moisture
- insulated against sound transmission
- insulated against thermal transmission
- fire-resistant.

External walls for new domestic dwellings are known as **cavity walls**. The purpose of the cavity is to create a break in the **capillary path** of moisture. Traditional cavity construction consists of an outer leaf of brickwork and an inner leaf of dense or aerated concrete block work. Older properties may not be constructed with cavity walls.

Figure 8.26 shows a typical detail through an external wall for a domestic dwelling. We'll use this diagram to illustrate the points above.

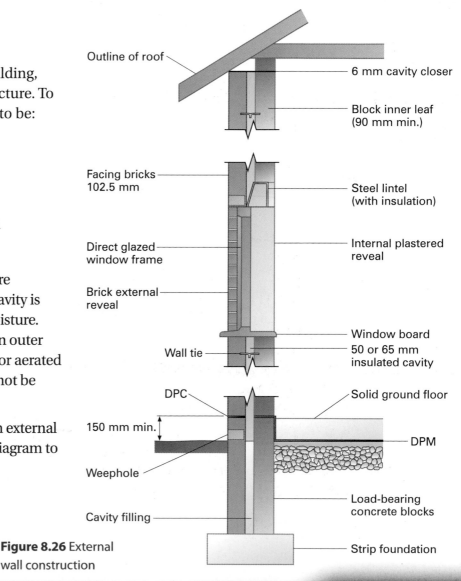

Figure 8.26 External wall construction

Stability

External cavity walls used for domestic dwellings are described as load bearing. That means they have to be strong and stable enough to carry their own weight plus the weight of the floors and roof and any additional loads placed on these components. The stability of the cavity wall is ensured by the strength of the bricks and blocks, the mortar used to join them together, and the use of **wall ties** to tie the two 'leaves' together.

Bricks

Bricks that are used in construction should conform to BS 3921. The size of a brick is 225 mm long, 100 mm wide and 65 mm thick. Bricks are manufactured using either clay or a sand and flint base mix known as calcium silicate. Clay bricks are mostly used to construct domestic dwellings and are classified as:

- common bricks, which are suitable for general work where their appearance is not important

- facing bricks, which are manufactured to give a high quality of surface finish

- engineering bricks, which are very dense and have low water absorption and a high compressive strength, making them suitable for carrying heavy loads.

There are three other ways in which bricks are classified:

- internal quality; used only for internal walls or partitions

- ordinary quality; used for external work above ground level in conditions which do not include severe exposure

- special quality; suitable for conditions of extreme exposure or below ground.

And they are also classified as solid, perforated or cellular, as shown in Figure 8.27.

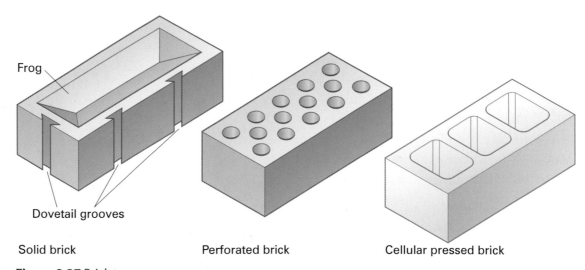

Solid brick   Perforated brick   Cellular pressed brick

Figure 8.27 Brick types

Blocks

Blocks are usually used on the inner leaf of the cavity wall. They're defined by BS 6073 as walling units larger than the sizes specified for bricks. Clay blocks are available, but in the main they're made from dense or aerated concrete. Sizes of blocks vary, but typical face dimensions are 440 mm × 215 mm, and 440 mm × 330 mm, and the width ranges from 75 mm to 200 mm, 100 mm being typical for a domestic dwelling.

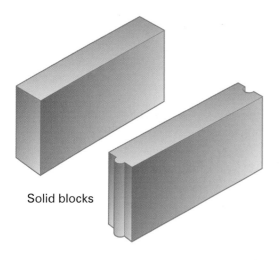

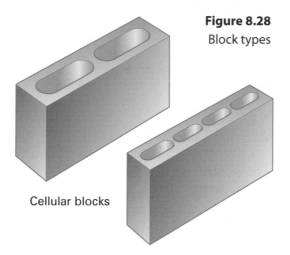

Figure 8.28
Block types

Solid blocks

Cellular blocks

Bonding of bricks and blocks

Bonding refers to the way that bricks and blocks are laid. The bond used for cavity wall construction is known as a stretcher bond. There are numerous other bonds used for solid walls, but these are beyond the scope of your Level 3 studies.

Figure 8.29 shows what the stretcher bond looks like for both the brickwork and the blockwork. The pattern of bonding ensures that the wall acts as a single structure and spreads the load towards the foundation.

Mortar mix requirements are covered by BS 5628 Part 3, which specifies the following compositions for internal and external use:

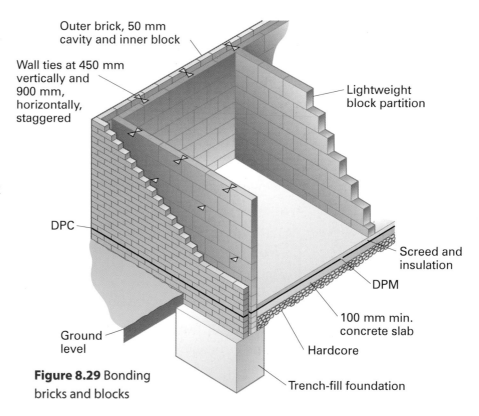

Outer brick, 50 mm cavity and inner block

Wall ties at 450 mm vertically and 900 mm, horizontally, staggered

Lightweight block partition

DPC

Screed and insulation

DPM

100 mm min. concrete slab

Ground level

Hardcore

Figure 8.29 Bonding bricks and blocks

Trench-fill foundation

Above DPC		Below DPC	
1:1:6	Cement: lime: sand by volume	1:4	Cement: sand by volume
1:6	Cement: sand by volume with plasticiser by volume	1:1½: 4¼	Cement: lime: sand
1:5	Masonry cement: sand by volume		

Table 8.2 Mortar mix compositions for internal and external use

Generally, the mortar is spread onto the last course of bricks laid, called the bed, and then mortar is applied to one end of the brick, which in turn is placed onto the mortar bed. This process is continued brick by brick. As the work proceeds, any excess mortar is cleaned off with the trowel. There are a number of options for finishing the mortar joint, as shown in Figure 8.30.

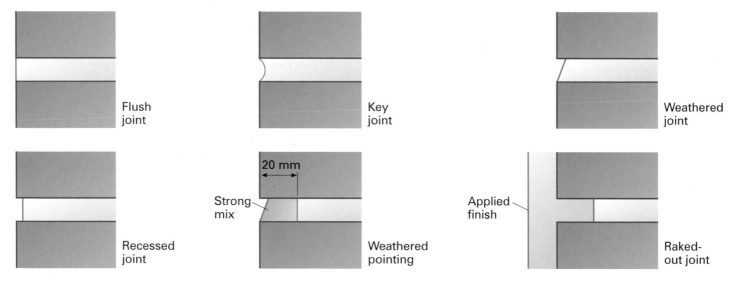

Figure 8.30
Mortar joints

Wall ties should comply with BS 1243. They can be made from **galvanised** steel and plastic, but the preferred material is now stainless steel. Typical patterns are shown in Figure 8.31.

The groove or twists in the ties are capillary drips and should be laid downwards.

For a domestic dwelling, wall ties should be positioned at minimum spaces of 450 mm vertically and 900 mm horizontally and staggered. Wall ties should also be included at the jambs of window and door openings for additional strength.

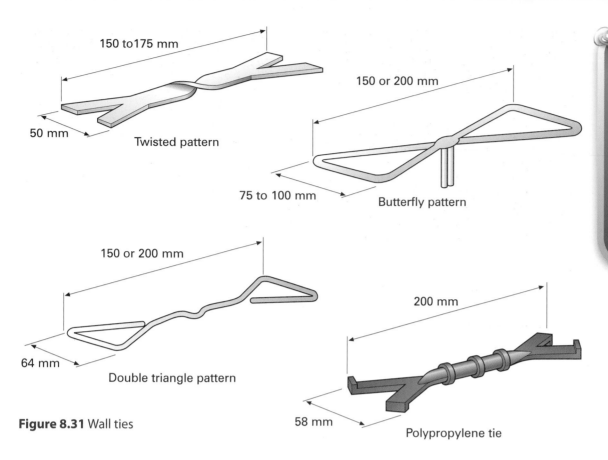

150 to175 mm

50 mm

Twisted pattern

150 or 200 mm

75 to 100 mm

Butterfly pattern

150 or 200 mm

64 mm

Double triangle pattern

200 mm

58 mm

Polypropylene tie

Figure 8.31 Wall ties

Definition

Lintel – a load-bearing concrete or metal strip placed above windows or doorways to take load from blockwork or brickwork above.

Jamb – the sides of a door frame

Openings in external walls

Openings in external walls need to be provided to allow for doors and windows. Figure 8.26 shows a typical detail of a section through a cavity wall to take a window frame. The actual window frame itself wouldn't be capable of withstanding the load above it, so this load is carried by a **lintel**.

The most common type of lintel is a purpose-made galvanised mild steel type, designed specifically for cavity walls. The lintel must be insulated to satisfy Building Regulations on energy conservation. Lintels are also required on masonry internal load-bearing walls.

The openings have to be sealed at the **jambs**, using insulated damp-proof course material and blockwork slip, or an insulated cavity closer. The joint between frame and brickwork is finished with sealant.

External walls are subject to two forms of moisture attack: rising damp and penetrating damp. Rising damp is prevented by incorporating a damp-proof membrane (DPM) and damp-proof course (DPC) on both leaves of the cavity wall. BS 743 recommends materials that can be used for a DPM, but the most common found on new-builds is polythene. In all cases the DPM should be:

* completely **impervious**

* durable

Moisture attacking brickwork

Find out

On a domestic dwelling, where do you think penetrating damp is most likely to occur?

- in relatively thin sheets so that it doesn't cause disfigurement to the building
- strong enough to support the loads placed upon it
- flexible enough to give way to any settlement of the building without failing.

Penetrating damp

Penetrating damp is caused by the transfer of moisture from the outside face of an external wall to the inside face of an internal wall. Cavity wall construction is a method of dealing with this, but there are instances in cavity wall construction where the cavity is bridged, such as window and door reveals. Where this occurs, the joint where the cavity is sealed at the inside wall is protected with a DPM. A DPM is also required above lintels, and where air bricks are used for suspended floor construction.

Sound transmission

A cavity wall constructed of brick and blockwork provides satisfactory sound insulation from outside noise. The Building Regulations (Part E) require that **party walls** (these are walls that separate semi-detached or terraced dwellings) 'shall resist the transmission of airborne sound'. Partitions and separating walls within dwellings should also provide adequate levels of sound insulation between rooms.

In terms of party walls, a cavity construction of 2×100 mm aerated concrete blocks with a 75 mm cavity, plaster or **dry-lined** finish will satisfy the Regulations. For domestic dwellings, most internal partitions are constructed from hollow walls, laminated or filled proprietary wall construction, the latter being manufactured to satisfy regulations.

For normal dwelling applications, covering the timber frame of a typical stud-partition construction with either 9.5, 12.5 or 15 mm thick plasterboard and a 5 mm plaster skim would be satisfactory – insulation material is commonly included between the timber studs to reduce noise levels.

Thermal transmission

The government is committed to reducing the emission of greenhouse gases, which are seen as a major contributor to global climate change. Building Regulations Part L1 specifies the requirements for ensuring that dwellings are constructed to the highest standards of thermal insulation. It also states what the energy efficiency of boilers and heating systems should be.

The Regulations state that for a dwelling heated with an appliance conforming to the relevant SEDBUK rating, the '**U**' **value** of the wall construction should be 0.35 W/m² K. Designers or builders therefore need to ensure that the external wall construction meets this requirement.

Brick and blockwork manufacturers make products that are as thermally efficient as practicable (in terms of thickness). But in order to satisfy Regulations, the wall construction usually needs additional insulation. This can be achieved in the ways shown in Figure 8.32.

Fire resistance of walls is covered under Building Regulations Part B. Because of the materials used to construct and finish walls, their fire resistance tends to be relatively high.

Internal partition walls

Internal partition walls fall into two main categories:

- solid
- hollow.

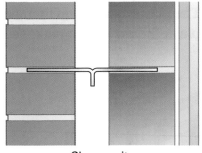

Clear cavity

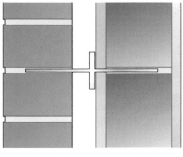

Partially filled cavity

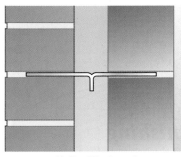

Fully filled cavity

Figure 8.32 Thermal insulation

Definition

U value – the rate of thermal transmittance from the inside of a building to the outside

Solid partitions

This type of construction usually consists of lightweight concrete blocks, 100 mm thick. While it was a popular method in the past for domestic dwellings, it has given way to more economical forms of construction such as laminated, filled and **stud partitions**. However, blockwork is often used in schools and other industrial, commercial and public buildings. You may also come into contact with it on maintenance work in older properties.

Laminated and filled partitions form another category of solid partition. This category includes a number of proprietary systems such as Gyproc laminated partitioning. This is produced by bonding layers of plasterboard to form a solid panel to the required thickness. The main consideration here is ensuring sound- and fire-resistance properties.

Cellular-filled partitioning is also included in this category. A typical example is the use of two sheets of plasterboard bonded to a cellular core such as cardboard.

Hollow partitions

For the domestic application, most new hollow walls are constructed using stud partitioning. A stud partition consists of vertical softwood sections, called studs, which are braced with horizontal cross pieces or noggins. These stiffen the structure and provide fixing points for the plasterboard, which is used to cover the timber frame.

In this detail, the slates are nailed at the head, although some are centre-nailed.

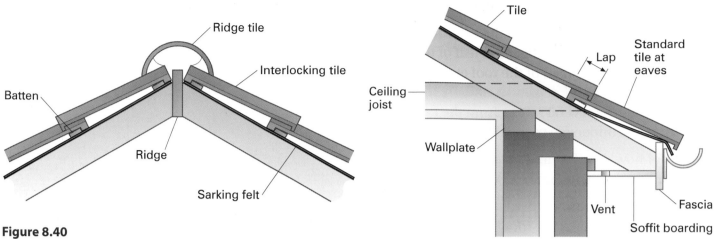

Figure 8.40
Interlocking
tile detail

The detail below shows interlocking tiles. It also provides details of how the roof is finished off at the ridge and eaves.

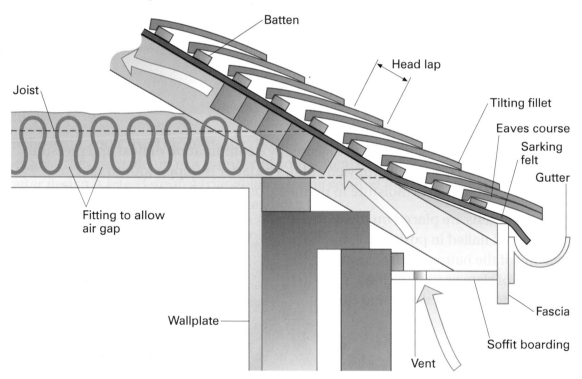

Figure 8.41 Roof
ventilation

Roof ventilation is a requirement of the Building Regulations to prevent the roof timbers from rotting as a result of the presence of moisture.

Timber-frame construction

Up to now you have dealt with what is known as 'traditional' building construction for domestic dwellings: with masonry external walls consisting of an external leaf of brickwork and an internal leaf of blockwork, separated by a cavity which in most cases will be insulated.

In timber-frame construction, the inner leaf of blockwork is replaced with a structural timber frame, which takes the load in the same way as the blockwork in a traditional construction. Most timber-frame dwellings are clad with an outside skin of brickwork, but other options such as timber or plastic weatherboarding or rendering can be used.

When first introduced, timber-framed construction suffered from a poor image, resulting mainly from bad detailing and poor workmanship during installation, which allowed moisture to enter the timber frames.

Quality control has now improved to a level where this form of construction is a popular alternative to traditional methods. Timber-framed panels are prefabricated in a factory under stringent quality-controlled conditions, and the panels are assembled as large units on site. Figure 8.42 shows what the structure of the panels looks like, and how provision is made for openings and first-floor joists. Figure 8.43 shows how a panel is constructed.

Find out

What do you think might be the advantages of timber-framed construction?

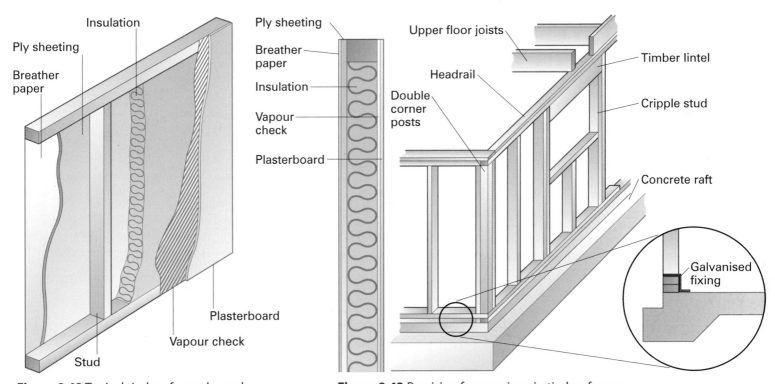

Figure 8.42 Typical timber-framed panel

Figure 8.43 Provision for openings in timber frames

Older building structures

Plumbers do have to work in buildings that were constructed many years ago. Generally speaking, the principles of construction have remained the same over the years, the changes being driven by the need to reduce over-specification and to improve the performance of buildings in terms of energy efficiency, resistance to moisture, and the comfort of the occupants.

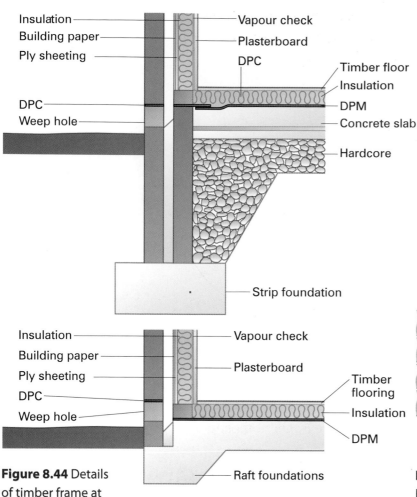

Insulation
Building paper
Ply sheeting

DPC
Weep hole

Vapour check
Plasterboard
DPC
Timber floor
Insulation
DPM
Concrete slab

Hardcore

Strip foundation

Insulation
Building paper
Ply sheeting
DPC
Weep hole

Vapour check
Plasterboard
Timber flooring
Insulation
DPM

Raft foundations

Figure 8.44 Details of timber frame at foundation level

Foundation details haven't really changed, but external and internal wall construction has. External walls used to be constructed in 9" (225 mm) solid brickwork, and bonded in a range of bonds based on laying the bricks in a stretcher bond for so many courses, then tying the two walls together with a course of 'headers'; for really old construction you could find limestone walls up to a metre thick with rubble infill. Figure 8.45 shows an example of solid brickwork in 'English bond'.

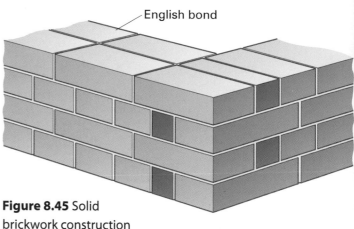

English bond

Figure 8.45 Solid brickwork construction

Two main problems of solid-wall construction are, firstly, the risk of moisture penetration over time, because of the moisture travelling through the solid material. The cavity wall with its 'gap' prevents that. The other is that the construction is limited in terms of its thermal insulation properties: it wouldn't meet the current Building Regulations requirement.

Internal walls used to be constructed in solid brick or concrete block, and some might be load-bearing, which means they could be carrying the weight of the first floor.

You may also encounter solid walls that didn't have a DPC installed when originally built. The chances are that they will have had one fitted later, a common process being to drill two holes in each brick, about 150 mm above ground level, and then to inject the brick with silicon to make the whole brick impervious.

Solid-floor construction has remained fairly consistent, although the DPM may not have been installed on very old properties. The principle of first-floor timber construction is the same, but the size of joists has been reduced in both depth and thickness, as generally they were over-specified. Similarly, the width of floorboards has been reduced, and more often than not chipboard sheets are now used.

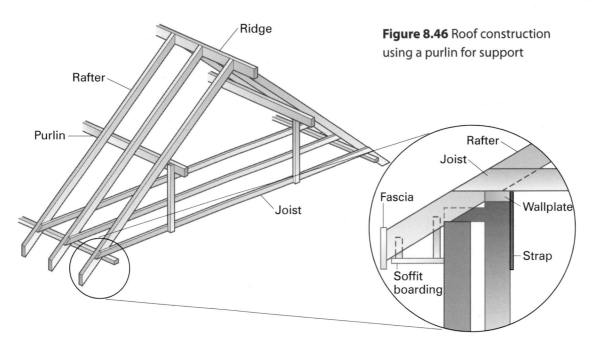

Figure 8.46 Roof construction using a purlin for support

Walls now tend to be dry-lined, replacing two-coat wet plastering and, in some cases, sand-and-cement rendering. This saves on labour costs and production time in waiting for the walls to dry out.

Prior to the introduction of trussed rafters, roofs were constructed using large section timber components. You may find that the roofs of some larger domestic dwellings are still constructed in this way.

Finally, internal fitments such as window boards now tend to be produced in MDF rather than traditional timber sections.

Provision within buildings for the accommodation of system components

There are Regulations governing where plumbing systems and components may be fitted, so as not to affect the stability of the building. Regulations cover the provision for installing system components concerning:

- pipework underfloor/in-floor
- chases and ducts
- roof spaces
- where services enter the building
- accessibility.

The requirements for installing system components are covered in the Water Regulations and the Building Regulations.

Inspecting buildings to confirm that provision for the system or components is suitable

This checklist for a bathroom installation applies equally to working in a new-build dwelling and a refurbishment job:

Checklist

- Make arrangements to gain access to the building to carry out the inspection.
- Confirm the customer's/client's requirements before commencing the work, including:
  - budget
  - fixtures and fittings (using a range of manufacturers' brochures)
  - best type of system to suit the customer's needs without compromising design and installation requirements, e.g. sealed/unvented systems versus open-vented, use of combination boilers, different control systems
  - preferred location of appliance and system components, and preferred flue option.
- Check the suitability of input services including:
  - cold water – whether the pressure and flow rate are adequate for the intended purpose
  - electricity supply, location of connections to system components, safety tests, isolation requirements, equipotential bonding (cross-bonding)
  - gas supply, location of incoming supply in relation to appliances; carry out tightness test prior to working on existing systems.
- Inspect the area where the work will be carried out, including:
  - checking the wall construction to determine specific fixing requirements (e.g. hollow walls)
  - the type of floor construction if running pipework below the surface; tools required for cutting traps
  - the suitability of load-bearing surfaces for appliances, any specific construction requirements, e.g. timber stands for CWSC
  - construction details if cutting chases, drilling for waste pipes or floors; any special tools required
  - any requirement to organise other trades to support the plumbing work – bricklayer, joiner, or electrician
  - working out pipe runs, and determining the fixing positions of appliances and components.
- Record the details clearly in a note book, including:
  - any specialist tools required, including testing equipment
  - list of appliances, materials, fittings and components for the installation
  - materials for making good if applicable

- unusual construction details that could affect the installation, e.g. restricted height in the roof space, solid 450 mm stone walling
- any specific installation details that may impact on the design of the system: remember to double check against the Water or Gas Regulations, Building Regulations or design information.
- Confirm the system design and performance requirements, including:
  - size of boiler and ventilation requirements
  - size of components, radiators, CWSC, HWSC, pump
  - pipe sizes for hot, cold and central-heating pipework
  - system layout.

Post inspection

After initial inspection the following should occur:

- Plumber confirms the system design specification and cost with the customer/client – first verbally, then in writing.
- Order materials from the merchants.
- Confirm a start date with the customer.
- Upon completion of the job, fully demonstrate the system operation to the customer/client, and leave all the manufacturers' customer information.

Plumbing information sources

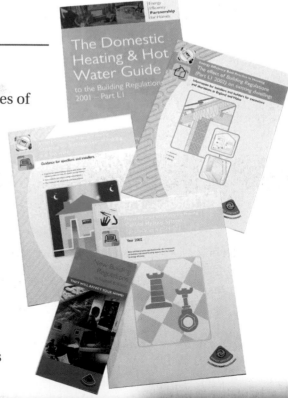

At the end of this section you should be able to:

- understand installation procedures for systems, the range and sizes of components, positions into which they are to be installed, and the preparatory work necessary before installation commences
- identify the persons with whom liaison should take place when carrying out the work
- understand the requirements for work by other trades, and how to negotiate these
- state the measures required to protect work in progress from damage by building operations
- recognise the information that provides the details required for a specific system installation
- access information to find the types of materials and components that are specified for a system

- confirm that the materials, components and designs specified for an installation meet the requirements of industry standards, including compliance with current legislation

- undertake basic job programming.

Plumbing information covers a range of sources, including plans, drawings and specifications, manufacturers' literature, Codes of Practice and legislative documentation. You will use this information in designing or specifying systems installations, so you need to know how to use it. You'll also use plans and drawings in construction, so you should be familiar with how to read them.

Information on the details required for a system installation

Plans and drawings take many forms. Here are some examples.

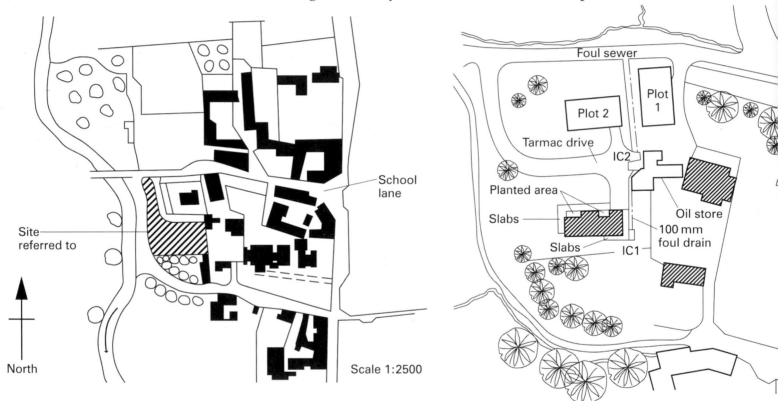

Figure 8.47 Block plan

Figure 8.48 Site plan

While a block plan doesn't provide details of a specific system installation, it is classified as part of the general documentation that's required for a new housing development, so it's worth a mention here.

A block plan records a number of uses of space. Developers will use it to show the proposed site's relationship to amenities such as schools, shops and leisure facilities, together with the transport infrastructure – meaning road and rail networks.

The site plan shows the proposed development in much greater detail. The road access to the site is shown, together with the position of each property. Drainage and sewerage requirements are also included.

A location drawing is used to show overall sizes, levels and references to assembly drawings. A location drawing could be:

- block plans
- site plans
- floor plans
- foundation plans
- roof plans
- section through the building
- elevations.

Assembly drawings are used to show how a building is erected on site, and will show a detailed section through each aspect of the construction.

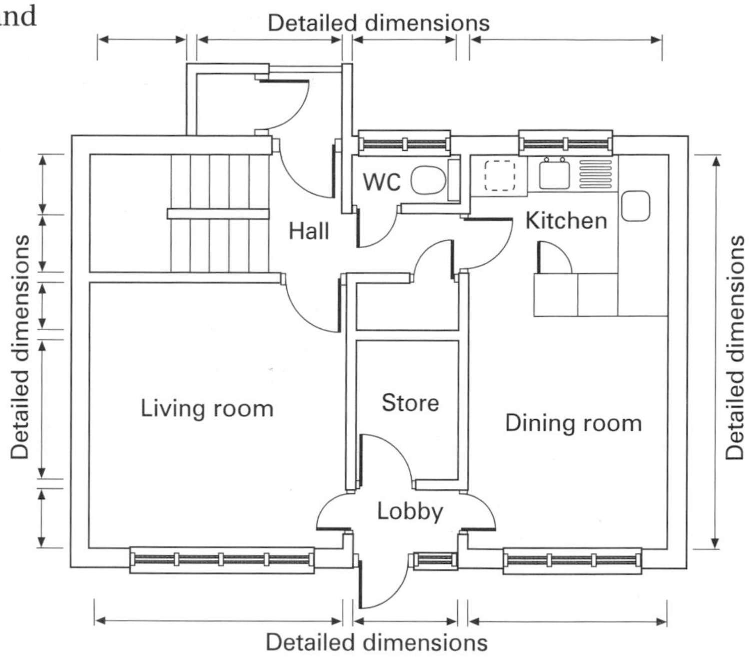

Installation drawings

Generally, on one-off installations, the pipe runs will be determined by the plumber, and the location of appliances and components agreed in consultation between the customer/client and the plumber. On some jobs, however, installation drawings may be provided, showing the position of appliances and components, pipe sizes and pipe runs.

Figure 8.49 Location drawing showing a ground-floor plan

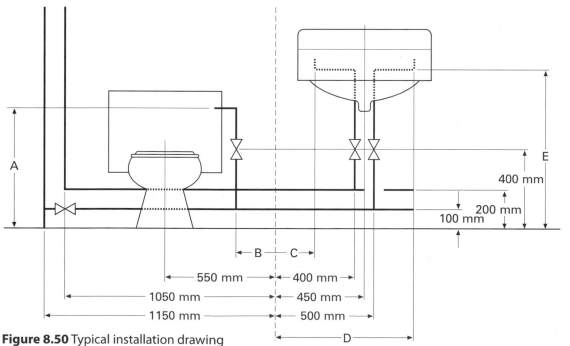

Figure 8.50 Typical installation drawing

Specifications

On larger installation contracts, a **bill of quantities (B of Q)** is produced to specify how the work will be carried out, as well as the quality and quantity of the materials. The B of Q is produced by measuring all the quantities, based on a drawing, referred to as 'taking off'. Table 8.3 is a typical example of an extract from a B of Q.

Item No.	Description	Quantity	Unit	Rate	Cost £
A	All sanitary ware supplied by a manufacturer . 'Hiline' pedestal-mounted wash basin in white vitreous china to BS 3402, 67 cm x 53 cm	10	Item	55.00	550.00
B	'Starly' bath in cast white acrylic sheet 170 cm x 70 cm, and slip-resistant base	10	Item	250.00	2500.00
C	'Space' close-coupled WC suite with washdown bowl and box flushing rim in vitreous china to BS 3402, 39 cm wide x 79 cm high	10	Item	150.00	1500.00
D	Table 15 mm copper tube grade X half-hard to EN 1057, supplied in 6 m lengths	120	M	1.10	132.00
E	Table copper integral solder ring fittings to EN 1254, 15 mm Tees Elbows Tap connectors Straight connectors	20 40 30 40	Item Item Item Item	1.20 0.80 1.80 0.40	24.00 32.00 54.00 16.00

Table 8.3 Extract from bill of quantities

The B of Q is produced by a quantity surveyor, usually on behalf of the architect, and its main purpose is to cost a job in great detail. It will also be used throughout the contract to control costs and provide milestones for contractors' payments. The architect, on behalf of the client, will use the B of Q before a contract starts as part of the tender documentation. It is sent out to contractors with the rate and cost columns blank for them to fill in when tendering for the work.

Material and component details

Material schedules

A material schedule contains information similar to the above, but would be used by a plumber on site as a working document to provide details of what materials would be specified for a particular dwelling. It is also unlikely to have any costings to it.

Component and appliance details

Generally speaking, most plumbers use manufacturers' instructions for specific details of components or appliances. This type of information is supplied with the component or appliance in the delivery packaging. If working on an existing appliance on a maintenance job (particularly boilers), you must have access to manufacturers' instructions. You should be able to obtain copies of instructions from most manufacturers.

Figure 8.51 shows a typical page from a manufacturer's installation instructions.

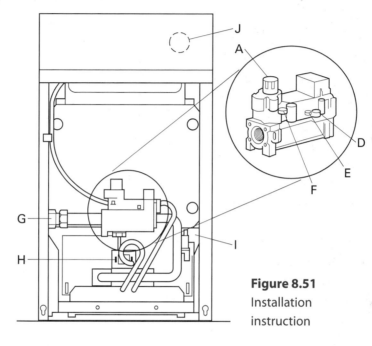

Figure 8.51
Installation instruction

A	Gas control knob	G	Gas service cock
D	Burner pressure test nipple	I	Piezo ignition button
E	Main burner pressure adjuster	J	Boiler thermostat knob
F	Inlet pressure test nipple	H	Flame window

Industry standards

There are two reasons for making sure that an installation meets industry standards:

- the materials are of a satisfactory standard
- the work is of a satisfactory standard and conforms to Regulations.

Materials used in plumbing installations should be to the relevant EN or BS number. BS also make recommendations on design and installation practice, e.g. BS 6700.

In addition to British Standards, the following legislation places statutory responsibilities on plumbers, and many of the requirements laid down are absolute, i.e. if not complied with severe penalties can follow:

- Water Supply (Water Fittings) Regulations 1999
- Gas Safety (Installation and Use) Regulations 1998

- Electricity Supply Regulations 1998 and Electricity at Work Regulations 1989
- Building Regulations 2000
- Health and Safety at Work Act 1974.

Basic programming

Work programmes

On a smaller job the plumber will have the work programme inside their head, based on an agreed start and finish date; but on larger contracts involving more dwellings the approach is more scientific, and a contract programme will be provided. This could consist of an overall programme for all site trades, as well as a separate programme for each trade, including plumbing.

Below is an example of a programme for a plumbing installation.

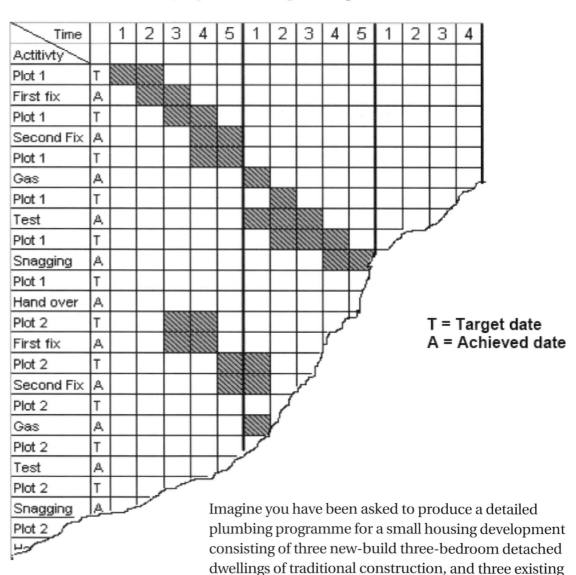

T = Target date
A = Achieved date

Imagine you have been asked to produce a detailed plumbing programme for a small housing development consisting of three new-build three-bedroom detached dwellings of traditional construction, and three existing three-bedroom detached dwellings which are to be completely stripped of their existing plumbing systems and refurbished.

The master bedroom will have en-suite facilities, and there will be a downstairs cloakroom. Plumbing will also be required for a dishwasher, washing machine and outside tap. Gas appliances include inset living-flame-type fire, boiler and hob.

Draw up the programme, to include:

- the points at which other trades/persons may be needed. This could include:
  - external ground works for the mains service
  - joiner for CWSC supports, and lifting floorboards on the existing dwellings
  - bricklayer for cutting holes and chases
  - building site manager for progress meetings or organising labour
  - water company approvals/approved installer
- when and what materials need to be ordered (think about having money tied up in materials, and having to store them)
- laying cold-water service pipe from external stop tap to dwelling
- first-fix carcassing for hot, cold water, heating and above ground discharge pipework
- first-fix carcassing for gas pipework
- second-fix heating appliances, components and controls
- second-fix sanitary appliances
- second-fix pipework
- testing for water tightness
- testing for gas tightness
- commissioning systems and components
- handing over system to client
- carrying out snagging.

Working around other trades, the contract period is six weeks for all the installation activities, followed by a two-week snagging period. Decide on an appropriate plumbing specification, in order to give an indication of installation time. Consider whether it will be:

- cold-water supply: direct or indirect?
- type of heating system, including number of radiators: fully pumped, sealed?
- type of gas boiler and controls, energy efficiency considerations
- unvented hot-water system or vented?
- type and quality of sanitary appliances; will you include a bidet?

Give some thought to the amount of time required to do the work – will one plumber be able to cope, or will it need more labour? A blank programme sheet is included on page 450 for this activity – photocopy it and have a go at completing it.

Programme sheet

Activity	Weeks					
	1	2	3	4	5	6
	Days					
	1 2 3 4 5	6 7 8 9 10	11 12 13 14 15	16 17 18 19 20	21 22 23 24 25	26 27 28 29 30

Variation orders

A variation order is a contractually binding document (usually an A4 sheet in triplicate) that allows an architect (or official company representative) to make changes to the design, quality or quantity of the building and/or its components. They're normally associated with larger contracts, and are issued for:

- additions, omissions or substitution of any work
- alterations to the kind or standard of materials
- changes to the work programme
- restrictions imposed by the client, such as:
  - access to the site or parts of the site
  - limitations of working space
  - limitations of working hours
  - specific order of work.

Here's an example of how the process works. If the client has requested that the architect change a first-floor construction, and this will result in the plumber having to change pipework runs significantly, then the plumber would request a **variation order** (probably via the construction site manager), which is written confirmation of a change to the contract.

The variation order will clearly state exactly what will change from the original specification, and will form the basis for claims for additional costs and/or time. In practice, the variation order is usually issued by the architect's representative on site, the **clerk of works**.

On smaller jobs there won't be an architect, so you may have to deal with issues like this yourself, although it is usual practice for you to pass it on to your employer to deal with. The main point is that you don't commit to any additional work without confirmation from the customer, and this needs to be in writing. In the event that your employer/supervisor can't deal with it, get confirmation (preferably on company documentation), including the customer's signature, to confirm the work, or get a signed note from them.

There are probably more disputes in the contractor/customer relationship over additional work/changes to the work programme than over anything else, so it needs to be properly documented.

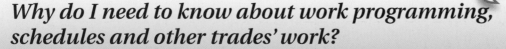

FAQ

Why do I need to know about work programming, schedules and other trades' work?

We are all working towards the same goal, i.e. completion of the project. It is important that we all work together efficiently and safely to this end.

Basic design principles for cold-water and hot-water systems

At the end of this section you should be able to:

- understand the design principles and pipe sizing for:
  - cold and hot-water systems
  - central-heating systems.

For systems in larger buildings (this could include larger detached domestic properties) that require several outlets over a greater area, or on several floors, the pipework will need to be sized correctly, ensuring that there is adequate pressure and flow at all the draw-off points, without any excessive system noise problems. Correct pipe sizing will ensure adequate flow rates at the appliances, preventing any associated problems caused by over-sizing or under-sizing. In this section we will look at the design principles required to help you achieve this.

Pipe-sizing procedure

You'll need access to BS 6700, which provides a series of tables and charts for pipe sizing. Pressure is not always measured in the same units on the charts, so here is a reminder about converting different values:

1 metre head = approx 10kPa (kN/m2) = approx 0.1 bar

Determining the flow rate

Generally, in most buildings, it is unlikely that all appliances installed will be in use at the same time. As the number of outlets increases, the probability of them all being used at the same time decreases. It is therefore more economical to design the system for peak flow rates, which are based on the probability theory, using 'loading units' instead of using possible maximum flow rates.

(a) A **loading unit** is a factor or number given to an appliance; it relates to the flow rate at the terminal fitting, to the period of time in use and frequency of use.

(b) Table 8.4 from BS 6700 gives the loading units for various appliances.

By multiplying the number of each type of appliance by its loading unit, then adding the results together, the total units for the installation will be found. The loading units can then be converted into litres per second by using the conversion chart in Figure 8.52.

In this example, a building with five floors has on each floor one bath, two WC cisterns, two wash basins with taps and one shower. At this stage, you're dealing with just the cold water; the hot water will be dealt with separately, as it comes from a distribution system (storage) – unless it's an unvented system (or a combi), in which case you'll need to look at sizing aspects of the hot and cold systems together.

Outlet fitting	Design flow rate (l/s)	Minimum flow rate (l/s)	Loading units
WC cistern – dual or single flush	0.13	0.01	2
WC trough cistern	0.15 per WC	0.01	2
Wash basin – ½" size	0.15 per tap	0.01	1½ to 3
Spray tap or spray mixer	0.05 per tap	0.03	–
Bidet	0.20 per tap	0.01	1
Bath tap – ¾" size	0.30	0.20	10
Bath tap – 1" size	0.60	0.40	22
Shower head	0.20 hot or cold	0.10	3
Sink tap – ½" size	0.20	0.10	3
Sink tap – ¾" size	0.30	0.20	5
Washing machine – ½" size	0.20 hot or cold	0.15	3
Dishwasher – ½" size	0.15	0.10	3
Urinal flushing cistern	0.004 per position	0.002	–

Table 8.4 Loading units for different appliances

Bath	=	1 × 10	=	10 units
WC cistern	=	2 × 2	=	4 units
Wash basin	=	2 × 1½	=	3 units
Shower	=	1 × 3	=	3 units
Total loading units			=	20 units

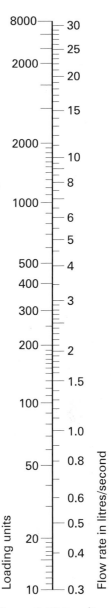

Figure 8.52 Loading unit calculator

By using the conversion chart in Figure 8.52, the figure can be converted to litres per second; this gives **0.45 l/s for the section of pipework**. Therefore, the five floors being supplied have a total loading unit rating of 20 × 5 = 100, converted to litres per second, gives 1.25 l/s.

(a) Continuous flows: appliances such as automatic flushing cisterns must be considered as having a continuous flow rate, and instead of applying the probability theory for using loading units the full design flow rate for the outlet fitting must be used.

(b) Design flow rate: the design flow rate for a pipe is the sum of the flow rate determined from 'loading units' and the 'continuous flows'.

Effective pipe length

Because valves and fittings create a resistance to the passage of water, you must convert the resistance created by them to an equivalent length of straight pipe run:

Effective pipe length = actual pipe length + equivalent pipe length (valves and fittings)

This calculation also benefits from a ready-made chart. Table 8.5 shows the equivalent pipe lengths for various types of fittings and pipe sizes for use on copper, plastic and stainless steel pipework.

Bore of pipe (mm)	Equivalent pipe length			
	Elbow (m)	Tee (m)	Stop valve (m)	Check valve (m)
12	0.5	0.6	4.0	2.5
20	0.8	1.0	7.0	4.3
25	1.0	1.5	10.0	5.6
32	1.4	2.0	12.0	6.0
40	1.7	2.5	16.0	7.9
50	2.3	3.5	22.0	11.5
65	3.0	4.5	–	–
73	3.4	5.8	34.0	–

Table 8.5 Equivalent pipe lengths for various types of fittings and pipe sizes

Note that:

1 For tees the change of direction only should be considered.

2 The pressure loss through gate valves can be ignored.

From the chart in Table 8.5 you can see that using a 20 mm stop valve is equivalent to adding another 7 m of pipe run.

Pressure loss through outlets

To size your pipework you'll also need to know the pressure loss across any outlet fittings – BS 6700 provides you with standard data for some common fittings; for more specialist fittings such as shower valves, the manufacturer will be able to provide this information.

For the system to work, the pressure at the inlet to the tap should be more than the pressure loss across it from inlet to outlet. Pressure loss through taps can be calculated using Table 8.6.

Nominal size of tap	Flow rate (l/s)	Loss of pressure (kPa)	Equivalent pipe length (m)
G½" basin	0.15	5	3.7
G½" basin	0.20	8	3.7
G¾"	0.30	8	11.8
G1"	0.60	15	22.0

Table 8.6 Calculating pressure loss through taps

The pressure loss through float-operated valves is worked out using another scale.

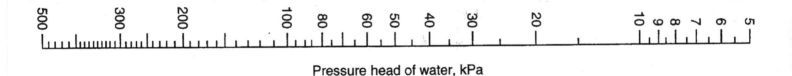

Pressure head of water, kPa

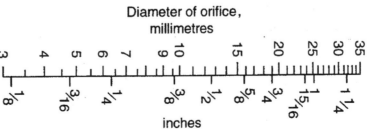

Diameter of orifice, millimetres

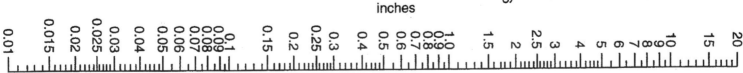

Flow through orifice, litres per second

Figure 8.53

Pipe size – cold water supply pipework

Section 1–2

Always start from the main, then on through the system

Assume a pipe diameter of 15 mm (standard for domestic properties) – column (4)

Total loading units on section = all appliances = 20 units (place in column 2), this equates to 0.45l/s using the conversion of loading units chart on page 83 (place in column 3).

Before we go any further the next point to establish is the head loss per metre and the velocity (using the chart on page 87 and our assumed pipe size (15mm) and design flow rate of 0.45l/s) – this establishes a head loss of 10kPa/m (column 6) and a velocity of 3.4m/s (column 5)

At this point it has been established that the velocity is above our maximum limit so we go to the next pipe size up to remedy the problem and re-enter the details to that point

The drop or rise in head is determined by the change in vertical height in the system from point 1 to 2 in the section which is 2m, this equates to a change of 20kPa, as point 2 is higher in the system it will be a drop in pressure, so a negative figure is shown in column 7.

The available head is therefore the pressure at point 1 less the drop or rise to point 2 = 500 kPa (main) – 20 kPa = 480kPa (column 8)

The actual pipe length is shown in column 9.

The effective pipe length (column 10) is the actual plus any fittings allowance from the table on page 85, so effective length is 8 + 0.8m for the 20mm elbow (note: valves are included separately at column12)

Pipe head loss (column 11) is calculated by multiplying column 10 x column 6

The head lost through the stop valve is inserted at column 12 – 20mm stop valve = 7.0m or 70kPa - page 85

Total head loss (column 13) is column 11+12

The available head (column 14) is column 8 – column 13 (on main runs columns 15,16 & 17 do not require completing)

1	2	3	4	5	6	7	8	9	10	11	12	13	14	15	16	17
Pipe reference	Flow rate		Pipe size (mm)	Velocity (m/s)	Head loss (kPa/m)	Drop– Rise+ (kPa)	Available head (7+14) (kPa)	Pipe length		Head loss			Residual head			
	Total (LU)	Design (l/S)						Actual (m)	Effective (m)	Pipe (10x6) (kPa)	Valves (kPa)	Total (11+12) (kPa)	Available (8-13) (kPa)	Fitting type	Required (kPa)	Surplus (kPa)
1-2	20	0.45	15	3.4	10	-20	480	8.0								
1-2	20	0.45	22	1.5	1.5	-20	480	8	8.8	13.2	70	83.2	396.8			

Table 8.7

Section 2-3

We're going to assume a diameter of 15mm again (column 4)

The loading units are 16.5 (column 2) and the flow rate for the section is 0.4l/s (chart page83) into column 3, this equates to a head loss of 7.5kPa/m (column 6) and a velocity of 3.0m/s (column 5) right on the limit but we're going to run with it this time.

The head drop/ rise across the points is 0 (the pipe is installed level) (column 7)

The available head box for this section (column 8) is the available residual head from column 14 in section 1-2 less the drop/rise shown in this section at column 7 = 459.8 – 0 = 459.8

The actual length is 2.0m (column 9), the effective pipe length is the tee change of direction for 15mm = 0.6m (column 10). Total pipe head loss is column 6 x column 10. The head loss for the stop valve and check valve on a 15mm pipe is 4.0 + 2.5 = 6.5 m or 65kPa – (page 85 and inserted into column 12).

The remaining sections on the main run are progressed in the same manner so these have been included here, we will now go on and look at what happens with a section to an appliance connection as this is slightly different.

1	2	3	4	5	6	7	8	9	10	11	12	13	14	15	16	17
Pipe reference	Flow rate		Pipe size (mm)	Velocity (m/s)	Head loss (kPa/m)	Drop– Rise+ (kPa)	Available head (7+14) (kPa)	Pipe length		Head loss			Residual head			
	Total (LU)	Design (l/S)						Actual (m)	Effective (m)	Pipe (10x6) (kPa)	Valves (kPa)	Total (11+12) (kPa)	Available (8-13) (kPa)	Fitting type	Required (kPa)	Surplus (kPa)
2-3	16.5	0.4	15	3.0	7.5	0	396.8	2.0	2.6	19.5	65	84.5	312.3			
3-4	13.5	0.36	15	2.6	7.0	0	312.3	1.5	1.5	10.5	0	10.5	301.8			
4-5	12	0.33	15	2.5	6.0	0	301.8	0.5	0.5	3.0	0	3.0	298.8			

Table 8.8

Section 5-6

This is similar but slightly different at the end.

Assume the pipe diameter to be 15mm (column 4)

Loading units for the WC are 2 (column 2) and the flow rate required is 0.05l/s (column 3)

The head loss (column 6) is 0.2kPa and the velocity is approx 0.4 m/s (column 5) – from chart on page 87.

The pressure drop/ rise from point 5 to 6 is a drop of 0.6 x 10 = 6kPa. The available head box for this section (column 8) is the available residual head from column 14 in section 4-5 less the drop/rise shown in this section at column 7 = 298.8 – 6 = 292.8.

The actual pipe length is 1.25m (column 9). The effective pipe length 1.25 + 0.5m allowance for a 15mm elbow (from table page 85) equates to 1.75m (column 10).

Column 6 is multiplied by column 10 to give the pipe head loss (column 11). There are no valves so column 12 is 0.

The available residual head loss (column 14) is column 8 – column 13.

Now we have established the head at the inlet to the float valve, we need to ensure that this is greater than the pressure loss across it, so going back to our original plan we need to establish the pressure loss across a 3mm orifice float valve at 0.05l/s, this is done by using the chart on page 86 = 43kPa – detail inserted in column 16.

The surplus kPa (column 17) which must always be a positive figure or else the sizing process will have to be reviewed is column 14 minus column 16.

We can now finish and check the other final sections in a similar manner remembering that we use the available head figure as the start point of the section and that we must include the change of direction for a tee. The pressure loss across taps are shown on page 85. The remaining sections have been shown.

1	2	3	4	5	6	7	8	9	10	11	12	13	14	15	16	17
	Flow rate		Pipe size (mm)	Velocity (m/s)	Head loss (kPa/m)	Drop–Rise+ (kPa)	Available head (7+14) (kPa)	Pipe length		Head loss				Residual head		
Pipe reference	Total (LU)	Design (l/S)						Actual (m)	Effective (m)	Pipe (10x6) (kPa)	Valves (kPa)	Total (11+12) (kPa)	Available (8-13) (kPa)	Fitting type	Required (kPa)	Surplus (kPa)
5-6	2	0.05	15	0.4	0.2	-6	292.8	1.25	1.75	0.4	0	0.4	292.4	Float valve	43	249.4
3-9	3	0.20	15	1.4	2.4	-10	302.3	1.0	1.0	2.4	0	2.4	299.9	0.2l/s tap	8	291.9
4-8	1.5	0.15	15	1.2	1.5	-7.5	294.3	0.75	0.75	1.2	0	1.2	293.1	0.15l/s tap	5	288.1
5-7	10	0.30	15	2.3	5	-5	298.8	0.5	0.5	2.5	0	2.5	296.3	0.2l/s tap	8	288.3

Table 8.9

You are required to size the indirect cold-water pipework from the cold-water storage cistern shown. The details of the system are similar to those you've previously seen:

- bath 0.30 l/s (10 loading units)

- basin – 0.15 l/s (1.5 loading units)

- WC – 0.05 l/s (2 loading units) – 3 mm orifice

- sink – 0.20 l/s (3 loading units)

- a **gate valve** is included in the pipework – but you can ignore this for pressure loss

- the head from the base of the cistern to the lowest point on the pipework is 4 metres.

Hint: section and label the pipework first, and remember that this time you start off gaining pressure from 4.0 metres head at the outlet of the cistern.

Use the calculation sheet on the following page to carry out the calculation.

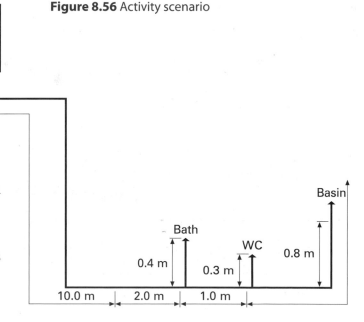

Figure 8.56 Activity scenario

Cold-water storage cisterns

Traditionally, cold water in domestic dwellings was stored to provide a reserve in the event of cold mains failure. However, over recent years there has been a steady decline in the use of indirect systems, mainly because of the increase in the use of combination boilers and unvented hot-water storage vessels that don't require a cold-water storage vessel.

British Standard 6700 gives a recommended minimum storage capacity for domestic dwellings of approximately 230 litres for indirect systems. Cisterns that are supplying cold water only should have a minimum capacity of approximately 115 litres.

1	2	3	4	5	6	7	8	9	10	11	12	13	14	15	16	17
Pipe reference	Flow rate		Pipe size (mm)	Velocity (m/s)	Head loss (kPa/m)	Drop–Rise+ (kPa)	Available head (7+14) (kPa)	Pipe length		Head loss			Available (8-13) (kPa)	Residual head		
	Total (LU)	Design (l/S)						Actual (m)	Effective (m)	Pipe (10 x 6) (kPa)	Valves (kPa)	Total (11 + 12) (kPa)		Fitting type	Required (kPa)	Surplus (kPa)

Figure 8.57 Pipe size chart

In larger premises the cold-water storage capacity will depend on the following factors:

- building type and use
- number of occupants
- number and types of fittings
- rate and pattern of use
- likelihood of an interruption or breakdown of the mains supply.

These factors have been taken into consideration, and Table 8.14 gives recommended guidance for the storage of water in a variety of different property types.

Type of building or occupation	Minimum storage (l)
Hostel	90 per bed space
Hotel	200 per bed space
Office premises – with canteen facilities – without canteen facilities	45 per employee 40 per employee
Restaurant	7 per meal
Day school – nursery – primary – secondary – technical	15 per pupil 20 per pupil
Boarding school	90 per pupil
Children's home/residential nursery	135 per bed space
Nurses' home	120 per bed space
Nursing/convalescent home	135 per bed space

Table 8.10 Minimum storage recommended for different property types

The minimum cold-water storage shown in Table 8.10 also includes water used to supply hot-water outlets.

As an example, a boarding school with 200 residential pupils has opened a new day nursery annexe. The nursery contains provision for a further 40 pupils. Determine the amount of cold-water storage required.

Storage capacity = number of pupils × storage per pupil

Day school nursery = 40 pupils × 15 litres per pupil = 600 litres

Boarding school = 200 pupils × 90 litres per pupil = 18,000 litres

Add the two figures together and the total storage capacity is:

600 + 1800 = 18,600 litres.

Remember

The cold-feed cistern must have a capacity at least equal to that of the hot storage vessel it's supplying

Hot-water storage vessels

The minimum hot-water storage capacity recommendations for domestic dwellings is given in the British Standard 6700 as:

- 35–45 litres per occupant, unless the heat source provides a faster heat recovery rate
- 100 litres for systems heated by solid-fuel boilers
- 200 litres for systems heated by off-peak electricity.

When sizing hot-water storage vessels, BS 6700 also recommends that the following factors are taken into consideration:

- pattern of use
- rate of heat input (see Table 8.11)
- heat recovery period of the vessel
- stratification of the stored water.

Table 8.11 can be used to assist with identifying the storage vessel size in relation to the heat input available for typical domestic properties.

Heat input to hot water (kW)	Dwelling with 1 bath		Dwelling with 2 baths	
	With stratification (l)	With mixing (l)	With stratification (l)	With mixing (l)
3	109	122	166	260
6	88	88	140	200
10	70	70	130	130
15	70	70	120	130

Table 8.11 Heat input available for typical domestic properties

Stratification is when hot water in the storage vessel, floats on a layer of cold-feed water, allowing the hotter water in the vessel to be drawn off without the incoming cold-feed water mixing. This in turn contributes to less frequent reheating of the storage vessel, saving in heating costs and energy.

Calculate the capacity of a hot-water storage vessel, assuming you cannot use the detail in Table 8.11. The storage capacity in any situation depends on the rate of heat input to the stored water and its pattern of use. BS 6700 provides a formula for calculating the required storage capacity, which is:

- The time, M (in minutes), taken to heat a quantity of water through a specified temperature rise

  $$M = VT / (14.3\ p)$$

This is explained by:

- V = volume of water heated in litres
- T = temperature rise in °C
- p = rate of heat input to the water in kW.

The formula can be applied to any pattern of use, irrespective of stratification taking place in the stored water. It also ignores heat loss from the storage vessel due to the relativity short reheat period after draw-off has taken place.

The following example explains the application of the formula; for convenience the figures have been rounded off.

Example

A small domestic dwelling has one bath installed. The maximum requirements are: one bath (60 litres of water at 60°C plus 40 litres of cold water) with an additional 10 litres of hot water at 60°C for kitchen use, followed by a second bath fill after 25 minutes.

That makes: a draw-off of 70 litres at 60°C, followed 25 minutes later by a requirement of 100 litres at 40°C. This can be achieved by mixing hot water at 60°C with cold water at 10°C.

Assume that good stratification is obtained, for example by using a top-entry immersion heater (stratification will prevent water mixing). The heat input rate from the immersion heater into the vessel is 3 kW. So to heat 60 litres of water from 10°C to 60°C using a 3 kW heat input takes:

$$M = (60 \times 50) / (14.3 \times 3)$$

Therefore M = 70 min.

So, the second bath that is required 25 minutes later has to be provided from storage. In 25 minutes the water heated to 60°C is:

$$V = M (14.3) / T$$

$$V = (25 \times 14.3 \times 3) / 50$$

$$V = 21 \text{ litres}$$

Therefore the minimum storage capacity to meet the requirement is:

$$70 + 60 - 21 = 109 \text{ litres}$$

Design principles for central-heating systems

The sizing of pipework and hence the pump in a domestic system will often use rule-of-thumb principles. However, it's always best to know how to do it properly, as you may well encounter a bigger job where rule of thumb doesn't apply, or a job that's a bit different. Radiator sizing, and hence boiler sizing, is a different story though, and this needs to be done properly on every job. It's easiest to start with the radiators.

The sizing of heat emitters (panel radiators)

Heat emitters must be capable of providing sufficient heat output to maintain a comfortable room temperature in the room in which they are installed. It would be very uneconomical to install under-sized heat emitters as this would give rise to higher fuel running costs, poor system efficiency and, more importantly, complaints from the customer. But if they are over-sized you will still struggle with efficiency problems – so they need to be sized correctly.

There are several different methods that can be used to establish the correct size of heat emitter for a room. They are:

- a central-heating calculator, e.g. 'mears wheel'
- manufacturer's heat emitter computer program, e.g. 'Stars Myson heat loss calculator'
- mathematical calculation.

To use any type of heat loss calculator the first principle is to know what you're doing. If you can understand the mathematical calculation you'll be able to use any type of calculator.

The calculation

In sizing a radiator in a property you are identifying the rate of heat that will be lost from that room, so you must know that heat is lost in two ways:

- due to air change and natural ventilation (air circulation in the room due to the number of air changes that occur in the room per hour)
- through the building fabric (this is primarily affected by the level of insulation in the building structure: high-insulation qualities lower the heat loss. A key measure of the insulation properties of a building component is the U value).

So, when starting to size a radiator in a room, first work out the heat loss using two different calculations.

Rate of heat loss due to ventilation

This is:

Heat loss (W) = Room volume (m³) × temperature difference (°C) × number of air changes per hour × a constant figure of 0.33

The number of air changes to a room are identified from tables in BS 5449. The air change rate table will be part of the same table that quotes room temperatures, so we'll come across this later.

Now look at the formula used for heat loss through the building fabric. This is:

Heat loss (W) = Surface area (m²) × temperature difference (°C) × U value (W/m/°C)

Part L of the Building Regulations (England and Wales) requires that 'U values are calculated to BS EN 6946'. Before looking at U values in depth, we'll look at temperature difference:

- In the case of a component that makes up the building fabric, the difference is measured across that fabric, so that could be inside a lounge to the outside (an external wall), between the lounge and the kitchen (an internal wall) and between the lounge and the next-door property (a party wall).

- In the case of ventilation, it's always taken as the difference between the required temperature in the room and the outside fresh-air temperature.

The external (outside the building) temperature is usually identified as –1°C, although a lower figure of –3°C is often used for more exposed locations.

Be careful when working out the temperature difference. For example: a lounge is at a temperature of 21°C and the outside temperature is –1°C. The temperature difference = 21 – (–1) = 22°C (minus a minus and it becomes a positive).

Internal design temperatures are generally in accordance with British Standard 5449 Section 3 1990. These are shown in Table 8.12.

Recommended internal air temperature and air change rates		
Room	Temperature (°C)	Air changes (per hour)
Living room	21	1.5
Dining room	21	1.5
Bedsitting room	21	1.5
Kitchen	18	2.0
Bedroom	18	1.0
Hall/landing	16	2.0
Bathroom	22	3.0
Toilet	18	3.0

Table 8.12 Internal design temperatures in accordance with BS 5449 Section 3 1990

Be careful when looking at the number of air changes required in rooms with solid-fuel open fires, as they need to increase greatly; refer to BS 5449 for details.

Where a building or dwelling adjoins another – e.g. a semi-detached house – assume a 10°C temperature difference.

The U value

A U value is termed as the thermal transmittance rate from the inside to the outside of a building, through the intermediate elements of constructions. The U value is defined as the energy in watts per square metre (W/m²) of construction for each degree of Kelvin temperature difference between the inside and outside of the building (W/m²K). Note that various textbooks and the table below refer to this as W/m²/°C. U value tables can be found in a variety of system design guides, and if you're going to do design calculations then you'll need access to U value tables.

Table 8.13 gives approximate U values through building fabric – but you'll need something better than this to do the job properly.

Construction	W/m²/°C	Construction	W/m²/°C
External solid wall	2.0	Ground floor – solid	0.45
External cavity wall	1.0	Ground floor – wood	0.62
External cavity wall (filled)	0.5	Intermediate floor – heat flow up	1.7
External timber wall	0.6	Intermediate floor – heat flow down	1.5
Internal wall	2.2	Flat roof	1.5
Window – single glazed	5.7	Pitched roof (100 mm insulation)	0.34
Window – double glazed	3.0	Pitched roof (no insulation)	2.2
Internal wall – solid block	2.1		

Table 8.13 U values through building fabric

By using the floor plan of the detached bungalow in Figure 8.58, work out the heat emitter requirement of the lounge.

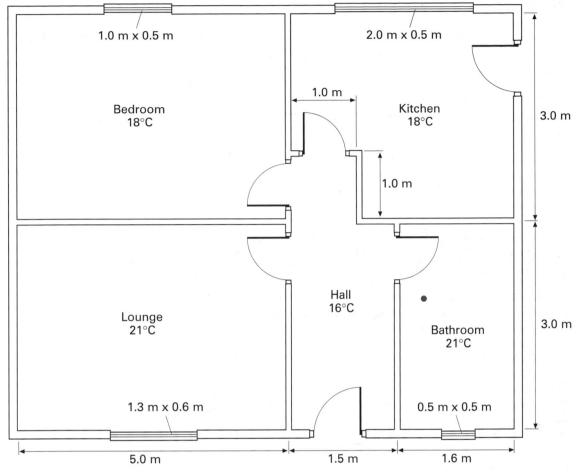

Figure 8.58 Bungalow floor plan

Notes:

- all dimensions to the bungalow are in metres
- the bungalow has solid brick external walls and a solid floor
- the windows are single-glazed, with double-glazed doors
- the roof insulation is 100 mm; the bungalow has a pitched roof
- the height of the rooms is 2.4 metres
- the internal walls are solid block.

This is the calculation procedure to work out the lounge heat loss. The key points are:

- the window or door heat loss is done first. The total area of glazing must be deducted from the wall area for the external heat loss calculation
- internal doors are treated as wall surface.

Fabric-loss element	Area: length x breadth (m²)	Temperature difference	U value W/m³/°C	Heat loss (W)
Window	1.3 x 0.6 = 0.78	x 22	x 5.7	97.8
External walls	8 x 2.4 = 19.2 − 0.76 = 18.42	x 22	x 2.0	810.5
Internal wall – bedroom	5.0 x 2.4 = 12	x 3	x 2.1	75.6
Internal wall – hall	3.0 x 2.4 = 7.2	x 5	x 2.1	75.6
Floor	5.0 x 3.0 = 15.0	x 22	x 0.45	148.5
Roof	5.0 x 3.0 = 15.0	x 22	x 0.34	112.2
			Total	1320.2
Ventilation loss				
Volume	**Air change**	**Temperature difference**	**Factor**	
5 x 3 x 2.4 x	1.5 x	22 x	0.33 = 392.0	

Table 8.14 Heat loss calculation

The total heat loss for the room is:

1320.2 + 392.0 = 1712.2 watts.

However, you've only calculated the amount of heat that will be lost from the room, and in cold weather conditions the amount of heat shown would not be sufficient to raise the temperature in a room in a reasonable timescale, so we need to add a percentage margin to the total room heat loss for intermittent heating of between 10 and 20 per cent, dependent on the system controls and the size of the property. With good controls and a small property, the percentage will be higher. Here, assume 15 per cent. So the heat loss is 1712.2 watts × 1.15 (15% add on) = 1969.0 watts.

You're now at a point when you can begin to select a radiator, and you'll need a radiator catalogue for this.

Radiator selection

The figures quoted in the manufacturer's catalogue are to a test standard. The pipework to them is connected flow at the top and return at the bottom – opposite ends of the radiator; if bottom opposite-end connections are used, which is the norm in domestic properties, then a correction factor needs to be applied to the figures in the catalogue. There's a table for the factor: this is known as f2.

Before you can move on, there's another correction factor – f1. The radiator is tested in a room with a difference between the mean water temperature in the radiator and the air temperature in the room of 60°C. You will probably notice in our bungalow that we use different temperatures, and therefore a further correction factor has to be applied.

The mean water temperature is the average of the flow and return water temperatures. Typically in most systems, the flow will be 80°C and the return will be 70°C. So the mean water temperature will be 80 + 70 divided by 2 = 75°C. You then need to deduct the room temperature from the mean water temperature to find the difference between mean water and air temperature.

In the case of the lounge it's 75 – 21 = 54°C; this is then applied to Table 8.16 to determine the f1 correction factor.

Top and bottom opposite-end connections	1.00
Bottom opposite-end connections with blind nipple	0.97
Bottom opposite-end connections	0.90

Table 8.15 Radiation correction factor – f2

Temp. diff. °C	f1	Temp. diff. °C	f1
40	0.605	56	0.918
41	0.624	57	0.938
42	0.643	58	0.958
43	0.662	59	0.979
44	0.681	60	1.000
45	0.700	61	1.020
46	0.719	62	1.041
47	0.738	63	1.062
48	0.758	64	1.062
49	0.778	65	1.104
50	0.798	66	1.125
51	0.818	67	1.146
52	0.838	68	1.168
53	0.858	69	1.189
54	0.878	70	1.211
55	0.898	71	1.232

Table 8.16 Temperature difference factor – f1

So to work out the radiator size, multiply f1 × f2 to arrive at an overall correction factor. For our lounge at 21°C with both corrections, f1 is 0.878 and f2 is 0.90:

$$0.878 \times 0.90 = 0.79$$

The size of the radiator required is therefore the total room heat loss including the intermittent use margin, divided by the overall correction factor = 1969 watts / 0.79 = 2492 watts. The nearest size radiator above this figure may be selected from the catalogue to suit space requirements.

Boiler sizing

Find out

Carry out this assignment: first, see if you can devise some simple tables for recording the sizing procedure. Now carry out the heat-loss calculation for the bedroom and the bathroom in the bungalow, and select appropriate radiators for each

Total up the sum of all the heat losses from the rooms and add an allowance for pipework heat losses and hot water.

Note: A pipe heat-loss allowance will typically be added where a fair proportion of pipework is installed under suspended timber floors or in roof spaces. If it is all surface mounted in the room then no allowance needs to be added; typically 10 per cent will be added.

The hot-water heat requirement is usually based on providing 1 kW of boiler heat output for every 50 litres of hot water stored.

When determining the heat requirement for the radiator circuit use the figure calculated for the room heat loss, together with any percentage applied for intermittent use, and *not* the figure used for the radiator size after correction factors have been applied, as this would over-size the boiler.

As an example, the heat loss including intermittent heat margin from four rooms is:

- lounge – 1.7 kW
- bedroom – 0.8 kW
- bathroom – 1.2 kW
- kitchen – 1.1 kW.

This gives a total radiator load of 4.8 kW. The hot-water storage capacity of the cylinder is 100 litres, which is 2 kW output.

The total heat load is 4.8 + 2.0 × 1.1 (10 per cent for the pipe losses) = 7.5 kW heat output from the boiler; this can now be selected from the manufacturer's catalogue.

Pipe sizing

Now you've looked at radiator and boiler sizing, take a look at pipe sizing. In undertaking this task, determine the size of the pump required and the setting it should be placed on during commissioning.

To determine the size of pipework for a system you need to be able to identify the required flow rate down the pipe to get the desired amount of heat from the radiators – this is usually measured in kg/s. In moving that water flow through the system you have to overcome the frictional resistance of the pipework and fittings that make up the circuit, so you need to apply pressure; therefore use a pump. That pressure will be greater at the beginning of a particular circuit than it is at the end. This is because the resistance will reduce as you get further down the circuit.

To undertake pipe sizing the first thing to know is the flow rate through a particular section of pipework. Take the lounge radiator that you worked out earlier, which totalled (with an allowance for intermittent heating) 1969 watts. The flow rate is calculated:

Heat in kW = flow rate (kg/s) × the specific heat capacity of water, which is a constant figure of 4.2 × temperature difference flow pipe to return pipe (normally 10°C)

So, moving the figures around in the equation:

$$\text{The flow rate} = \frac{\text{heat in kW}}{10 \times 4.2} = \frac{\text{kW}}{42}$$

Remember, this will change if the temperature difference between flow and return temperatures is different. So for the lounge radiator the flow rate to the pipework immediately supplying that radiator is:

$$\frac{1.969\text{kW}}{42} = 0.047 \text{ kg/s}$$

If there is exposed pipework don't forget to add the allowance to the radiator output.

The next thing you need to know is the length of pipework throughout the circuit. You need a very simple layout drawing of a system to draw the flow and return pipes as single lines, but remember that the length of each circuit will be the length of both flow and return pipes together, so it's doubled.

Most plumbers usually use a chart for pipe sizing to make it easier. However, before you look at one take a look at the principles of how it's done, starting with Section 1.

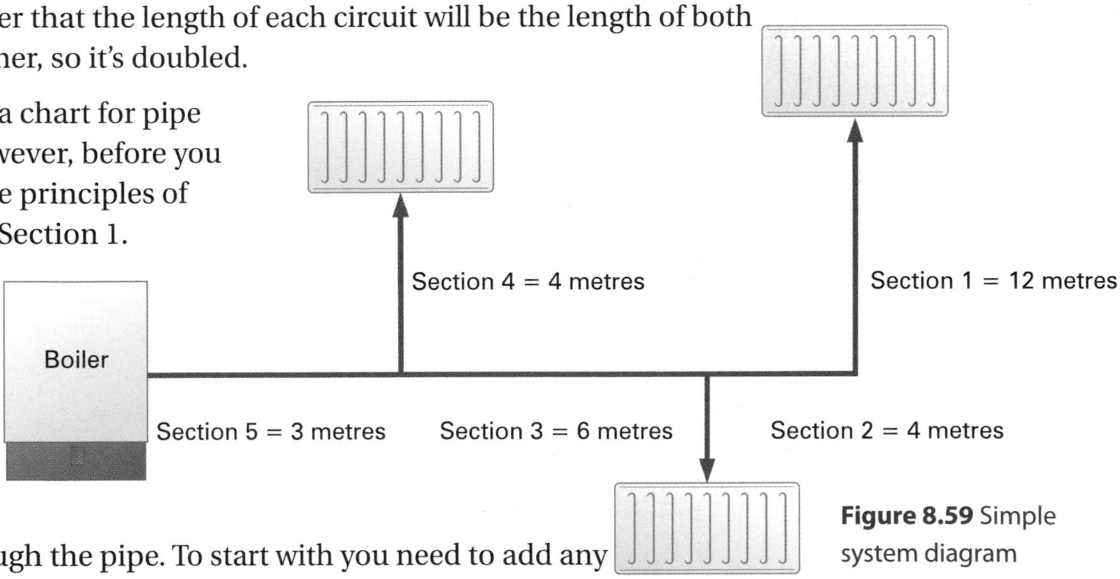

Section 4 = 4 metres

Section 1 = 12 metres

Boiler

Section 5 = 3 metres   Section 3 = 6 metres   Section 2 = 4 metres

Figure 8.59 Simple system diagram

First find the flow rate through the pipe. To start with you need to add any mains pipe heat losses to the radiator output. So section 1 carries 3 kW × 1.1 (10 per cent mains pipe heat loss) = 3.3 kW.

Therefore the flow rate in that section for a 3.3 kW load = $\frac{3.3 \text{ kW}}{42}$ = 0.08 kg/s

Now, the length of pipe run is 12 metres, but that doesn't include any pressure loss due to fittings. Here, use a percentage addition of 33 per cent if the pipework has an average number of changes of direction; if a lot of changes are used then the figure should rise to 50 per cent. This will give an overall effective pipe length of 12.0 m × 1.33 (33 per cent for fittings) = 16.0 m.

Use the chart in Figure 8.60 to determine the pressure loss per metre run of pipe. To start with, you need to select a pipe size that you think might be able to meet the requirement. However, the pipe selected needs to fit within a maximum velocity reading:

- 1.0m/second for standard small-bore pipework, and

- 1.5m/second for microbore pipework.

Figure 8.60 Pressure-loss chart

Pressure loss (rn/rn)	8 mm kg/s	10 mm kg/s	15 mm kg/s	22 mm kg/s	28 mm kg/s	35 mm kg/s	Velocity rn/s
0.008		0.0108	.0380	0.109	0.227	0.400	0.50
0.009		0.0114	0.040	0.117	0.235	0.424	
0.010	0.0064	0.0122	0.042	0.124	0 250	0448	
0.011	0 0067	0 0129	0.044	0.131	0.263	0.475	
0.012	0.0071	0.0135	0.047	0.137	0.277	0.499	
0.013	0.0074	0.0141	0.049	0.144	0.289	0.523	
0.014	0.0077	0.0147	0.052	0.150	0.302	0.543	
0.015	0.0081	0.0154	0.054	0.156	0.314	0.564	
0.016	0.0084	0.0159	0.056	0.161	0.325	0.594	
0.017	0.0086	0.0165	0.058	0.167	0.336	0.604	0.75
0.018	0.0089	0.0171	0.060	0.172	0.348	0.623	
0.019	0.0092	0.0176	0.061	0.178	0.359	0.645	
0.020	0.0095	0.0182	0.063	0.183	0.369	0.669	
0.021	0.0098	0.0185	0.065	0.188	0.380	0.686	
0.022	0.0101	0.0192	0.067	0.193	0.390	0.704	
0.024	0.0106	0.0203	0.070	0.203	0.408	0.735	
0.026	0.0111	0.0212	0.073	0.212	0.428	0.773	
0.028	0.0116	0.0221	0.076	0.221	0.446	0.805	1.00
0.030	0.0120	0.0230	0.080	0.230	0.464	0.838	
0.032	0.0125	0.0238	0.082	0.238	0.482	0.869	
0.034	0.0129	0.0245	0.085	0.247	0.500	0.898	
0.036	0.0133	0.0253	0.088	0.255	0.518	0.925	
0.038	0.0138	0.0261	0.091	0.263	0.533	0.952	
0.040	0.0142	0.0268	0.094	0.270	0.548	0.982	
0.042	0.0146	0.0276	0.096	0.278	0.564	1.010	1.25
0.044	0.0150	0.0283	0.099	0.286	0.578	1.035	
0.046	0.0154	0.0290	0.101	0.293	0.592	1.048	
0.048	0.0158	0.0298	0.104	0.300	0.608	1.075	
0.050	0.0162	0.0305	0.106	0.307	0.622	1.100	
0.052	0.0167	0.0312	0.018	0.314	0.637	1.123	
0.054	0.0170	0.0320	0.111	0.321	0.651	1.150	
0.056	0.0173	0.0326	0.113	0.328	0.665	1.178	
0.058	0.0177	0.0332	0.115	0.334	0678	1.194	1.50
0.060	0.0180	0.0339	0.117	0.340	0.691	1.215	
0.062	0.0184	0.0345	0.120	0.347	0.705	1.235	
0.064	0.0187	0.0351	0.122	0.353	0.718	1.253	
0.066	0.0190	0.0358	0.124	0.359	0.724	1.272	
0.068	0.0193	0.0364	0.126	0.364	0.736		
0.070	0.0196	0.0370	0.128	0.370	0.750		
0.072	0.0200	0.0377	0.130	0.375	0.762		
0.074	0.0203	0.0382	0.132	0.381	0.774		
0.076	0.0206	0.0388	0.134	0.386	0.785		
0.078	0.0208	0.0394	0.136	0.391	0.797		
0.080	0.0211	0.0400	0.138	0.397	0.808		
0.082	0.0215	0.0406	0.140	0.402	0.819		
0.084	0.0217	0.0411	0.142	0.407	0.830		
0.086	0.0220	0.0417	0.144	0.412	0.841		
0.088	0.0223	0.0423	0.146	0.417	0.851		
0.090	0.0226	0.0429	0.148	0.422	0.862		
0.092	0.0229	0.0433	0. 149	0.426	0.872		
0.094	0.0231	0.0439	0.151	0.431			
0.096	0.0234	0.0445	0.153	0.435			
0.098	0.0237	0.0450	0.155	0.440			
0.100	0.0240	0.0455	0.156	0.445			
0.102	0.0243	0.0460	0.158	0.449			
0.104	0.0245	0.0465	0.160	0.453			
0.106	0.0247	0.0469	0.162	0.458			
0.108	0.0250	0.0474	0.164	0.462			
0.110	0.0253	0.0479	0.165	0.466			
0.112	0.0256	0.0484	0.167	0.471			
0.114	0.0258	0.0488	0.169	0.475			
0.116	0.0261	0.0493	0.170	0.479			
0.118	0.0264	0.0498	0.172				
0.120	0.0266	0.0502	0.174				
0.130	0.0279	0.0523	0.181				
0.140	0.0291	0.0548	0.189				
0.150	0.0302	0.0568	0.197				
0,160	0.0314	0.0588	0.204				
0. 170	0.0326	0.0608	0.211				
0.180	0.0336	0.0628					
0.190	0.0347	0.0648					
0.200	0.0357	0.0668					

The velocity readings are the stepped scale from the right-hand side to the left-hand side of the chart. So try an 8 mm pipe size first for the 0.08 kg/s. Look for 0.08 kg/s under the section of the table for 8 mm pipe – notice that it's not even on the table, indicating that it's not suitable.

Try 15 mm pipe. You can see that 0.080 kg/s is equal to 0.030 loss/metre run of pipe from the right-hand scale, and the velocity is between 0.5 and 0.75 m/s, so this is acceptable.

So the resistance in the section of pipe is the head loss/metre run × the effective pipe length. The head loss is:

0.030×16.0 metres of pipe = <u>0.48 metres</u>

Now look at Section 3.

Section 3 carries the heat loads of sections 1 and 2. You already know the flow rate through section 1 as 0.080 kg/s. You therefore need to know the heat load in section 2:

$$\frac{2.1 \times 1.1 \text{ (10 per cent mains loss)}}{42} = 0.055 \text{ kg/s}$$

The total load on the section is therefore sections 1 and 2 = 0.08 + 0.055 = 0.135 kg/s.

The effective pipe length for the section is 6 m × 1.33 (fittings resistance) = 7.98 m.

Back to the chart. Look at 15 mm first. 0.135 is between 0.75 and 1.0 m/s and is acceptable; the pressure loss/metre run of pipe to the nearest figure above is 0.078 m/m run of pipe.

The pressure loss across the section of pipe is therefore 7.98m × 0.078 m/m = 0.62 metres.

The remaining pipe sections are shown completed in the form of Table 8.17, which makes the calculation process easier.

Section	Section heat requirement (kW)	Mains loss (%)	Total heat loss (kW)	Flow rate (kg/s)	Pipe size (mm)	Actual pipe run (m)	Fittings resistance (%)	Effective pipe length (m)	Pressure loss per metre run of pipe (m/m)
1	3.0	1.1	3.3	0.08	15	12	1.33	16	0.030
2	2.1	1.1	2.31	0.055	15	4	1.33	5.32	0.016
3	–	–	–	0.135	15	6	1.33	7.98	0.078
4	2.5	1.1	2.75	0.065	15	4	1.33	5.32	0.021
5	–	–	–	0.2	22	3	1.33	3.99	0.024

Table 8.17 Pipe-sizing calculations

Pump sizing

Now we come to the pump – and you need this not only to specify the right size of pump but also to commission the system. To size the pump you need to know the required flow rate. This one's straightforward – it's the heat load on the section from the boiler to the first branch, so in your worked example it is the flow rate on section 5. From the chart this is – **0.2 kg/s**.

Size the pump for the head of pressure to be generated to overcome the pipe losses in the circuit. Don't be tempted to add up all the pipework resistances on all the pipework sections – **this is wrong**.

The pump requirement is based on the individual pipework circuit with the highest pressure loss. This circuit must have only one radiator on it – it is known as the **index circuit**.

From the drawing there are three possible circuits:

- Section 5-3-1
- Section 5-3-2
- Section 5-4.

The section with the highest pressure loss is the index circuit.

Section 5-3-1 = (from the chart) 0.10 + 0.62 + 0.48 = 1.2 m

Section 5-3-2 = (from the chart) 0.10 + 0.62 + 0.08 = 0.8 m

Section 5-4 = (from the chart) 0.10 + 0.11 = 0.21 m

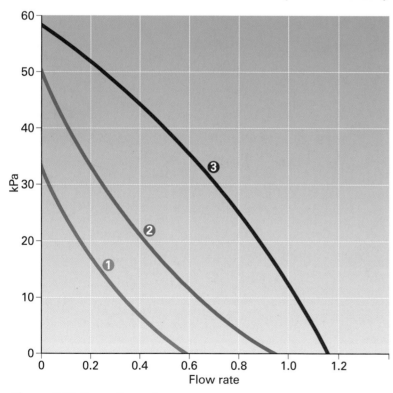

You can see that the index circuit is therefore section 5-3-1, which has a pressure loss of 1.2 m head. It's at this point that you need to establish whether the boiler generates a significant pressure loss through it by consulting manufacturer's information (many low-water-content boilers do), so add this on to the pressure loss through the pipework.

If the head loss through a boiler is 2.0 m and the pipe loss is 1.2 m head, then the pump should be sized at 3.2 m head, delivering a flow rate of 0.2 kg/s.

You now need to consult a pump manufacturer's catalogue to select your pump and determine any speed setting that it may be placed on. Look at the example in Figure 8.61.

To convert 1.2 metres to kPa = 3.2 × 10 = 32 kPa. Speed setting 2 will meet the requirement.

Figure 8.61 Pump flow rate

High flow rate (boilers)

Many newer low-water-content boilers (combis in particular) have a higher flow rate requirement through them than the heating load that the circuit requires, so with most boilers you need to check the manufacturer's minimum water flow rate requirement through the **heat exchanger**.

If you had used a boiler that required a minimum water flow rate of 0.4 kg/s through the exchanger, it can be seen from the worked example that the system only required 0.2 kg/s – a shortfall. This is where the system bypass comes in. 0.2 kg/s would be required to be circulated around a **bypass circuit** to ensure the minimum flow through the boiler.

Checking back to the pipe-size chart, that bypass would have to be sized at 22 mm. The pump would also now need to deliver 0.4kg/s at 3.2 m head, heading towards speed setting 3.

Find out

Carry out this assignment on pipe and pump sizing, by using Figure 8.62 to size this system:

- temperature drop – 10 degrees
- mains pipework loss – 10 per cent
- fittings loss – 33 per cent
- boiler head loss – 1.5 m

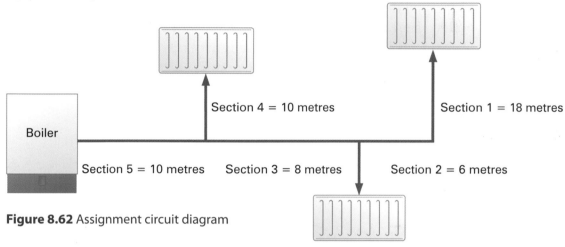

Figure 8.62 Assignment circuit diagram

Design principles for above ground discharge systems

At the end of this section you should be able to:

- understand the design principles and pipe sizing for:
  - above ground discharge systems.

You have already covered Above Ground Discharge Systems (AGDS) and pipework installation requirements, including the following systems:

- primary ventilated stack
- ventilated discharge branch
- secondary modified ventilated stack
- stub stack.

The driving force for the design and installation of AGDS is Building Regulations Part H1, and the general requirements of Part H1 are that the foul-water system:

- conveys the flow of water to a foul-water outfall (a foul or combined sewer, a cesspool, septic tank or settlement tank)
- minimises the risk of blockage or leakage
- prevents foul air from the drainage system from entering a building under working conditions
- is ventilated
- is accessible for cleaning blockages.

The majority of installations you are likely to encounter, as a domestic plumber, will be primary ventilated stacks. The design requirements for these systems are governed by Part H1 and as such this is reflected by manufacturers' pipework and fitting product sizes and designs. So, for sanitary pipework systems for domestic applications, there's virtually nothing to design as it's all done for you, e.g. size of basin pipework given to you and the maximum length of run. Where it differs a bit, though, is with the rainwater system, and you need to know the basics of this.

Sizing gutters

There are two factors to take into account when specifying the size of a gutter:

- the effective maximum roof area to be drained
- gutter flow capacity.

Effective roof area

This is calculated by using the formula:

$$(B + C \div 2) \times \text{length of roof} = \text{effective roof area in m}^2$$

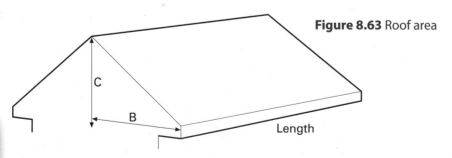

Figure 8.63 Roof area

Gutter flow capacity

This is dependent on three factors:

- system type
- outlet position
- fall of the gutter.

The information contained in Table 8.18 calculates the correct gutter flow capacity for a specified system profile, outlet position and fall, and is based on a rainfall intensity of 75 mm per hour. If a roof's dimensions are length = 10 m, B = 4 m and C = 2 m, the effective roof area would be:

$(4 + 2 \div 2) \times 10 = 30$ m²

Use Table 8.18 to specify the gutter for a 1:600 fall. A 76 mm half-round gutter with the outlet at the centre of the gutter run would be adequate, and would have a gutter flow capacity of 0.75 litres/sec.

	Outlet at end of gutter run			
	Gutter flow capacity (litres per second)		Maximum roof area (m²)	
System profile	Gutter level	1:600 fall	Gutter level	1:600 fall
76 mm half round	0.27	0.38	16	23
112 mm half round – Clickfit	0.83	1.17	40	56
112 mm square section	1.08	1.52	52	72
150 mm half round	1.83	2.56	88	123
115 mm high capacity	2.05	2.87	98	137
Ogee	2.34	2.50	112	122
Outlet at centre of gutter run				
76 mm half round	0.54	0.75	32	46
112 mm half round	1.67	2.34	80	112
112 mm square section	2.17	3.04	104	144
150 mm half round	3.67	5.13	176	246
115 mm high capacity	3.97	5.56	190	266
Ogee	4.58	4.84	217	236

Table 8.18 Gutter flow capacity

Knowledge check

1   List three installation requirements for a cold-water storage cistern.

2   Give three Water Regulations requirements for all fittings.

3   Give three examples of when Building Regulations approval is required.

4   State three types of floor construction used on domestic dwellings.

5   An external wall should be 'stable'; state two other key design features.

6   What type of rafters are used for most new roof construction?

7   Because of the limited strength of timber-framed construction, it can only be used under certain conditions. What are they?

8   What are the two main aspects in making sure that an installation meets industry standards?

9   Give three examples of documentation that covers installation practice.

Glossary

adventitious ventilation Air from gaps around doors and windows.

aerated burner Burner which has air mixed with gas before it leaves burner ports.

aggregates Various sizes of chippings, as used in concrete.

air brick Specially designed brick which allows air for proper combustion of gas, oil and solid fuelled appliances.

air vent Non-adjustable grille to give a free passage of air at all times.

anti-legionella valve Valve which incorporates an isolation and drain down to protect against legionella.

anti-vacuum valve Valve designed to open a pipe to atmosphere should the internal pressure drop.

appliance compartment Enclosure (not being a habitable space) specifically designed or adapted to house one or more gas appliances only.

approved contractors Contractors who have been deemed to comply with the requirements.

Approved Documents Guidance booklets published in association with the Building Regulations.

approved inspectors Persons appointed by the regulating authority.

artexing Decorative finish to ceilings or walls.

atmospheric sensing device Safety control which shuts off the gas if there is a lack of oxygen going to the burner (also known as an oxygen-depletion system).

back-flow Water flowing in a direction contrary to the direction intended.

back pressure Reversal of flow in a pipe caused by an increase in pressure in the system.

back siphonage Backflow caused by siphonage of water from a cistern or appliance back into the pipe which feeds it.

baffle Plate to slow down combustion products, thereby extracting more heat before products are discharged.

balanced compartment Sealed enclosure (not being a habitable space) specifically designed or adapted to house one or more open-flued

gas appliances only, but which takes its air supply from a place outside the enclosure that is adjacent to the flue discharge point.

balanced-flue appliance Room-sealed appliance which draws its combustion air from a point at which the combustion products are discharged, the inlet and outlet being so disposed that wind effects are the same on the outlet and inlet.

balanced pressure valve Mixing valve designed to operate on equal water pressures.

bayonet connector 'Plug in' end of a flexible hose as used on gas cookers.

beam and block Construction method for a suspended pre-cast concrete floor.

bill of quantities (B of Q) Document which lists all material for a contract.

bi-metal strip Two different metals which expand at different rates to bend and activate a switch.

bonding Term used to describe electrical earthing arrangement.

bottle trap Neater trap than a tubular type, used where appearance is important.

branched flue system Shared open-flued system serving appliances situated on two or more floors.

builder's opening Enclosure constructed by the builder to accommodate fireplace components such as back boilers.

bypass circuit Open circuit which always allows some water to flow through the boiler, even if all other devices are closed.

calorific value Amount of heat produced when a unit quantity of fuel is burned.

capillary path Route that moisture takes during capillary action.

carbon monoxide (CO) Gas formed as a result of incomplete combustion – poisonous.

catchment space Purpose-made space or void below the flue spigot to allow for the collection of soot, rubble or debris.

cathodic protection Action taken to protect a cathode against galvanic action, such as the placing of a sacrificial anode inside a hot-water vessel, cistern or tank, or on pipelines.

cavitation Eroding of a metal by constant water turbulence.

cavity wall External wall of building with an inner and outer leaf separated by a cavity.

chimney liner Rigid or flexible pipe inserted in a chimney to form a flue.

chimney plate Permanent plate or label fixed in a secure and accessible position in the building, giving details of the chimney or flue installation.

circulating head Distance in height between centre of boiler and centre of cylinder, which affects gravity circulation.

clerk of works Person who monitors quality of work done for benefit of customer.

close coupling Arrangement for feed and vent in open-vented systems.

closure plate Metal plate used to seal off a fireplace opening when fitting a gas fire.

clout nails Short nails used in felt roof work and for securing ceilings.

commissioning Completing an installation, checking for faults, putting the system in use, ensuring that it operates safely and efficiently and is to the customer's satisfaction.

commissioning report Written document which the plumber compiles upon commissioning an installation, and which is left with the customer.

common flue system Shared open-flued system serving two or more appliances installed in the same room or space.

compartment Enclosed space (not a habitable room) especially adapted to house a gas appliance.

complete combustion Occurs when all fuel is burnt.

composite valve Valve that can provide a number of functions in one unit e.g. line strainer, check valve and expansion valve.

condensate drain Fixture in a flue or appliance (resistant to corrosion from condensate formed from the products of combustion), where condensate can be drained.

condensate pipe Pipe, which may also be part of the flue pipe (resistant to corrosion from condensate formed from the products of combustion), which is leak free, and along which condensate may flow.

condensing appliance Appliance designed to make use of the latent heat from the water vapour in the combustion products by condensing the water vapour within the appliance.

connecting flue pipe Component used for connecting an appliance outlet to the flue within the chimney or flue system.

control relay Possible requirement when wiring valves in a control system.

cooling-off period Time during which the customer can cancel the order.

cost of labour calculation Standard method for working out cost of time taken to do a job.

credit meter Instrument that records the gas used; a periodic bill/account is sent to the customer.

damp-proof membrane (DPM) Moisture barrier used in floors.

data plate Plate fixed to an appliance which gives details of pressures, gas rates etc.

decorative fuel-effect gas appliance Open-flued appliance designed to simulate a solid-fuel open fire for decorative purposes, and intended to be installed so that the products of combustion pass unrestricted from the firebed to the chimney or flue.

discharge pipe Term given to a waste-water pipe.

dot and dab Method of fixing wall boarding to walls.

downstream Part of the installation after a certain point e.g. installation pipe is downstream of meter.

DPC Damp-proof course.

draught break Opening into any part of an open-flued system, including that part integral with the appliance.

draught-diverter Device for preventing conditions in a secondary flue from interfering with the combustion performance of an appliance.

dry-lined Term given to a wall covered with boarding.

duct Purpose-made enclosure for containing a pipe or pipes.

elastomeric Ability to flex under temperature movement.

emergency control valve Clearly labelled valve for shutting off the mains supply of gas in an emergency.

equipotential bonding Electrical bond between the outlet of a gas meter, the main water stop valve and the earth terminal.

equivalent height The effective height of a flue after calculating the restrictions caused by flue pipe and fittings.

estimate Approximate price for a job.

expansion-relief valve Valve which opens to relieve excess pressure in a system.

fanned draught Flue system which incorporates a fan.

flame-failure device (FFD) Device to stop gas entering a burner if no ignition source is available.

flame lift Term given to a flame which has too high a speed of mixture input to burner.

flame-supervision device Control that stops gas to a burner if it detects that a flame is absent.

float-operated valve Valve for closing off supply when a cistern reaches a pre-determined level.

flue break Opening in the secondary flue such as a draught-diverter.

flueless appliance Installation designed for use without a flue e.g. cooker.

flux paste Substance used to assist in preventing oxidisation when soldering.

gable end Side of building where roof angles up to the ridge.

galvanic action Electrolytic corrosion of dissimilar metals in a damp or wet environment. It can cause severe corrosion of pipes and fittings, and can occur between pipes of dissimilar materials such as lead and copper.

galvanised Protective zinc coating to metal.

gate valve Valve with an opening wedge/gate which gives full-bore flow.

gravity primaries Pipes conveying hot water from boiler to cylinder by gravity.

gully Drainage water trap to which discharge pipes are connected.

hardcore blinding Rubble used to underlay concrete in the construction of a solid concrete floor.

heat exchanger Device that transfers heat from one source to another e.g. a cylinder coil.

herringbone strutting Wooden cross pieces between joists to prevent them from twisting.

honeycomb walls Type of wall construction in which gaps are provided for timber floor joists.

impervious Not allowing passage of water.

inclined manometer A device used for measuring flue draught.

index circuit Circuit with the largest heating load/resistance.

installation pipework Any pipework to appliances.

jamb Sides of a door frame.

job order sheet Written confirmation of job to be carried out.

job record Document completed by the plumber on completion of a job, and which is kept by the customer.

kettling Term given to boiler noises when scaling/air is present.

light back Caused when gas/air pressure is too low.

limited-capacity relief valve 'Valve within a valve': it allows excess pressure to discharge to the atmosphere, and is triggered when an UPSO/OPSO valve to a closed position.

lint arrester A fibre-based pad or mesh filter to stop particles of dust and fluff from getting to the airways of the burner.

lintel Supporting beam above a window or door opening.

loading unit Factor used in pipe sizing.

lower flammability level (LFL) Lowest level at which a fuel/air mixture will burn.

macerator Special unit which cuts up and pumps waste from a WC to a remote drain/pipe.

Material change in use A change in how a premises is used.

micro-switch On a motorised valve, an auxiliary switch which powers the electrical supply to the boiler and pump.

mortar Mixture of sand and cement.

motorised valve Valve which is operated by an electrical device such as a room stat/programmer.

multi-functional control valve Series of components incorporated in one gas valve.

neutral point Point where a circulating pump may be positioned in a heating circuit to avoid positive or negative pressure in a system.

no-flow conditions When no demand for water is placed on the system.

non-potable Water not fit for drinking.

non-return/single check valve Valve allowing water to flow only one way.

not to current standards (NCS) Term used in classification of faults to gas installations.

nuisance discharge Unwanted discharge of material likely to damage the environment or cause inconvenience.

open flue Flue that discharges products to outside and takes its air from the building.

open-flued appliance Appliance designed to be fitted to a flue pipe system.

oversite concrete Layer of concrete used to seal the earth under the ground floor of a house.

oxygen-depletion device (ODD) Safety device which shuts down appliance if lack of oxygen occurs.

party wall Dividing wall between two properties.

pathogenic organisms Micro-organisms such as bacteria, viruses or parasites which are capable of causing illness, especially in humans, e.g. salmonella, vibrio cholera, campylobacter. These generally form in living creatures and can be released into the environment, for example in faecal matter, animal wastes or body fluids.

permanent live Live feed, even when programmer is off, such as to give a decorative light to a gas fire even when heat is off.

piezo-electric igniter Instrument that produces a spark by application of mechanical stress to produce electrical polarisation and a high voltage in certain types of crystal.

polyethylene Type of plastic.

pressure-flushing cistern WC flushing device that utilises the pressure of water within the cistern supply pipe to compress air and increase the pressure of water available for flushing a WC pan.

pressure-flushing valve Self-closing valve supplied with water directly from a supply pipe or a distributing pipe which when activated will discharge a pre-determined flush volume.

pressure gauge Gauge which shows pressure in system.

pressure-jet burner Burner in which combustion air is supplied under pressure.

pressure-limiting valve Valve which prevents design pressures being exceeded.

pressure-reducing valve Valve to reduce high pressures.

primary meter Meter that is connected to the gas service for the purpose of charges by the supplier of gas.

primary ventilated Term given to the main waste discharge stack.

printed circuit board (PCB) Thin board on to which electronic components are fixed by solder. Component leads may pass through holes in the board or can be surface mounted.

programmer Time switch with extra switching functions.

pump-overrun thermostat Thermostat to keep pump running until boiler has cooled down sufficiently.

purging Process of getting air out of a gas pipe on first filling.

quotation Formal statement of the estimated cost of a job or service.

reduced-pressure (RPZ) valve A verifiable backflow preventer activated by pressure reduction.

Regulator The ultimate authority on water Regulations. The Secretary of State in England and the National Assembly of Wales in Wales.

retention ports Ports on a gas burner to help prevent flame lift.

room sealed Air for combustion taken from outside.

sacrificial anode Metal that is gradually dissolved, such as magnesium, is placed in a system, therefore giving longer life to other metals.

screed finished layer of a solid cement floor.

secondary flue Part of the flue from the draught break or draught-diverter to the terminal.

SE duct Duct rising vertically through a building opening at the base to bring in combustion air and to remove products of combustion from special room-sealed appliances.

service pipe Pipe from the distribution main or gas storage vessel to the customer's property; usually ends at the emergency control valve.

service valve Valve for shutting off water supply for maintenance or service.

siphonage Transfer of liquid from a vessel via a pipe that goes above that vessel to a lower one, which receives the liquid.

sparge outlet Flush outlet to a urinal which sprays water onto it.

specification Document which describes the types of materials and quality of work to be used.

spider Manifold used to distribute flow and return pipework to radiators.

Statutory Instrument Legal document relating to regulations.

stoichiometric mixture Mixture of gas and air in the correct ratio, as determined by the theoretical air requirement.

stop valve Valve for shutting off, for the purpose of maintenance or service, the flow of water in a pipe connected to a water fitting.

stratification Formation of layers of water in a hot-water storage vessel.

stub stack Short discharge soil pipe used in limited applications.

stud partition Hollow-boarded internal wall.

supply pipe Pipe that carries water from the water main to a single property.

supply valve Customer's valve to control water to premises.

temperature-relief valve Valve which will open at a set temperature to avoid a dangerous situation occurring.

terminal Component fitted at the flue outlet to help products of combustion disperse and prevent downdraught; also prevents any material from entering the flue.

test pressure Internal water pressure of not less than 1½ times the maximum pressure that the installation or relevant part is designed to withstand in operation.

thermal envelope Area of a building that is enclosed within the walls, floor and roof, which is thermally insulated in accordance with the Building Regulations.

thermal insulation Material to slow down the passage of heat.

thermostatic mixing valve Safe mixing valve which prevents excessive temperatures.

thermostatic radiator valve Radiator valve which is controlled by the room temperature to give automatic control.

throughflow expansion vessel Device designed to compensate for the expansion of water through heat and allows the water to constantly flow through it to prevent blockages.

tongue and groove Term given to spigot and socket of floorboards or of trench blocks used in wall construction.

trace heating Electric heating element in the form of a cable attached to water pipework to provide frost protection.

trap Pipe fitting, or part of a sanitary appliance, that retains liquid to prevent the passage of foul air.

trussed roof Roof supported by a framework of rafters, ties and struts, and specifically trussed rafters.

U duct Vertical duct in the form of a 'U' open at roof level. One leg of the duct carries combustion air to special room-sealed appliances while the other leg conveys products to roof level and out to atmosphere.

unbalanced supply Delivered from two different pressures.

upper flammability level (UFL) Highest level/concentration at which a fuel/air mixture will burn.

utility services Services such as electricity, water and gas.

U value Thermal transmission rate from the inside to the outside of a building through the intermediate elements of construction.

variation order Written record of a customer's request to have work done/alterations and be charged accordingly.

vent pipe Open pipe on a system which goes to atmosphere.

venturi Device which causes a differential pressure in a pipe to activate a switch such as on a gas water heater.

vitiated Air which does not have the correct amount of oxygen in it.

vitiation-sensing device Device for sensing a lack of oxygen available for complete combustion of a gas appliance.

wallplate Timber on which the ceiling joists rest.

wall ties Ties, normally of galvanised steel, that hold together the inner and outer structures of a cavity wall.

warning pipe Pipe whose outlet is located in a position where the discharge of water can be readily seen.

waste-disposal unit Electrically operated device fitted to the waste pipe of a kitchen sink for grinding up food waste.

water-level indicator Instrument used to measure water levels in standpipes or wells.

water undertaker Legal term for the water companies that supply domestic water.

whole-site protection Used to protect one building from another (it was formerly termed 'secondary protection').

Wobbe number Indication of the heat produced by a burner for a particular gas.

workmanlike manner Working in line with appropriate British and European Standards, to a specification approved by the Regulator or the water undertaker.

written confirmation Signed document authorising work to be done.

zone protection Used to protect one part (zone) of a building from another part.

Appendix: Summary of backflow-prevention devices

Type	Description of backflow-prevention arrangements and devices	Fluid category for which suited	
		Back-pressure	Back-siphonage
AA	Air gap with unrestricted discharge above spillover level	5	5
AB	Air gap with weir overflow	5	5
AC	Air gap with vented submerged inlet	3	3
AD	Air gap with injector	5	5
AF	Air gap with circular overflow	4	4
AG	Air gap with minimum size circular overflow determined by measure or vacuum test	3	3
AUK1	Air gap with interposed cistern (e.g. a WC suite)	3	5
AUK2	Air gaps for taps and combination fittings (tap gaps) discharging over domestic sanitary appliances such as a washbasin, bidet, bath or shower tray shall be not less than the following –	x	3
AUK3	Size of tap or combination fitting — Distance of tap outlet above appliance spillover level Not exceeding G½ Exceeding G½ but not exceeding G¾ — 20 mm — 25 mm Exceeding G¾ — 70 mm Air gaps for taps or combination fittings (tap gaps) discharging over any higher risk domestic sanitary appliances where a fluid risk category 4 or 5 is present, such as (a) Any domestic or non-domestic sink (b) Any appliance in premises where a higher level of protection is required such as some appliances in hospitals or other health care premises The air gap shall be not less than 20 mm or twice the diameter of the inlet pipe to the fitting, whichever is the greater	x	5
DC	Pipe interrupter with permanent atmospheric vent	x	5

Notes:

1   X indicates that the backflow-prevention arrangement or device is not applicable or not acceptable for protection against back pressure for any fluid category within water installations in the UK.

2   Arrangements incorporating type DC devices shall have no control valves on the outlet side of the device, they shall be fitted not less than 300 mm above the spillover level of a WC pan, or 150 mm above the sparge pipe outlet of a urinal, and discharge vertically downward.

3   Overflows and warning pipes shall discharge through, or terminate with, an air gap, the dimension of which should satisfy a Type AA air gap.

Table A.1 Non-mechanical backflow-prevention devices acceptable under the Regulations

Type	Description of backflow-prevention arrangements and devices	Fluid category for which suited	
		Back-pressure	Back-siphonage
BA	Verifiable backflow preventer with reduced pressure zone (RPZ valve)	4	4
CA	Non-verifiable disconnector with difference between pressure zones not greater than 10%	3	3
DA	Anti-vacuum valve (or vacuum breaker)	x	3
DB	Pipe interrupter with atmospheric vent and moving element	x	4
DUK1	Anti-vacuum valve combined with verifiable check valve	2	3
EA	Verifiable single check valve	2	2
EB	Non-verifiable single check valve	2	2
EC	Verifiable double check valve	3	3
ED	Non-verifiable double check valve	3	3
HA	Hose union backflow preventor. Only permitted on existing hose union taps in house installations	2	3
HC	Diverter with automatic return (normally integral with domestic appliance applications only)	x	3
HUK1	Hose union tap which incorporates a verifiable double check valve. Only permitted for replacement of existing hose union taps in house installations	3	3
LA	Pressurised air inlet valve	x	2
LB	Pressurised air inlet valve with check valve downstream	2	3

Notes:

1 X indicates that backflow-prevention device is not acceptable for protection against back-pressure for any fluid category within water installations in the UK.

2 Arrangements incorporating a type BD device shall be fitted not less than 300 mm above the spill-over level of the appliance and discharge vertically downwards.

3 Types DA and DUK1 shall have no control valves on the outlet of the device and be fitted on a minimum 300mm type A upstand.

4 Relief outlet ports from types BA and CA backflow-prevention devices shall terminate with a type AA air gap.

Table A.2 Mechanical backflow devices acceptable under the Regulations

Non-mechanical backflow prevention devices

Type AA air gaps

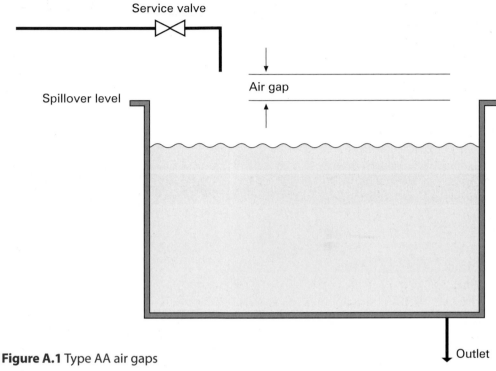

Figure A.1 Type AA air gaps

Back-pressure	Back-siphonage
5	5

Application
Cold water storage vessel containing fluid that is a serious health risk; such sinks in domestic or industrial premises or cisterns for industrial use, agricultural use or water for fire fighting purposes.

Table A.3 Type AA air gaps

Type AB air gaps

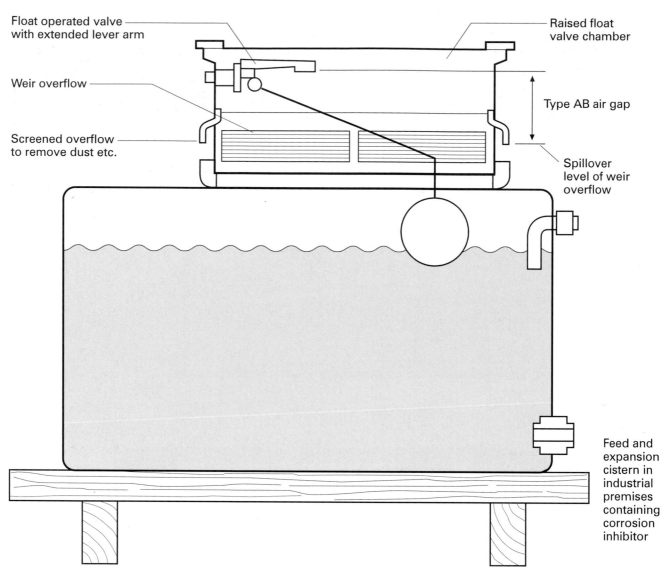

Figure A.2 Type AB air gaps (with weir overflow)

Back-pressure	Back-siphonage
5	5

Application
The example in the drawing shows a method of supplying water to a feed-and-expansion cistern feeding a heating system in an industrial/commercial premise. It's also suitable for supplying high-quality stored water to applications such as dentistry. Air gap 20 mm or twice the diameter of the inlet pipe whichever is the greater.

Table A.4 Type AB air gaps

Type AC air gaps

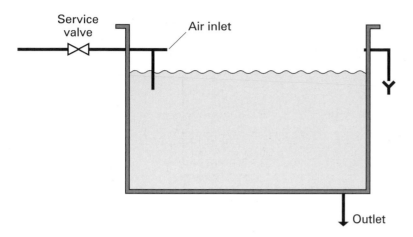

Figure A.3 Type AC air gaps (with submerged inlet and circular overflow discharging to a tundish)

Back-pressure	Back-siphonage
3	3
Application	
An example of this arrangement could be a silencer pipe in a storage cistern.	

Table A.5 Type AC air gaps

Type AD air gaps

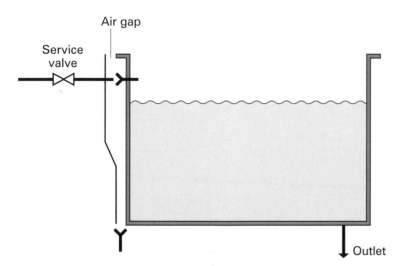

Figure A.4 Type AD air gap (air gap with injector and overflow)

Back-pressure	Back-siphonage
5	5
Application	
Device often in-built into modern washing machines.	

Table A.6 Type AD air gap

Type AF air gaps

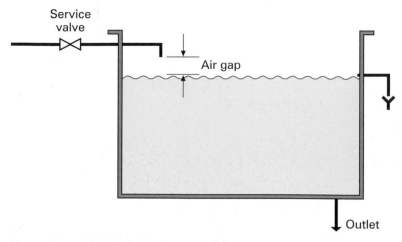

Figure A.5 Type AF air gap (air gap with circular overflow)

Back-pressure	Back-siphonage
4	4
Application	
Standard cold water storage cistern application with properly sized overflow and incorporating tundish.	

Table A.7 Type AF air gap

Type AG air gaps

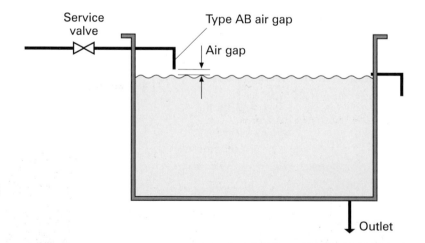

Figure A.6 Type AG air gap (type B air gap to BS 6281)

Back-pressure	Back-siphonage
3	3
Application	
Cold water storage cisterns for domestic use and interposed cisterns type AUK1.	

Table A.8 Type AG air gap

Type AUK1 air gaps

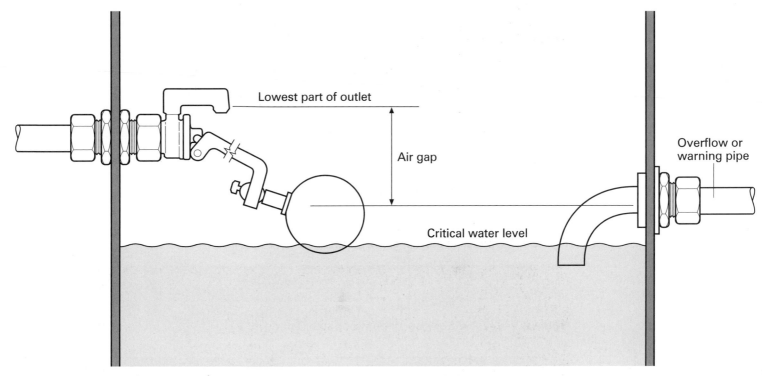

Figure A.7 Interposed cistern

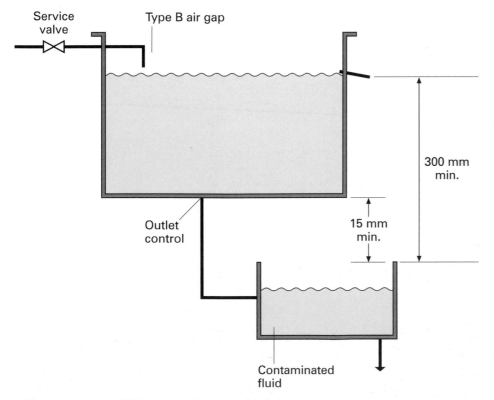

Figure A.8 Type AUK1 air gap (air gap with interposed cistern)

Back-pressure	Back-siphonage
3	5
Application	
WCs and vessels containing highly contaminated fluids.	

Table A.9 Type AUK1 air gaps

Type AUK2 air gaps

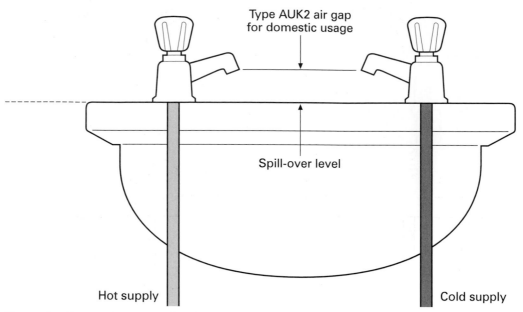

Figure A.9 Type AUK2 air gap for domestic applications

Back-pressure	Back-siphonage
x	3

Application
Air gaps for taps and combination fittings over domestic sanitary appliances – baths, basins, bidets or shower trays.

Air gap required –	
Not exceeding G½	20 mm
Exceeding G½ but not exceeding G¾	25 mm
Exceeding G¾	75 mm

Table A.10 Type AUK2 air gaps

Type AUK3 air gaps

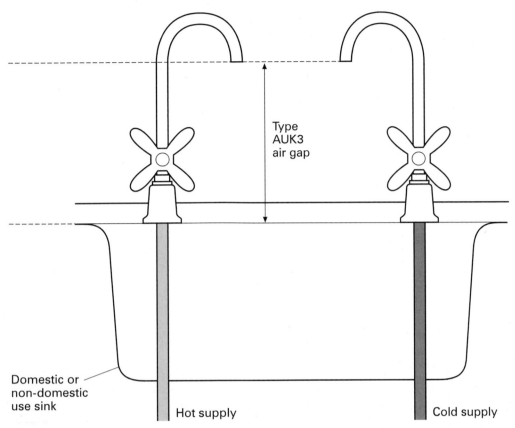

Figure A.10 Type AUK3 air gap for non-domestic and high-risk applications

Back-pressure	Back-siphonage
x	5

Application

Air gaps for taps or combination fittings discharging over any high risk sanitary appliances where a fluid category 4 or 5 risk is present e.g. sinks, hospital applications etc.

Air gap required –
Not less than 20 mm, or twice the diameter of the inlet pipe to the fitting, whichever is the greater.

Table A.11 Type AUK3 air gaps

Type DC pipe interrupter

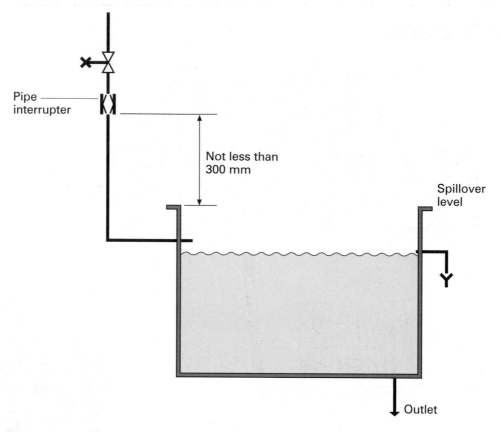

Pipe interrupter

Not less than 300 mm

Spillover level

Outlet

Figure A.11 Type DC pipe interrupter (with permanent atmospheric vent)

Back-pressure	Back-siphonage
x	5

Application
Flush pipes to WCs and urinals fed from pressurised systems and including pressure flushing valves.

Table A.12 Type DC pipe interrupter

Mechanical backflow-prevention devices

Type BA RPZ valve

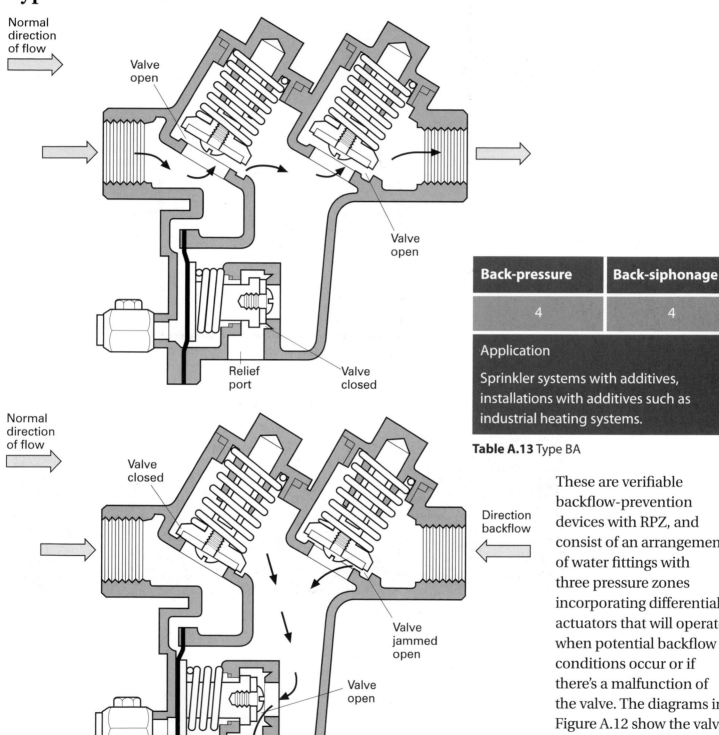

Normal direction of flow

Valve open

Valve open

Valve open

Relief port

Valve closed

Normal direction of flow

Valve closed

Valve jammed open

Valve open

Direction backflow

Relief port

Figure A.12 Type BA (verifiable backflow preventer with reduced pressure zone – RPZ valve)

Back-pressure	Back-siphonage
4	4

Application

Sprinkler systems with additives, installations with additives such as industrial heating systems.

Table A.13 Type BA

These are verifiable backflow-prevention devices with RPZ, and consist of an arrangement of water fittings with three pressure zones incorporating differential actuators that will operate when potential backflow conditions occur or if there's a malfunction of the valve. The diagrams in Figure A.12 show the valve operating in its correct operating state and what happens if the valve detects a backflow condition.

Type CA non-verifiable disconnector

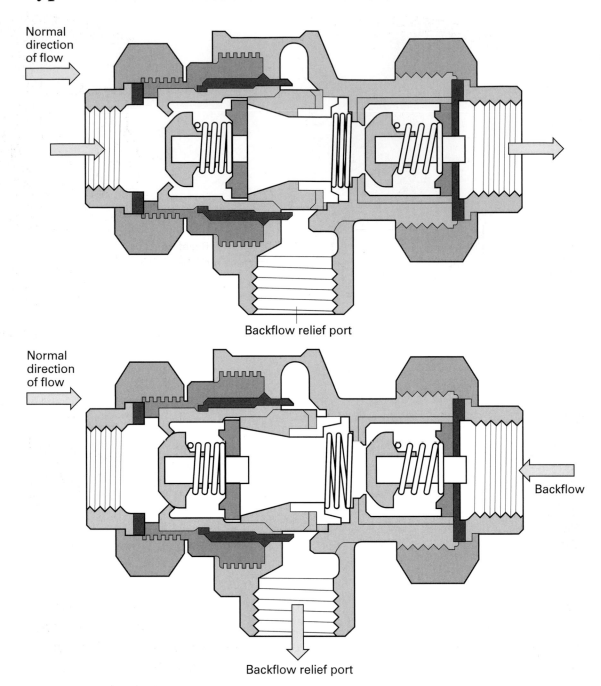

Normal direction of flow

Backflow relief port

Normal direction of flow

Backflow

Backflow relief port

Figure A.13 Type CA (non-verifiable disconnector)

Back-pressure	Back-siphonage
3	3

Application
Sprinkler systems with additives, installations with additives such as industrial heating systems.

Table A.14 Type CA

Type CA is a non-verifiable mechanical backflow-prevention device that provides disconnection by venting the intermediate pressure zone of the device to the atmosphere when the difference of pressure between the intermediate zone and the upstream zone isn't greater than 10 per cent of the upstream pressure. The diagrams in Figure A.13 show the valve first in its normal operating condition and then under backflow conditions. A type AA air gap should be provided between the relief outlet port and the top of the allied tundish.

Type DA anti-vacuum valve or vacuum breaker

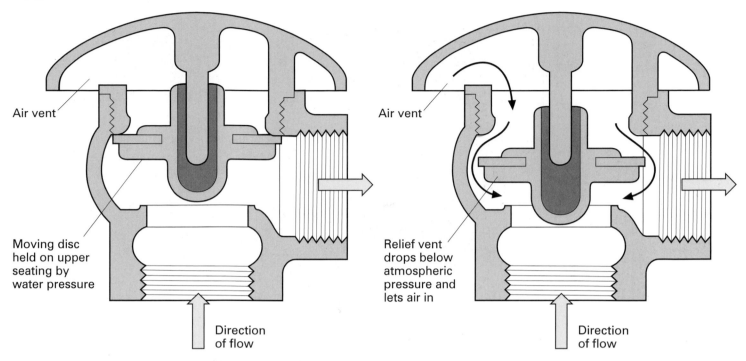

Figure A.14 Type DA anti-vacuum valve or vacuum breaker

Back-pressure	Back-siphonage
x	3
Application	
Air break to supply to storage cistern with submerged inlet below the water line.	

Table A.15 Type DA

An anti-vacuum valve or vacuum breaker mechanical backflow-prevention device has an air inlet which is closed when water in the device is at or above atmospheric pressure, but opens to admit air if a vacuum occurs at the inlet of the device. The device must be fitted on an upstand so that the outlet isn't less than 300 mm above the free discharge point or spill-over level of the appliance, and must have no valve or restriction on its outlet.

Type DB pipe interrupter with atmospheric vent and moving element

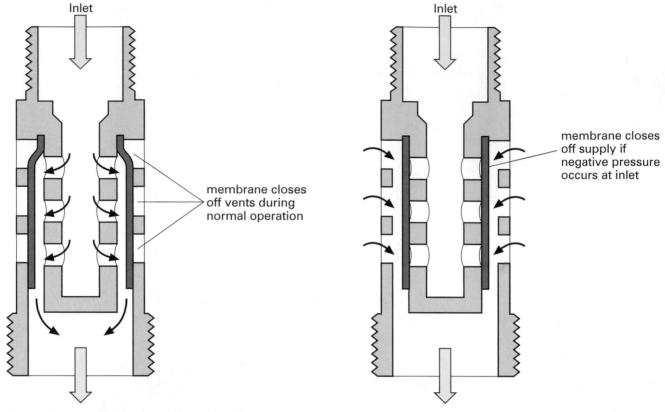

Figure A.15 Type DB (pipe interrupter with atmospheric vent and moving element)

Back-pressure	Back-siphonage
x	4

Application
Garden irrigation systems and porous hoses.

Table A.16 Type DB

Type DB is a moving element mechanical backflow-prevention device, with the air inlet closed by a moving element when the device is in normal use, but which opens and allows air to enter if the water pressure upstream of the device falls to atmospheric pressure. The device must be installed so that the flow water is in a vertical downward direction.

Type DUK 1 anti-vacuum valve combined with verifiable check valve

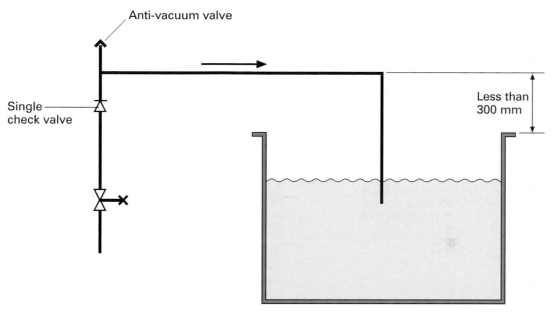

Figure A.16 Type DUK1 (anti-vacuum valve combined with verifiable check valve)

Back-pressure	Back-siphonage
2	3
Application	
Special applications not usually encountered in plumbing systems.	

Table A.17 Type DUK1

Type DUK1 is a mechanical backflow prevention device comprising an anti-vacuum valve with a single check valve located upstream.

Type EA verifiable single check valve

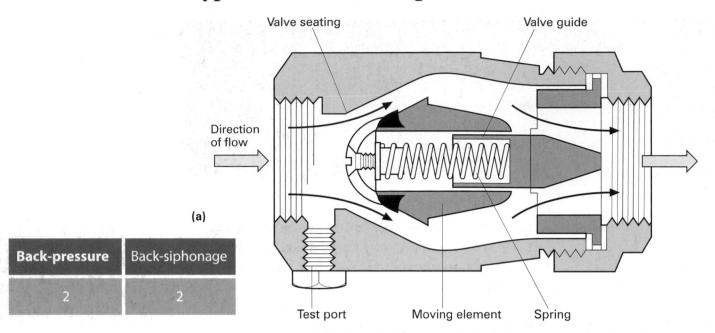

Back-pressure	Back-siphonage
2	2

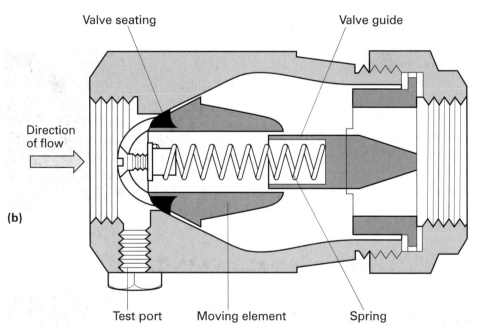

Figure A.17 Type EA (verifiable single check valve) (a) In normal flow conditions (b) In backflow conditions

Type EA are verifiable single check valves (verifiable because they have a test port), mechanical backflow-prevention devices that will allow water to flow from upstream to downstream but not in the reverse direction. The drawings in Figure A.17 show the valve in its correct operating state and then in the event of a malfunction. Type EB (non-verifiable single check valve) is similar to type EA, except the valve does not have a test port.

Type EC verifiable double check valve

Back-pressure	Back-siphonage
3	3
Application	
Filling of mains-fed domestic central heating systems, hose union tap connections in domestic properties.	

Table A.18 Type EC

Type EC are verifiable double check valve mechanical backflow-prevention devices consisting of two verifiable single check valves in series, which will allow water to flow from upstream to downstream, but not in the reverse direction. They consist of two valve assemblies situated in a valve body.

Type ED non-verifiable double check valve

This is similar to type EC, except the valve does not have a test port.

Back-pressure	Back-siphonage
3	3
Application	
Protecting against risk of wholesome water coming into contact with warmed water – unvented hot-water system taps with hot and cold water mixing in valve body.	

Table A.19 Type ED

Type HA hose union backflow preventer

Back-pressure	Back-siphonage
2	3
Application	
Only suitable for application on existing unprotected hose union taps and not new installations.	

Table A.20 Type HA

Type HA are hose union mechanical backflow-prevention devices, for fitting to the outlet of a hose union tap, and consist of a single check valve with outlets that open if the water flow ceases.

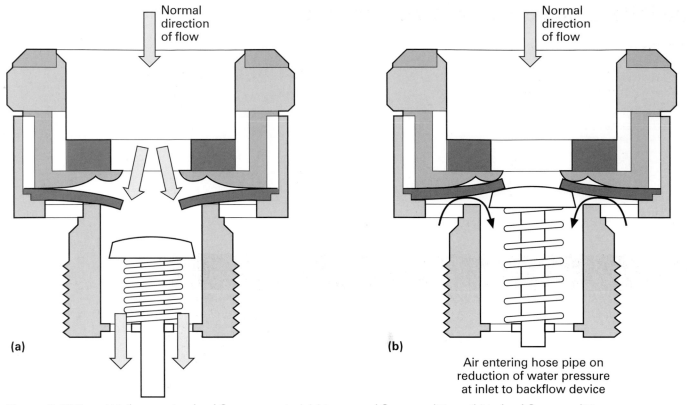

Figure A.18 Type HA (hose union backflow preventer) (a) In normal flow conditions (b) In backflow conditions

Type HC diverter with automatic return

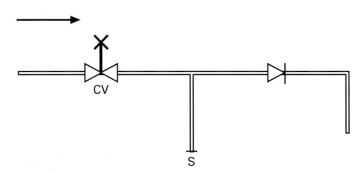

Figure A.19 Type HC (diverter with automatic return)

Back-pressure	Back-siphonage
x	3
Application	
Bath/shower combination tap assemblies.	

Table A.21 Type HC

Type HC are mechanical backflow diverters with automatic return devices which are activated if there's a reduction of water pressure at the inlet of the backflow-prevention device.

Type HUK1, non-verifiable disconnector

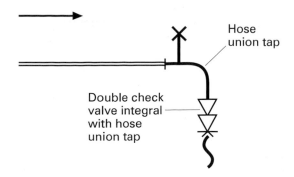

Figure A.20 Type HUK1 (hose union tap with integral double check valve)

Back-pressure	Back-siphonage
3	3
Application	
Only permitted for replacement of an existing hose union tap.	

Table A.22 Type HUK1

Type HUK1 are hose union tap devices, incorporating a verifiable double check valve.

Type LA pressurised air inlet valve

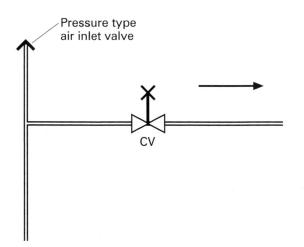

Figure A.21 Type LA (pressurised air inlet valve)

Back-pressure	Back-siphonage
x	2
Application	
Special applications not usually encountered in plumbing systems.	

Table A.23 Type LA

Type LA are pressurised air inlet valves, incorporating an anti-vacuum valve or vacuum breaker very similar to type DA, but are suitable for conditions where the water pressure at the outlet of the device under normal conditions of use is greater than atmospheric pressure.

Type LB pressurised air inlet with check valve

Back-pressure	Back-siphonage
2	3
Application	
Special applications not usually encountered in plumbing systems.	

Table A.24 Type LB

These are mechanical pressurised air inlet valves combined with a single check valve downstream. Their use is limited to locations where operational waste is acceptable e.g. gardens or similar.

Index

Let the web do the work!

Why not visit our website and see what it can do for you?

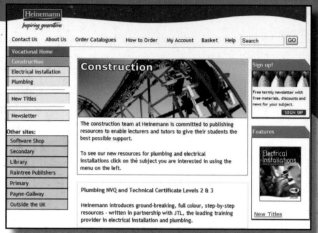

Free online support materials

You can download free support materials for many of our new construction products. We even offer a special e-alert service to notify you when new content is posted.

Lists of useful weblinks

Our site includes lists of other websites, which can save you hours of research time.

Online ordering – 24 hours a day

It's quick and simple to order your resources online, and you can do it anytime – day or night!

Find your consultant

The website helps you find your nearest Heinemann consultant, who will be able to discuss your needs and help you find the most cost-effective way to buy.

It's time to save time – visit our website now!

www.heinemann.co.uk/vocational

 01865 888068  01865 314029  orders@heinemann.co.uk 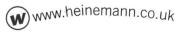 www.heinemann.co.uk

Inspiring generations

L155 R